Chemistry and Applications of Benzimidazole and its Derivatives

Edited by Maria Marinescu

Published in London, United Kingdom

IntechOpen

Supporting open minds since 2005

Chemistry and Applications of Benzimidazole and its Derivatives
http://dx.doi.org/10.5772/intechopen.81426
Edited by Maria Marinescu

Contributors
Kantharaju Kamanna, Aravazhi Amalan Thiruvalluvar, Jayabharathi Jayaraman, Sivaraman Rosepriya, Gopalsamy Vasuki, Yousef Najajreh, Pavel Vassiliev, Alexander A. Spasov, Vera A. Anisimova, Olga N. Zhukovskaya, Jørn H. Hansen, Richard Fjellaksel, Vincent Njar, Puranik Purushottamachar, Senthilmurugan Ramalingam, Ema Horak, Robert Vianello, Ivana Murković Steinberg, Ana Beloqui, Pilar Ocon, Daniel Herranz, Shivaputra Patil, Renukadevi Patil, Olivia Powrozek, Binod Kumar, William Seibel, Kenneth Beaman, Gulam Waris, Neelam Sharma-Walia, Maria Marinescu

First published in London, United Kingdom, 2019 by IntechOpen
IntechOpen is the global imprint of INTECHOPEN LIMITED, registered in England and Wales, registration number: 11086078, The Shard, 25th floor, 32 London Bridge Street
London, SE19SG – United Kingdom
Printed in Croatia

British Library Cataloguing-in-Publication Data
A catalogue record for this book is available from the British Library

Additional hard and PDF copies can be obtained from orders@intechopen.com

Chemistry and Applications of Benzimidazole and its Derivatives
Edited by Maria Marinescu
p. cm.
Print ISBN 978-1-78984-552-5
Online ISBN 978-1-78984-553-2
eBook (PDF) ISBN 978-1-83962-241-0

For EU product safety concerns:
IN TECH d.o.o., Prolaz Marije Krucifikse Kozulić 3, 51000 Rijeka, Croatia, info@intechopen.com or visit our website at intechopen.com.

Meet the editor

Maria Marinescu received her PhD in 2010 from the University of Bucharest, under the supervision of Prof. Mihaela Hillebrand, studying the chemistry and the electronic structure of the 1,2,3,4,5,6,7,8-octahydroacridine compounds, as key structures for new biomolecules and hydrogen-bonding receptors. Dr. Marinescu is currently an assistant professor at the University of Bucharest, Department of Organic Chemistry, Biochemistry and Catalysis. Her current research interests are in organic synthesis of various heterocycles with applications in medicinal and materials chemistry, density functional theory studies in organic molecules related with their biological properties, and applications of organic materials in art and archaeology.

Contents

Preface **XI**

Section 1
Introduction to Chemistry and Applications of Benzimidazoles **1**

Chapter 1 **3**
Introductory Chapter: Short Insight in Synthesis and Applications of Benzimidazole and Its Derivatives
by Maria Marinescu

Section 2
Chemistry of Benzimidazoles **13**

Chapter 2 **15**
Catalytic Intermolecular Functionalization of Benzimidazoles
by Jørn H. Hansen and Richard Fjellaksel

Chapter 3 **31**
X-Ray Crystal Structure Analysis of Selected Benzimidazole Derivatives
by Aravazhi Amalan Thiruvalluvar, Gopalsamy Vasuki, Jayaraman Jayabharathi and Sivaraman Rosepriya

Chapter 4 **51**
Synthesis and Pharmacological Profile of Benzimidazoles
by Kantharaju Kamanna

Section 3
The Therapeutic Potential of Benzimidazoles **71**

Chapter 5 **73**
Antidiabetogenic Features of Benzimidazoles
by Alexander A. Spasov, Pavel M. Vassiliev, Vera A. Anisimova and Olga N. Zhukovskaya

Chapter 6 **89**
Development of Benzimidazole Compounds for Cancer Therapy
by Puranik Purushottamachar, Senthilmurugan Ramalingam and Vincent C.O. Njar

Chapter 7 105
Benzimidazoles: From Antiproliferative to Multitargeted Anticancer Agents
by Yousef Najajreh

Chapter 8 139
Bisbenzimidazoles: Anticancer Vacuolar (H^+)-ATPase Inhibitors
by Renukadevi Patil, Olivia Powrozek, Binod Kumar, William Seibel, Kenneth Beaman, Gulam Waris, Neelam Sharma-Walia and Shivaputra Patil

Section 4
Benzimidazole Derivatives in Chemistry of Materials 157

Chapter 9 159
Optical Sensing (Nano)Materials Based on Benzimidazole Derivatives
by Ema Horak, Robert Vianello and Ivana Murković Steinberg

Chapter 10 181
Benzimidazole as Solid Electrolyte Material for Fuel Cells
by Daniel Herranz and Pilar Ocón

Chapter 11 203
Supramolecular Assembly of Benzimidazole Derivatives and Applications
by Ana Beloqui

Preface

Finding new strategies for synthesizing benzimidazole derivatives and functionalizing the benzimidazole core has proved to be important due to the compound's various applications in medicine, chemistry, and other areas. The development of new domains, such as the chemistry of optically active benzimidazole compounds, is imperative because of the compounds' particular pharmacological properties and their use as catalysts or as ligands.

The multitude of benzimidazole derivatives marketed as drugs has led to intensive research in the field for the discovery of new biologically active structures. The general applications of benzimidazole derivatives in materials chemistry, electronics, technology, dyes, pigments, and agriculture open up new research horizons. This book guides the rational design of benzimidazole derivatives synthesis with certain applications. It is organized as follows:

Section I: **Introduction to Chemistry and Applications of Benzimidazoles**

Chapter 1 "**Introductory Chapter: Short Insight in Synthesis and Applications of Benzimidazole and Its Derivatives**" summarizes the general and special methods of synthesis of benzimidazole derivatives as well as a scheme of applications of these compounds in the areas of medicine, pharmacology, and chemistry, among others.

Section II: **Chemistry of Benzimidazoles**

Chapter 2 "**Catalytic Intermolecular Functionalization of Benzimidazoles**" presents an updated review on functionalizations of nitrogen and at position C–2 in the benzimidazole scaffold. Functionalization of positions C–4 through C–7 is more challenging as these are less activated. A key success factor for reactions is the various catalytic systems.

Chapter 3 "**X-Ray Crystal Structure Analysis of Selected Benzimidazole Derivatives**" describes the X-ray crystal structure analysis of six selected benzimidazole derivatives. The detailed structural analyses indicate that the planes of benzimidazole units are associated in different modes through hydrogen bonds.

Chapter 4 "**Synthesis and Pharmacological Profile of Benzimidazoles**" summarizes the synthetic methods of benzimidazole derivatives and discusses therapeutic use of benzimidazole in conditions like diabetes, cancer, parasitic diseases, viruses, and allergies, as well as for pain management.

Section III: **The Therapeutic Potential of Benzimidazoles**

Chapter 5 "**Antidiabetogenic Features of Benzimidazoles**" examines the effect of a new cyclic aminobenzimidazole derivative, diabenol, which contains a guanidine group and combines pharmacological effects of biguanides and sulfonylurea derivative of gliclazide. Preclinical and clinical studies are also discussed.

Chapter 6 "**Development of Benzimidazole Compounds for Cancer Therapy**" discusses the synthesis and clinical pharmacology of three anticancer compounds: bendamustine, selumetinib, and galeterone. Based on impressive preclinical data, the University of Maryland, Baltimore, was granted an exclusive license for the development of galeterone for the treatment of patients with castration-resistant prostate cancer.

Chapter 7 "**Benzimidazoles: From Antiproliferative to Multi-targeted Anticancer Agents**" focuses on multi-targeted benzimidazole derivatives reported during the last decade, especially dovitinib and AT9283, reported to exhibit potent in vitro and in vivo activities against a group of kinases and non-kinases. Dovitinib, an anti-neoplastic benzimidazolylquinolinone, inhibits multiple growth factor receptor tyrosine kinases important for tumor angiogenesis and tumor growth and is effective against sixteen colorectal cancer cell lines.

Chapter 8 "**Bisbenzimidazoles: Anticancer Vacuolar (H+)-ATPase Inhibitors**" presents the research on several bisbenzimidazoles as proton pump inhibitors with good antiproliferative activity and very good cytotoxicity over most cancer cell lines.

Section IV: **Benzimidazole Derivatives in Chemistry of Materials**

Chapter 9 "**Optical Sensing (Nano)Materials Based on Benzimidazole Derivatives**" reviews benzimidazole derivatives as important materials for multi-functional building blocks in optical chemical sensors, materials for optical sensing, fluorescent sensors, metal organic frameworks, and copolymers.

Chapter 10 "**Benzimidazole as Solid Electrolyte Material for Fuel Cells**" discusses the application of polybenzimidazoles as economical materials for fuel cells with excellent thermal and mechanical properties as well as chemical resistance in acidic and alkaline media.

Chapter 11 "**Supramolecular Assembly of Benzimidazole Derivatives and Applications**" focuses on the chemical and physical properties of benzimidazole and its derivatives used for the synthesis of supramolecular materials, including metal-organic frameworks, metal-coordination polymers, and smart nanocontainers.

I am immensely grateful to IntechOpen and Author Service Manager Ms. Ivana Barac, who offered me the opportunity to be the editor of this book.

I would also like to thank the authors for their valuable contributions.

I hope this book will serve as a guide for students and researchers interested in making advancements in benzimidazole chemistry and applications of these compounds in medicinal chemistry and other areas.

Dr. Maria Marinescu
Faculty of Chemistry,
University of Bucharest,
Romania

Section 1

Introduction to Chemistry and Applications of Benzimidazoles

Chapter 1

Introductory Chapter: Short Insight in Synthesis and Applications of Benzimidazole and Its Derivatives

Maria Marinescu

1. Introduction

Benzimidazole is well known as an important pharmacophore among heterocyclic compounds due to the remarkable medicinal and pharmacological properties of its derivatives [1–3]. Among these currently marketed benzimidazole drugs to treat several diseases, we can mention bendamustine, selumetinib, galeterone, and pracinostat as antitumor agents; pantoprazole, lansoprazole, esomeprazole, and ilaprazole as proton pump inhibitors; bezitramide as an analgesic; mebendazole, albendazole, thiabendazole, and flubendazole as antihelminthics; ridinilazole as antibacterial; astemizole and bilastine as antihistamines; enviradine, samatasvir, and maribavir as antivirals; and candesartan and mibefradil as antihypertensive [1, 4–7]. Recent research recommends benzimidazole derivatives as potential EGFR and erbB2 inhibitors [8, 9], DNA/RNA binding ligands [10, 11], antitumor agents [12–14], anti-Alzheimer agents [15, 16], antidiabetic agents [17, 18], antiparasitic agents [10, 19], antimicrobial agents [20, 21], antiquorum-sensing agents [12], and antimalarial agents [19]. Intensive studies have demonstrated the use of the benzimidazole scaffold as key pharmacophore in clinically approved analgesic and anti-inflammatory agents [22]. Chiral benzimidazole derivatives were found to be $Na_V1.8$ (voltage-gated sodium channels) blockers, which play a key role in the transmission of pain signals, with excellent preclinical in vitro ADME and safety profile [23]. Other benzimidazole derivatives have been shown to be anti-HIV-1 agents through the protection of APOBEC3G protein [24]. Benzimidazoles grafted with aromatic nuclei have been noted as antioxidant agents [25]. A correlation of the grafted organic functions on the benzimidazole scaffold has been found with their therapeutic potential [26]. Thus, carboxylic acids, carbamates, and amidines have been shown to be effective anticancer drugs [26–28], benzimidazole esters were reported as antifungal agents [29], and 2-aminobenzimidazole derivatives possesses very good antimicrobial activity [30].

Structure-activity relationship (SAR) studies have shown that 1,2,5,6-substituted benzimidazoles with various substituents are analgesic and anti-inflammatory agents [22]. Also, SAR studies were accomplished for antiviral, anticancer, antihelminthic, antimicrobial, antimycobacterial, antidiabetic, antiprotozoal, antipsychotic, antidepressant, and antioxidant benzimidazole derivatives [1, 31–33].

IntechOpen

2. Synthesis of the benzimidazole derivatives

Benzimidazole synthesis reported by Hoebrecker in 1872 has greatly improved and diversified over last decades precisely because of its very diverse applications which will be discussed in the third part of this chapter. Classical synthesis was improved in terms of reaction conditions: catalysts, solvents or solvent-free, heating source, microwaves or ultrasound, and of course, nonpollutant or 'green' conditions. In the following, we will make (1) a very short presentation of classical syntheses and (2) an introduction to benzimidazole syntheses by rearrangement reactions.

2.1 Classical syntheses of benzimidazoles

Synthesis methods of the benzimidazoles have been extensively summarized in previous studies, published by Wright [34] and Preston [35]. Actually, all classical syntheses of benzimidazoles represent modifications to two of the classic reactions [26]: (i) the Phillips-Ladenburg reaction, coupling 1,2-diaminobenzenes with carboxylic acids (see **Figure 1**) and (ii) Weitenhagen reaction, coupling of 1,2-diaminobenzenes with aldehydes and ketones (pathway 3) *via* benzimidazoline **3**. In the case of the Phillips-Ladenburg reaction, esters, acid anhydrides, acid chlorides, and lactones (pathway 1) can be used instead of the acids, and benzimidazoles were generated *via* amide **1** cyclization or amides, nitriles, amidines, guanidines and benzimidazoles were resulted *via* cyclization of amidine **2** (pathway 2). The Phillips synthesis of benzimidazoles uses 4 N hydrochloric acid or glacial acetic acid, but various methods applied today use sulfuric acid or polyphosphoric acid. Reaction temperatures are high, reaching 250–300°C.

2.2 Synthesis of benzimidazoles *via* rearrangement of quinoxalinones

The limitations of classical synthesis, especially with respect to the synthesis of heterocyclic substituted benzimidazoles, have led to other methods [36].

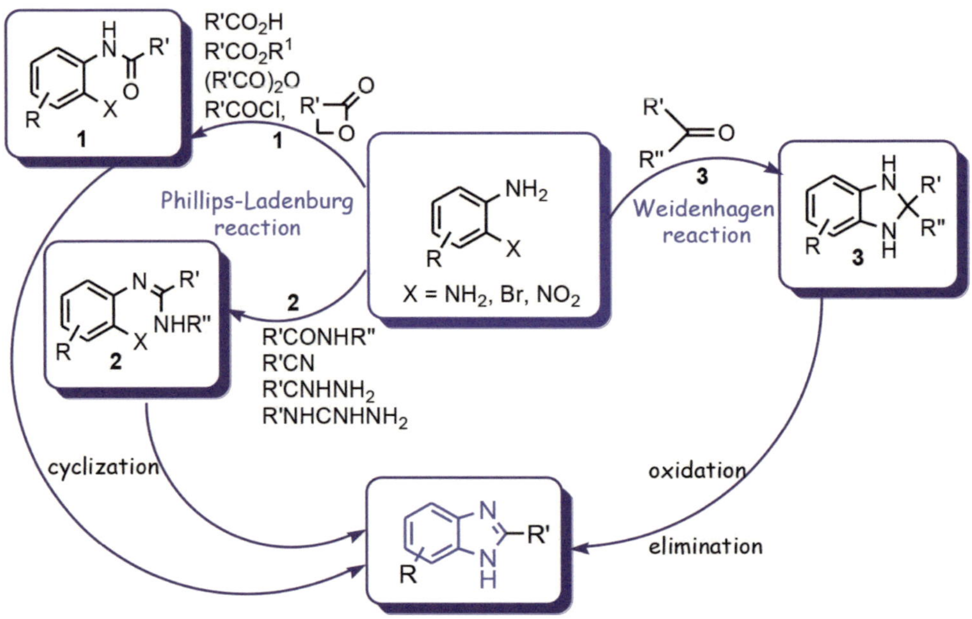

Figure 1.
Classical methods for synthesis of benzimidazoles.

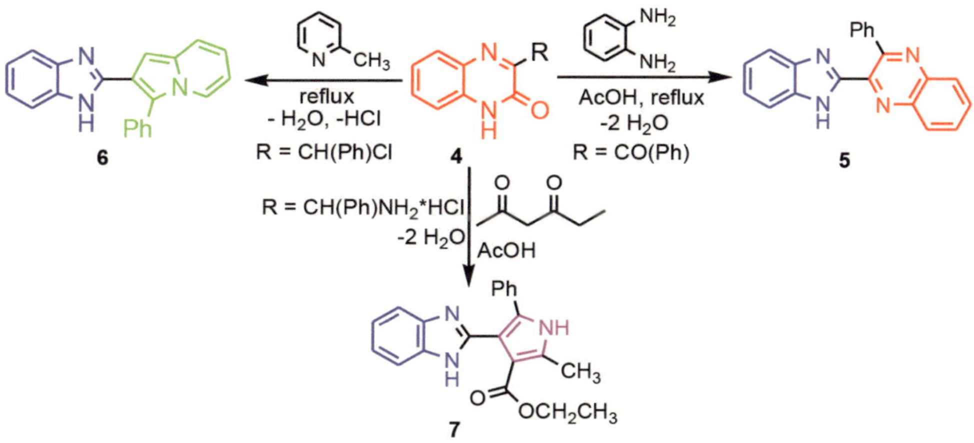

Figure 2.
Synthesis of 2-heteroaryl benzimidazoles by rearranging the quinoxalinones.

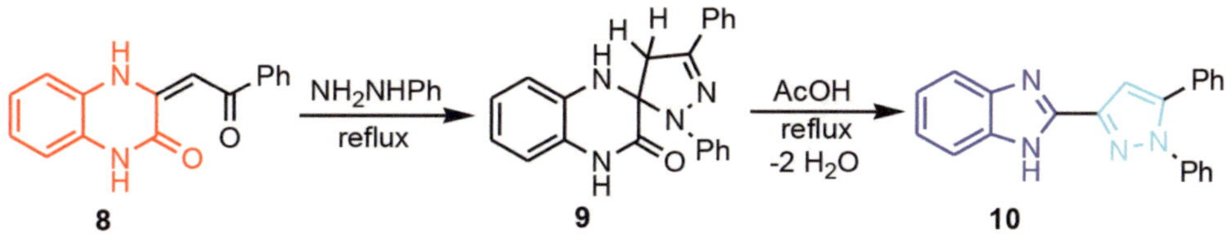

Figure 3.
Synthesis of 2-(1,5-diphenyl-1H-pyrazol-3-yl)-1H-benzo[d]imidazole **10**.

Rearrangements of quinoxalinones represent the most advantageous methods of synthesis currently reported [26, 36]. Hereinafter, some newer syntheses of benzimidazole derivatives are presented by quinoxalinone rearrangements. These new syntheses represent a combination of rearrangements, multicomponent reactions, and tandem sequences [26].

Thus, synthesis of benzimidazoles by the Hinsberg reaction implies condensation between 1,2-diaminobenzene and quinoxalin-2-one **4** to afford 2-benzimidazolylquinoxaline **5** in a 97% yield (see **Figure 2**). 2-(Indolizinyl)benzimidazoles **6** were obtained in high yields using a Chichibabin reaction, by refluxing quinoxalin-2-one **4** with α-picoline [37].

2-(Pyrol-3-yl)benzimidazole **7** was synthesized by a Knorr reaction between α-aminoketone of quinoxalinone **4** and ethyl acetoacetate [37].

Reaction of phenylhydrazine with 3-arylacylidene-3,4-dihydroquinoxalin-2(1H)-one **8** in boiling acetic acid implies the formation of spiro-compound **9**, which rearranges into pyrazolylbenzimidazole **10** (see **Figure 3**) [26].

3. Applications of benzimidazole derivatives in other fields than medicinal and pharmaceutical chemistry

There are a large number of published scientific papers that refer to the synthesis, properties, and applications of benzimidazoles. Thus, if we search the keyword "benzimidazole" on Science Direct, we get 26,386 results, of which 915 are published in the last 4 months.

Particular attention has been paid to improving the synthesis of chiral benzimidazoles, a relatively young branch of chiral chemistry, due to their importance in the field of therapeutic agents [38]. Also, chiral benzimidazoles were used as organocatalysts in Diels-Alder reaction, asymmetric aldol type reactions, asymmetric Michael addition, or enantioselective α-chlorination reactions as well as in palladium and rhodium benzimidazole complexes used as catalysts in Mizoroki-Heck [39] and Suzuki-Miyaura coupling reactions or in reduction reactions [40].

But recent research shows that benzimidazole scaffold is important not only for its therapeutic applications but also for its different uses in (nano) materials chemistry as optical chemical sensors [41], with special applications in medicine, environmental science, and chemical technology and has obvious advantage over other sensing devices, such as ease of operation and low cost (see **Figure 4**).

Supramolecular assemblies with interesting properties and with a wide range of applications like adsorbent materials, thermostable polymers, nanocontainers for small molecules, or liquid crystals for electronic conduction make up another use of benzimidazole and its derivatives [42–45].

Polybenzimidazole (PBI) derivatives: solid electrolyte for fuel cells [46], fibers [47], thin coatings [48], protective coatings for aerospace applications [49], or for the removal of uranium, thorium, and palladium from aqueous medium [50] are intensively studied in recent years. With an experience of 32 years, PBI Performance Products from Charlotte, North Carolina, is the leader in firefighter safety in Europe, USA, and the Middle East. PBI fabrics protect firefighters in a number of fire services, being renowned for their proven protection from heat and flame [51]. Another use of polybenzimidazoles is as PBI-based mixed matrix membranes with exceptional high water vapor permeability and selectivity [52].

In addition, the organic compounds are the most preferred for future photonic technology. Thus, several benzimidazoles with very good non-linear optic (NLO) properties, from very small molecules, such as 2-mercaptobenzimidazole, 2-phenyl benzimidazole, and 2-hydroxybenzimidazole [53], till molecules with more complicated structures [54], were studied.

Benomyl and carbendazim are recommended as benzimidazole fungicides having low toxicities in low doses and also are not carcinogenic, mutagenic, or teratogenic [55].

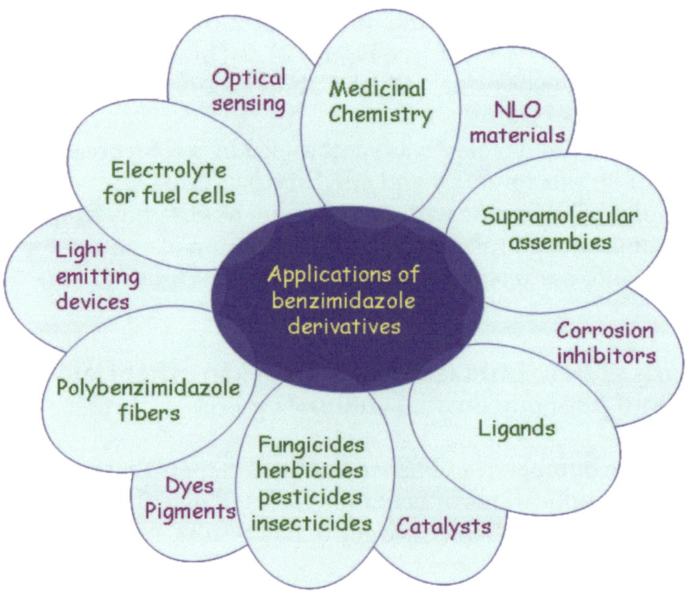

Figure 4.
Applications of benzimidazole derivatives.

The literature shows the conditions of using common benzimidazole pesticides and reported the use of benzimidazoles as herbicides and insecticides [56].

More and more research is being reported on the use of benzimidazoles as corrosion inhibitors for various metals (Cu, Fe, and Zn) under acidic conditions [57–58].

Other authors have shown that benzimidazole is a versatile and essential chromophore for organic dyes with photophysical, electrochemical, and photovoltaic properties due to the position of donors, acceptors, and π-linkers in the benzene ring [59]. A broad range of nuances in watercolor painting and electrophotographic developer toner has been made over three decades using benzimidazol-2-one derivatives, highly appreciated for their durability and light resistance [36]. Benzimidazole proved to be an essential core for organic light emitting devices (OLEDs) with superior phosphorescence, thermal properties, and morphological stabilities [60].

4. Conclusion

Benzimidazole occupies a central place in the class of heterocyclic compounds used in pharmaceutical and medicinal chemistry. The chemistry and applications of benzimidazole and its derivatives are in continuous development, especially in the last decades. In the coming years, we expect new synthesis strategies and more exciting applications to meet world market requirements.

Conflict of interest

There is no 'conflict of interest' in writing this chapter.

Author details

Maria Marinescu
Faculty of Chemistry, University of Bucharest, Bucharest, Romania

*Address all correspondence to: maria.marinescu@chimie.unibuc.ro

References

[1] Bansal Y, Silakari O. The therapeutic journey of benzimidazoles: A review. European Journal of Medicinal Chemistry. 2012;**20**(21):6208-6236. DOI: 10.1016/j.bmc.2012.09.013

[2] Akhtar W, Khan MF, Verma G, Shaquiquzzaman M, Rizvi MA, Mehdi SH, et al. Therapeutic evolution of benzimidazole derivatives in the last quinquennial period. European Journal of Medicinal Chemistry. 2017;**126**:705-753. DOI: 10.1016/j.ejmech.2016.12.010

[3] Rajesakhar S, Maiti B, Balamurali MM, Chanda K. Synthesis and medicinal applications of benzimidazoles: An overview. Current Organic Synthesis. 2017;**14**(1):40-60. DOI: 10.2174/15701794136661608181 51932

[4] Njar VC, Brodie AM. Discovery and development of galeterone (TOK-001 or VN/124-1) for the treatment of all stages of prostate cancer. Journal of Medicinal Chemistry. 2015;**58**(5):2077-2087. DOI: 10.1021/jm501239f

[5] Moniruzzaman RS, Mahmud T. Quantum chemical and pharmacokinetic studies of some proton pump inhibitor drugs. American Journal of Biomedical Sciences & Research. 2019;**2**(1):3-8. DOI: 10.34297/AJBSR.2019.02.000562

[6] Scholten WK, Christensen AE, Olesen AE, Drewes AM. Quantifying the adequacy of opioid analgesic consumption globally: An updated method and early findings. American Journal of Public Health (AJPH). 2019;**109**(1):52-57. DOI: 10.2105/AJPH.2018.304753

[7] Tahlan S, Kumar S, Ramasamy K, Lim SM, Shah SAA, Mani V, et al. Design, synthesis and biological profile of heterocyclic benzimidazole analogues as prospective antimicrobial and antiproliferative agents. BMC Chemistry. 2019;**13**(50):1-15. DOI: 10.1186/s13065-019-0567-x

[8] Celik I, Ayhan-Kilcigil G, Guven B, Kara Z, Gurkan AAS, Karayel A, et al. Design, synthesis and docking studies of benzimidazole derivatives as potential EGFR inhibitors. European Journal of Medicinal Chemistry. 2019;**173**:240-249. DOI: 10.1016/j.ejmech.2019.04.012

[9] Akhtar MJ, Siddiqui AA, Khan AA, Ali Z, Dewangan RP, Pasha S, et al. Design, synthesis, docking and QSAR study of substituted benzimidazole linked oxadiazole as cytotoxic agents, EGFR and erbB2 receptor inhibitors. European Journal of Medicinal Chemistry. 2017;**126**:853-869. DOI: 10.1016/j.ejmech.2016.12.014

[10] Popov AB, Stolic I, Krstulovic L, Taylor MC, Kelly JM, Tomic S, et al. Novel symmetric bis-benzimidazoles: Synthesis, DNA/RNA binding and antitrypanosomal activity. European Journal of Medicinal Chemistry. 2019;**173**:63-75. DOI: 10.1016/j.ejmech.2019.04.007

[11] Ding Y, Chai J, Centrella PA, Gondo C, DeLorey JL, Clark MA. Development and synthesis of DNA-encoded benzimidazole library. ACS Combinatorial Science. 2018;**20**:251-255. DOI: 10.1021/acscombsci.8b00009

[12] El-Gohary NS, Shaaban MI. Synthesis, antimicrobial, antiquorum-sensing and antitumor activities of new benzimidazole analogs. European Journal of Medicinal Chemistry. 2017;**137**:439-449. DOI: 10.1016/j.ejmech.2017.05.064

[13] Kanwal A, Saddique FA, Aslam S, Ahmad M, Zahoor AF, Moshin NA. Benzimidazole ring system as a

privileged template for anticancer agents. Pharmaceutical Chemistry Journal. 2018;**51**(12):1068-1077. DOI: 10.1007/s11094-018-1742-4

[14] Yadav S, Narasimhan B, Kaur H. Perspectives of benzimidazole derivatives as anticancer agents in the new era. Anti-Cancer Agents in Medicinal Chemistry. 2016;**16**(11):1403-1425. DOI: 10.2174/1871520616666151103113412

[15] Chaves S, Hiremathad A, Tomas D, Keri RS, Piemontese L, Santos MA. Exploring the chelating capacity of 2-hydroxyphenyl-benzimidazole based hybrids with multi-target ability as anti-Alzheimer's agent. New Journal of Chemistry. 2018;**42**(20):16503-16515. DOI: 10.1039/c8nj00117k

[16] Fang Y, Zhou H, Xu J. Synthesis and evaluation of tetrahydroisoquinoline-benzimidazole hybrids as multifunctional agents for the treatment of Alzheimer's disease. European Journal of Medicinal Chemistry. 2019;**167**:133-145. DOI: 10.1016/j.ejmech.2019.02.008

[17] Aboul-Enein HY, Rashedy AAE. Benzimidazole derivatives as antidiabetic agents. Medicinal Chemistry. 2015;**5**(7):318-325. DOI: 10.4172/2161-0444.1000280

[18] Adegboye AA, Khan KM, Salar U, Aboaba SA, Kanwal CS, Fatima I, et al. 2-Aryl benzimidazoles: Synthesis, in vitro α-amylase inhibitory activity, and molecular docking study. European Journal of Medicinal Chemistry. 2018;**150**:248-260. DOI: 10.1016/j.ejmech. 2018.03.011

[19] Farahat AA, Ismail MA, Kumar A, Wenzler T, Brun R, Paul A, et al. Indole and benzimidazole bichalcophenes: Synthesis, DNA binding and antiparasitic activity. European Journal of Medicinal Chemistry. 2018;**143**:1590-1596. DOI: 10.1016/j.ejmech.2017.10.056

[20] Marinescu M, Tudorache GD, Marton GI, Zalaru CM, Popa M, Chifiriuc MC, et al. Density functional theory molecular modeling, chemical synthesis, and antimicrobial behaviour of selected benzimidazole derivatives. Journal of Molecular Structure. 2017;**1130**:463-471. DOI: 10.1016/j.molstruc.2016. 10.066

[21] Bansal Y, Kaur M, Bansal G. Antimicrobial potential of benzimidazole derived molecules. Mini-Reviews in Medicinal Chemistry. 2019;**19**(8):624-646. DOI: 10.2174/1389557517666171101104024

[22] Gaba M, Singh S, Mohan C. Benzimidazole: An emerging scaffold for analgesic and anti-inflammatory agents. European Journal of Medicinal Chemistry. 2014;**76**:494-505. DOI: 10.1016/j.ejmech.2014.01.030

[23] Brown AD, Bagal SK, Blackwell P, Blakemore DC, Brown B, Bungay PJ, et al. The discovery and optimization of benzimidazoles as selective $Na_V1.8$ blockers for the treatment of pain. European Journal of Medicinal Chemistry. 2019;**27**:230-239. DOI: 10.1016/j.bmc. 2018.12.002

[24] Pan T, He X, Chen B, Chen H, Gheng G, Luo H, et al. Development of benzimidazole derivatives to inhibit HIV-1 replication through protecting APOBEC3G protein. European Journal of Medicinal Chemistry. 2015;**95**:500-513. DOI: 10.1016/j.ejmech.2015.03.050

[25] Hameed A, Hameed A, Farooq T, Noreen R, Javed S, Batool S, et al. Evaluation of structurally different benzimidazoles as priming agents, plant defence activators and growth enhancers in wheat. BMC Chemistry. 2019;**13**(29):1-11. DOI: 10.1186/s13065-019-0546-2

[26] Mamedov VA. Recent advances in the synthesis of benzimidazol(on)es *via* rearrangements of quinoxalin(on)es. RSC Advances. 2016;**6**:42132-42172. DOI: 10.1039/C6RA03907C

[27] Cheong JE, Zaffagni M, Chung I, Xu Y, Wang Y, Jernigan FE, et al. Synthesis and anticancer activity of novel water soluble benzimidazole carbamates. European Journal of Medicinal Chemistry. 2018;**144**:372-385. DOI: 10.1016/j.ejmech.2017.11.037

[28] Bistrovic A, Krstulovic L, Harej A, Grbcic P, Sedic M, Kostrun S, et al. Design, synthesis and biological evaluation of novel benzimidazole amidines as potent multi-target inhibitors for the treatment of non-small cell lung cancer. European Journal of Medicinal Chemistry. 2018;**143**:1616-1634. DOI: 10.1016/j.ejmech.2017.10.061

[29] Si W, Zhang T, Li Y, She D, Pan W, Gao Z, et al. Synthesis and biological activity of novel benzimidazole derivatives as potential antifungal agents. Journal of Pesticide Science. 2016;**41**(1):15-19. DOI: 10.1584/jpestics.D15-037

[30] Nguyen TV, Peszko MT, Melander RJ, Melander C. Using 2-amino-benzimidazole derivatives to inhibit *Mycobacterium smegmatis* biofilm formation. MedChemComm. 2019;**10**(3):456-459. DOI: 10.1039/C9MD00025A

[31] Siddiqui M, Alam MS, Sahu M, Yar MS, Alam O, Siddiqui MJA. Antidepressant, analgesic activity and SAR studies of substituted benzimidazoles. Asian Journal of Pharmaceutical Research. 2016;**6**(3):170-174. DOI: 10.5958/2231-5691.2016. 00024.1

[32] Tahlan S, Ramasamy K, Lim SM, Shah SAA, Mani V, Narasimhan B. Design, synthesis and therapeutic potential of 3-(2-(1H-benzo[d]imidazol-2-ylthio) acetamido)-N-(substituted phenyl)benzamide analogues. Chemistry Central Journal. 2018;**12**(139):1-12. DOI: 10.1186/s13065-018-0513-3

[33] Xu M, Wang SL, Zhu L, Wu PY, Dai WB, Rakesh KP. Structure-activity relationship (SAR) studies of synthetic glycogen synthase kinase-3β inhibitors: A critical review. European Journal of Medicinal Chemistry. 2019;**164**:448-470. DOI: 10.1016/j.ejmech.2018.12.073

[34] Wright JB. The chemistry of the benzimidazoles. Chemical Reviews. 1951;**43**(3):397-541. DOI: 10.1021/cr60151a002

[35] Preston PN. Synthesis, reactions, and spectroscopic properties of benzimidazoles. Chemical Reviews. 1974;**74**(3):279-314. DOI: 10.1021/cr60289a001

[36] Mamedov VA, Khavizova EA, Syakaev VV, Gubaidullin AT, Samigullina AI, Algaeva NE, et al. The rearrangement of 1H, 1'H-spiro[quinoline-4,2′-quinoxaline]-2,3'(3H,4'H)-diones: A new and efficient method for the synthesis of 4-(benzimidazol-2-yl)quinolin-2(1H)-ones. Tetrahedron. 2018;**74**(45):6544-6557. DOI: 10.1016/j.tet.2018.09.035

[37] Mamedov VA, Zhukova NA, Sinyashin OG. Advances in the synthesis of benzimidazolones *via* rearrangements of benzodiazepinones and quinoxalin(on)es. Mendeleev Communications. 2017;**27**(1):1-11. DOI: 10.1016/j.mencom.2017.01.001

[38] Khose VN, John ME, Pandey AD, Karnik AV. Chiral benzimidazoles and their applications in stereodiscrimination processes. Tetrahedron: Asymmetry. 2017; **28**(10):1233-1289. DOI: 10.1016/j.tetasy.2017.09.001

[39] Said NR, Mustakim MA, Sani MMM, Baharin SNA. Heck reaction using palladium-benzimidazole catalyst: Synthesis, characterisation and catalytic activity. IOP Conference Series: Materials Science and Engineering. 2018;**458**(012019):1-7. DOI: 10.1088/1757-899X/458/1/012019

[40] Gunnaz S, Gokce AG, Turkmen H. Synthesis of bimetallic complexes bridged by 2,6-bis(benzimidazol-2-yl) pyridine derivatives and their catalytic properties in transfer hydrogenation. Dalton Transactions. 2018;**47**:17317-17328. DOI: 10.1039/c8dt03178a

[41] Horak E, Kassal P, Steinberg M. Benzimidazole as a structural unit in fluorescent chemical sensors: The hidden properties of a multifunctional heterocyclic scaffold. Supramolecular Chemistry. 2017;**30**(10):838-857

[42] Jiang JJ, Pan M, Liu JM, Wang W, Su CY. Assembly of robust and porous hydrogen-bonded coordination frameworks: Isomorphism, polymorphism, and selective adsorption. Inorganic Chemistry. 2010;**49**(21):10166-10173. DOI: 10.1021/ic1014384

[43] Agarwal RA, Aijaz A, Ahmad M, Sañudo EC, Xu Q, Bharadwaj PK. Two new coordination polymers with Co(II) and Mn(II): Selective gas adsorption and magnetic studies. Crystal Growth & Design. 2012;**12**(6):2999-3005. DOI: 10.1021/cg300217v

[44] Nath I, Chakraborty J, Verpoort F. Metal organic frameworks mimicking natural enzymes: A structural and functional analogy. Chemical Society Reviews. 2016;**45**(15):4127-4170. DOI: 10.1039/C6CS00047A

[45] Tan S, Wei B, Liang T, Yang X, Wu Y. Anhydrous proton conduction in liquid crystals containing benzimidazole moieties. RSC Advances. 2016;**6**(40):34038-34042. DOI: 10.1039/C6RA03375J

[46] Yuan S, Guo X, Aili D, Pan C, Li Q, Fang J. Poly(imidebenzimidazole)s for high temperature polymer electrolyte membrane fuel cells. Journal of Membrane Science. 2014;**454**(12):351-358. DOI: 10.1016/j.memsci.2013.12.007

[47] Yin C, Dong J, Zhang Z, Zhang Q, Lin J. Structure and properties of polyimide fibers containing benzimidazole and amide units. Journal of Polymer Science. 2015;**53**:183-191. DOI: 10.1002/polb.23606

[48] Nabavian S, Naderi R, Asadi N. Determination of optimum concentration of benzimidazole improving the cathodic disbonding resistance of epoxy coating. Coatings. 2018;**8**(12):471. DOI: 10.3390/coatings8120471

[49] HMS I, Bhowmik S, Benedictus R. Performance evaluation of poly-benzimidazole coating for aerospace application. Progress in Organic Coatings. 2017;**105**:190-199. DOI: 10.1016/j.porgcoat.2017.01.005

[50] Kumar VV, Kumar CR, Suresh A, Jayalakshmi S, Mudali UK, Sivaraman N. Evaluation of polybenzimidazole-based polymers for the removal of uranium, thorium and palladium from aqueous medium. Royal Society Open Science. 2018;**5**(171701):1-16. DOI: 10.1098/rsos.171701

[51] Mandal S, Gwoen S. Characterizing thermal protective fabrics of firefighters' clothing in hot surface contact. Journal of Industrial Textiles. 2016;**47**(5):1-18. DOI: 10.1177/1528083716667258

[52] Akhtar FH, Kumar M, Villalobos LF, Vovusha H, Shevate R, Schwingenschlögl U, et al. Polybenzimidazole-based mixed

membranes with exceptional high water vapor permeability and selectivity. Journal of Materials Chemistry A. 2017;**5**(41):21807-21819. DOI: 10.1039/C7TA05081J

[53] Muthuraja A, Kalainathan S. A study on growth, optical, mechanical, and NLO properties of 2-mercaptobenzimidazole, 2-phenylbenzimidazole and 2-hydroxy benzimidazole single crystals: A comparative investigation. Materials Technology. 2017;**32**(6):335-348. DOI: 10.1080/10667857.2016.1235080

[54] Tayade RP, Sekar N. Benzimidazole-thiazole based NLOphoric styryl dyes with solid state emission – Synthesis, photophysical, hyperpolarizability and TD-DFT studies. Dyes and Pigments. 2016;**128**:111-123. DOI: 10.1016/j.dyepig.2016.01.012

[55] Gupta PK. Toxicity of fungicides. In: Gupta RC, editor. Veterinary Toxicology. Basic and Clinical Principles. Hopkinsville, KY: Academic Press; 2018. pp. 569-580. DOI: 10.1016/B978-0-12-811410-0.00045-3

[56] Emler S, Scholze M, Kortenkamp A. Seven benzimidazole pesticides combined at sub-threshold levels induce micronuclei in vitro. Mutagenesis. 2013;**28**(4):417-426. DOI: 10.1093/mutage/get019

[57] Eldebss TMA, Farag AM, Shamy AYM. Synthesis of some benzimidazole-based heterocycles and their application as copper corrosion inhibitors. Journal of Heterocyclic Chemistry. 2019;**56**(2):371-390. DOI: 10.1002/jhet.3407

[58] Wang X, Yang H, Wang F. An investigation of benzimidazole derivative as corrosion inhibitor for mild steel in different concentration HCl solutions. Corrosion Science. 2011;**53**:113-121. DOI: 10.1016/j.corsci.2010.09.029

[59] Saltan GM, Dincalp H, Kiran M, Zafer C, Erbaş SC. Novel organic dyes based on phenyl-substituted benzimidazole for dye sensitized solar cells. Materials Chemistry and Physics. 2015;**163**:387-393. DOI: 10.1016/j.matchemphys.2015.07.055

[60] Zhao Y, Chao W, Qiu P, Li X, Wang Q, Chen J, et al. New benzimidazole-based bipolar hosts: Highly efficient phosphorescent and thermally activated delayed fluorescent OLEDs employing the same device structure. ACS Applied Materials & Interfaces. 2016;**8**(4):2635-2643. DOI: 10.1021/acsami.5b10464

Section 2

Chemistry of Benzimidazoles

Chapter 2

Catalytic Intermolecular Functionalization of Benzimidazoles

Jørn H. Hansen and Richard Fjellaksel

Abstract

This chapter describes contemporary strategies for selective catalytic intermolecular functionalization of the benzimidazole scaffold. Functionalization at nitrogen and position C-2 is well developed employing copper, palladium, rhodium, nickel, and cobalt catalysis. Direct CH activation is the predominant approach to C-2 functionalization. Nickel-based catalysts can activate C—O bonds in conjunction with C—H activation at benzimidazole which grants access to a very broad range of phenols and enols as convenient functionalization precursors in this chemistry. The remaining carbon positions of benzimidazoles are typically functionalized via a sequential halogenation/coupling strategy to ensure selectivity. A key success factor in enabling these chemistries has been the fine-tuning of catalyst-ligand combinations.

Keywords: benzimidazoles, catalysis, C—H activation, cross-coupling, late-stage functionalization

1. Introduction

Benzimidazoles are tremendously important heterocycles in chemistry. They play a vital part in modern medicinal chemistry due to the importance of the benzimidazole as a pharmacophore in natural products and pharmaceuticals [1]. Benzimidazoles have a central role in contemporary homogeneous catalysis, particularly as ligands in metal catalysis and as a source of *N*-heterocyclic carbenes [2]. Moreover, they are important components of organic materials, e.g., optoelectronic materials [3]. Thus, the generation of a broad range of structurally diverse benzimidazoles is of paramount importance for enabling novel applications and unique properties to emerge.

Substituted benzimidazoles are typically synthesized de novo using a range of methods [1]. This is by far the most common approach, and new methods emerge steadily [4]. However, large libraries of benzimidazoles are needed in medicinal chemistry, catalysis, and materials science in order to discover fine-tuned properties and to optimize these. Thus, de novo synthesis makes for a rather inefficient approach since the benzimidazole scaffold must be constructed for each new analogue needed. A more powerful strategy would be to start with available benzimidazoles and be able to do functionalization with desired groups directly onto the scaffold—so-called late-stage functionalization [5].

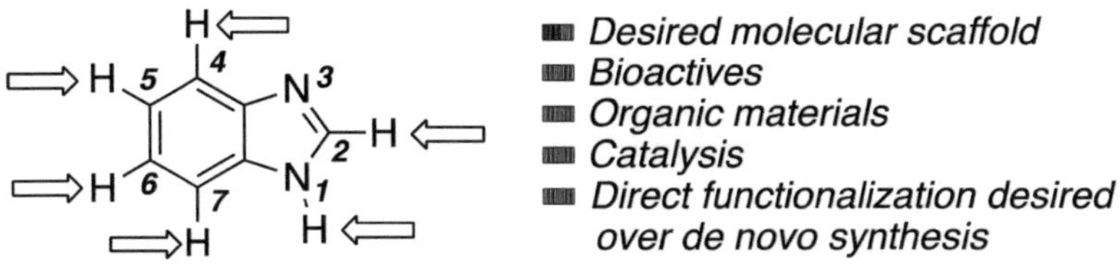

Figure 1.
The benzimidazole scaffold and its functionalization sites.

In this chapter the current catalytic strategies and methods for the intermolecular functionalization of benzimidazoles will be presented. These are surprisingly few, which illustrates the complexity involved when trying to selectively functionalize specific positions in the scaffold (**Figure 1**). The current strategies to solve this problem, and thus greatly streamline the synthesis of substituted benzimidazoles, will be presented herein with selected examples appearing since around 2003. We consider only *catalytic* strategies for *intermolecular* functionalization and aim to provide a good overview of state of the art in late-stage functionalization of benzimidazoles.

2. Catalytic functionalization chemistry

2.1 *N*-functionalization

Functionalization at nitrogen is perhaps the most straightforward. Classical bimolecular substitution can be used to alkylate in the presence of suitable bases. Nucleophilic aromatic substitutions or Ullmann couplings give rise to a number of *N*-arylated heterocycles, albeit with major limitations [1]. In recent years, catalytic methods and more sophisticated reaction conditions facilitate arylation and alkylation of this position with groups unavailable via classical chemistry [6].

A selectivity issue must be mentioned for unsymmetrical benzimidazoles with a free N—H. There is a rapid tautomeric equilibrium in which the proton shuffles between the two nitrogen sites. Thus, a particular tautomer must be "locked" in advance of the functionalization by substitution in order to obtain one distinct isomer.

Shieh and co-workers have reported an effective 1,4-diazabicyclo[2.2.2]octane (DABCO)-catalyzed N-benzylation reaction using dibenzyl carbonate (**2**) in the presence of a stoichiometric amount of the ionic liquid tetraoctyl ammonium bromide (**Figure 2**) [7]. The ionic liquid had a dramatic effect on both the reaction rate and yield, and the model benzylation of benzimidazole (**1**) afforded excellent 95% yield of **3**.

Buchwald et al. have reported one of the most general catalytic systems for efficient and mild *N*-arylations of benzimidazoles **4** (**Figure 3**) [8]. The bidentate ligand **6** in combination with dimethyl sulfoxide (DMSO) as solvent enabled this copper(I)-catalyzed coupling to occur even with unactivated aryl bromides **5** and afford 2-substituted benzimidazoles **7** in 71–98% yields. Notably, the coupling is also efficient with ortho-substituted aryl bromides.

The authors propose a mechanism for this transformation initiated by *N*-coordination of the benzimidazole to a Cu(I)Br-**6** complex followed by deprotonation of the N—H by the base. Subsequent oxidative addition of the aryl bromide to generate a cationic Cu(II) intermediate followed by reductive elimination releases the *N*-arylated benzimidazole 7 and regenerates the active Cu(I) catalyst. The method appears general and is a powerful tool for direct *N*-arylation of benzimidazoles [8].

Bao et al. have reported practical copper-catalyzed methods for *N*-vinylation [9]. In the most practical version, employing copper(I) oxide in the presence of a β-ketoester ligand precursor, they were able to couple electronically diverse *E*-bromostyrenes **8** with benzimidazole **1** to generate *N*-vinylated products **9** in 54–91% yields (**Figure 4**). Notably, the products retained the *E*-stereochemistry in the reaction with excellent >95:5 selectivity.

2.2 Functionalization at C-2

Catalytic functionalization at position C-2 is by far the most predominant in the literature. This is likely due to the more reactive nature of this particular C—H bond as it is situated between the two electron-withdrawing nitrogen sites. Although palladium, nickel, and rhodium play the major roles in this chemistry, some examples of copper-catalyzed C-2 functionalization exist. For example, copper(II) acetate in the presence of air has been employed for oxidative couplings of benzimidazoles at C-2 [10]. Furthermore, one exciting example of a copper(I) iodide (10 mol%)-catalyzed arylation with iodobenzene at C-2 of *N*-methylbenzimidazole was reported by Daugulis et al. to proceed in 89% yield [11]. Despite these promising results, there are no extensive studies of copper-catalyzed functionalizations at C-2 of benzimidazoles in particular.

2.2.1 Arylation and vinylation

An early example of palladium-catalyzed C-2 arylation by Bellina and Rossi involves the coupling of aryl iodides **10** with benzimidazole **1** under ligandless and base-free conditions [12]. They employed 5 mol% of palladium(II) acetate and a

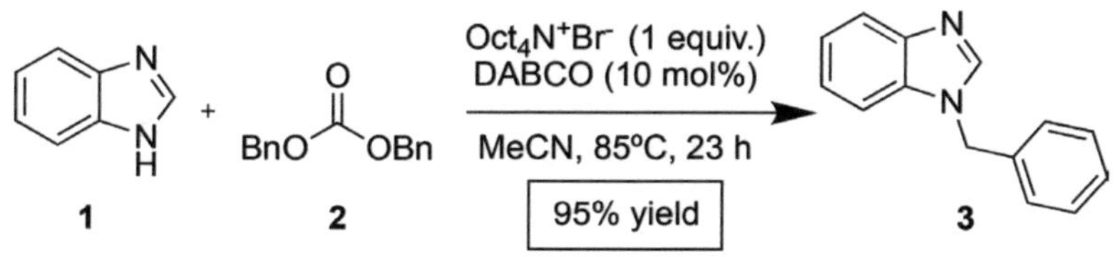

Figure 2.
Selective N-benzylation.

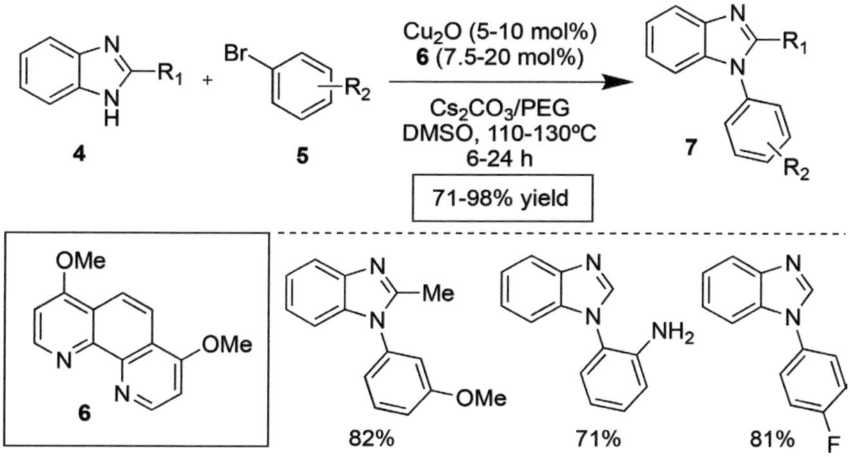

Figure 3.
Cu-catalyzed N-arylation.

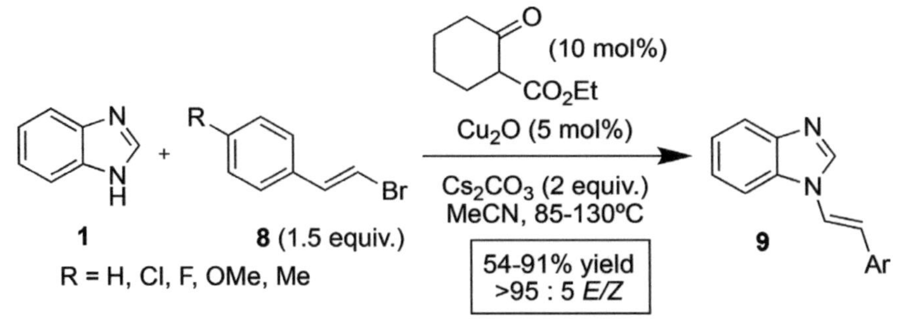

Figure 4.
Cu-catalyzed N-vinylation.

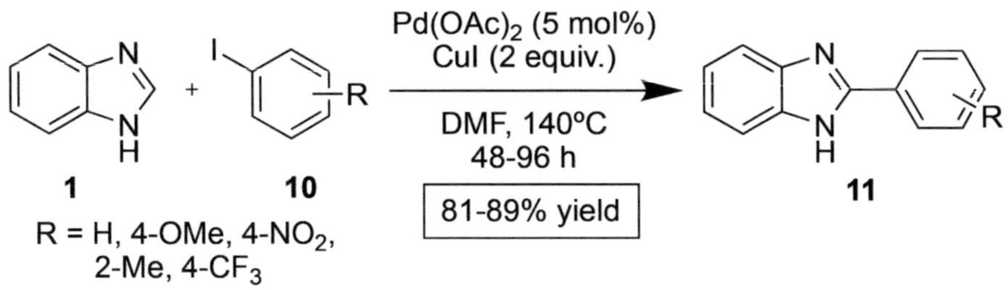

Figure 5.
Early Pd-catalyzed C-2 arylation under ligandless conditions.

superstoichiometric amount of copper(I) iodide in dimethylformamide (DMF) at high temperatures to afford 2-arylbenzimidazoles **11** in high purities and 81–89% yields. This demonstrated for the first time the possibility of using base-sensitive, unprotected N—H containing heterocycles without prior protection in this chemistry. A major drawback of the ligandless approach is relatively long reaction times (>48 hrs) at elevated temperatures in addition to the large amounts of copper(I) salt needed [12] (**Figure 5**).

The contemporary power of palladium-catalyzed coupling chemistry lies in ligand design [13]. The size and nature of the ligand play a crucial role in determining the possible pathways, selectivity, and kinetics, and, as such, optimization of ligand structure to suit the needs of the desired coupling reaction is key to modern catalytic reaction design. The number of ligands with vast spread in electronic and steric properties for palladium catalysis available today is large and expanding rapidly. This area lies at the forefront of modern catalysis research in organic chemistry [14].

A major step forward from the ligandless C-2 arylation reported above is the C—H arylation of *N*-substituted benzimidazoles **12** and aryl/heteroaryl chlorides **13** in the presence of the well-defined *N*-heterocyclic carbene-imidazole catalyst **14** (**Figure 6**). The reaction afforded moderate to good yields (56–97%) of a variety of C-2-arylated benzimidazoles **15** [15].

The first Ni-catalyzed C—H arylation and vinylation at C-2 of benzimidazoles were reported by Itami et al. in 2015 [16]. A major advance in the chemistry was the use of carbamate derivatives of phenols (**16**) or enols (**17**) as the source of aryl and vinyl groups, respectively (**Figure 7**). The catalytic system consists of nickel(II) triflate with potassium phosphate as base and bis-phosphine ligands; the latter are crucial. The use of a tertiary alcohol (AmylOH) as solvent was also crucial for this chemistry. 1,2-Bis(dicyclohexylphosphino)ethane (dcype) was the optimal ligand for the arylation chemistry and afforded C-2-arylated benzimidazoles **15** in 53–95%

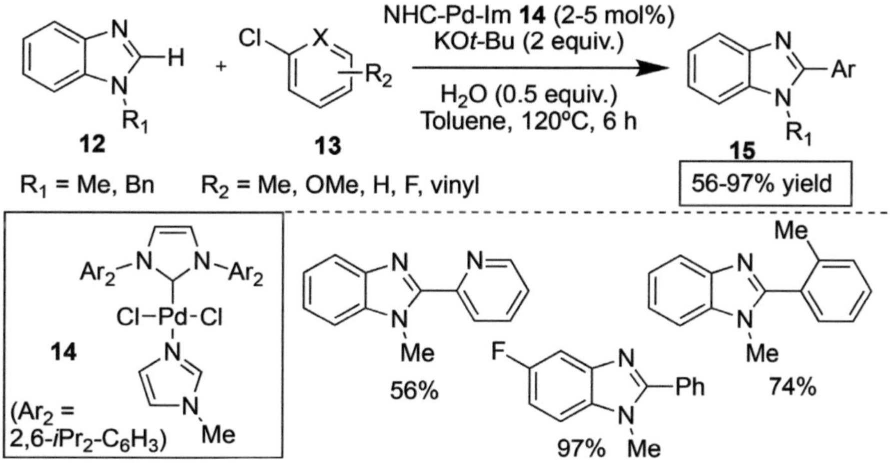

Figure 6.
Pd-catalyzed C-2 arylation with aryl chlorides.

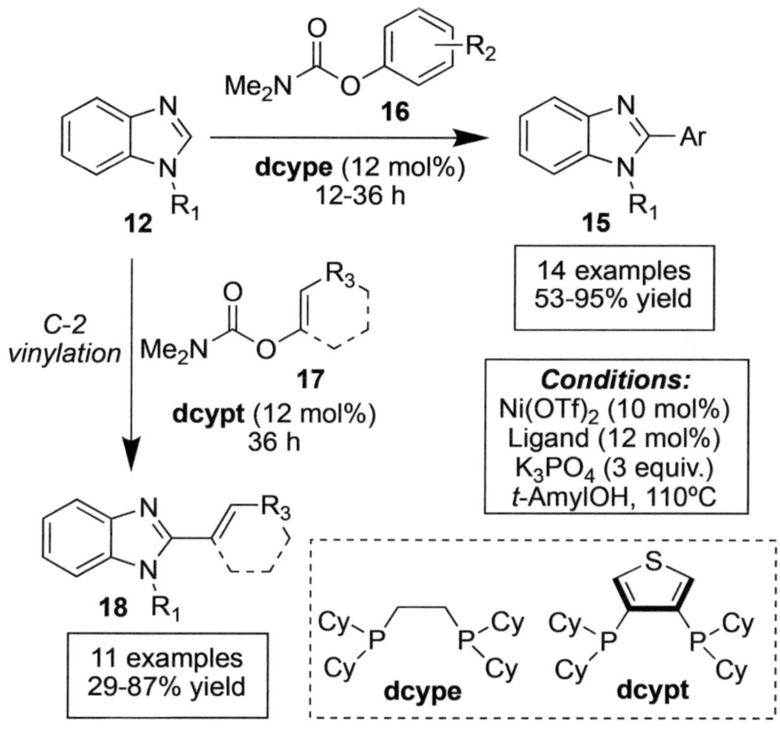

Figure 7.
Ni-catalyzed arylation and vinylation by domino C—H/C—O activation.

yield. This catalytic system could also facilitate arylations with aryl chlorides (**13**) in 64–86% yield. The potential synthetic power of this arylation approach was demonstrated by performing a functionalization of the nonsteroidal anti-inflammatory drug indomethacin with *N*-methylbenzimidazole at the chlorine site (26% yield). A different ligand, 1,2-bis(dicyclohexylphosphino)thiophene (dcypt), was employed as the optimal to achieve efficient C-2 vinylations with enol derivatives (**17**) to afford 2-vinylbenzimidazoles **18** in 29–87% yields [16].

The proposed mechanism for the nickel(II)-catalyzed approach involves reduction to a nickel(0) species by action of the bisphosphines or benzimidazole to initiate a Ni(0)–Ni(II) catalytic cycle [16]. Oxidative addition of the carbamate aryl C—O bond onto the activated Ni(0)-bisphosphine complex followed by C—H nickelation assisted by departure of the corresponding carbamic acid generates a key intermediate for reductive elimination of the product and regeneration of the active Ni(0) species [16].

The ability of nickel to undergo C—O activation enables the use of a range of practical and available substrates for functionalization. Wang and co-workers have recently disclosed C-2 arylations of benzimidazoles using methoxyarenes as functionalization agents in the presence of Grignard reagents [17]. Thus, the reaction system effects tandem C—O/C—H activation with subsequent coupling. The Grignard reagent was critical in order to minimize nonproductive couplings. A major demonstration of the applicability of this method is the reaction between steroidal hormone β-estradiol methoxy derivative **19** and *N*-methylbenzimidazole which afforded the C-2 steroid-functionalized benzimidazole **20** in very good 74% yield (**Figure 8**). The selectivity of aromatic methoxy group activation is striking, as the aliphatic methoxy group is left intact. The unusual dicarbene ligand carbodicarbene (CDC) was crucial for this reactivity and also demonstrates the importance of benzimidazoles as components of ligands for transition metal catalysis [17].

C-2 vinylation with alkynes as functionalization reagents has been reported, and these reactions occur under mild conditions with nickel or cobalt complexes in the presence of phosphine ligands. In a nickel-catalyzed process reported by Nakao et al., *N*-methylbenzimidazole reacts with internal alkynes to afford the C-2 vinylated products in 80–92% yields [18]. The cobalt-catalyzed vinylation of *N*-pyrimidylbenzimidazole **21** with alkyne **22** affords vinylation product **23** in 82% yield (**Figure 9**) in the presence of the phosphine ligand 2-[2-(diphenylphosphanyl) ethyl]pyridine (pyphos) and an equivalent of a Grignard reagent in tetrahydrofuran (THF) at ambient temperatures [19]. The *N*-pyrimidyl group is required for directing the chemistry to the C-2 site but can be easily removed from the scaffold after functionalization.

2.2.2 Alkylation

In the area of catalytic C-2 alkylations, rhodium(I) and nickel(0) complexes play a major role. Rhodium(I)-catalyzed *linear* C-2 alkylation was reported first by Bergman and Ellman [20]. In 2012, Shih et al. reported alkylations with full

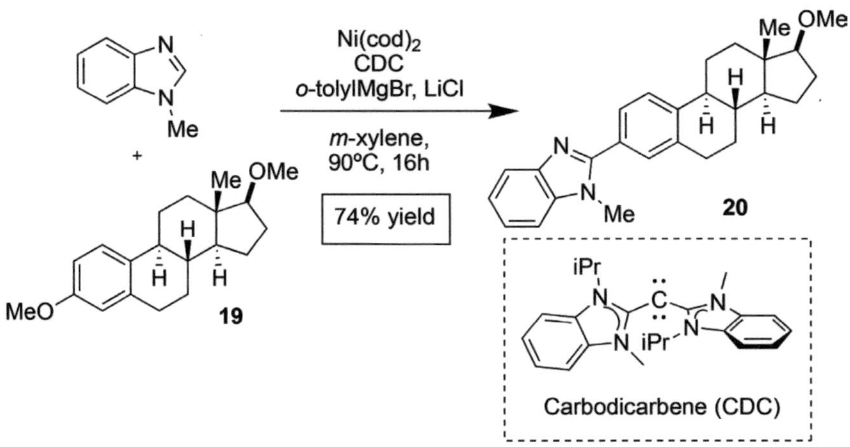

Figure 8.
Complex Ni-catalyzed C-2 arylation with estradiol dimethyl ether.

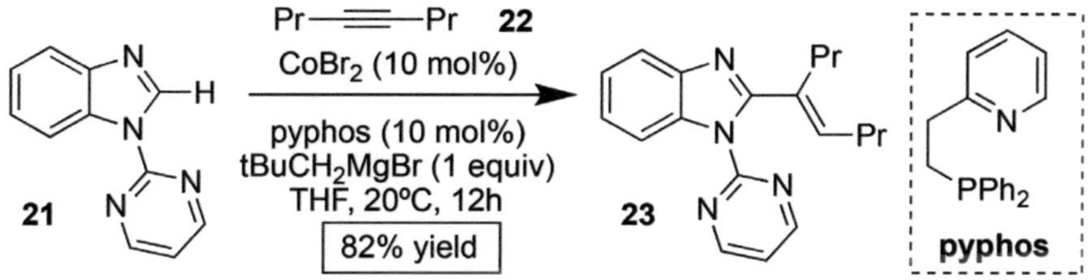

Figure 9.
A Co-catalyzed C-2 vinylation with alkynes.

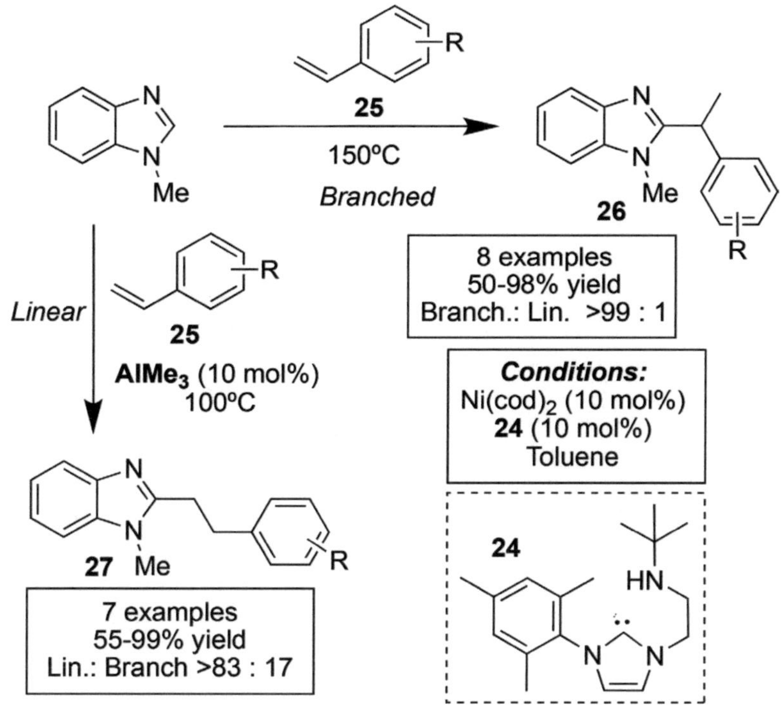

Figure 10.
$AlMe_3$ as a chemical switch for branched vs. linear alkylations at C-2.

control of linear versus branched selectivity using $AlMe_3$ as a chemical switch (**Figure 10**) [21]. In the presence of 10 mol% of $Ni(cod)_2$ (cod = cyclooctadiene) and the bidentate *N*-heterocyclic carbene (NHC) ligand **24** in toluene, the reaction between *N*-methylbenzimidazole and substituted styrenes **25** afforded exclusively branched alkylation products **26** in 50–98% yields. The addition of 10 mol% of $AlMe_3$ completely switched the selectivity toward linear alkylation products **27** in 55–99% yields. The branched product is electronically favored, but the linear product arises in the presence of $AlMe_3$ because the benzimidazole nitrogen at position 3 will generate a Lewis acid/base adduct causing a steric switch in the preferred binding orientation of the styrene during the catalytic cycle. Thus, the linear product is formed predominantly [21].

Obtaining branched selectivity in C-2 alkylations has been one of the major challenges in this chemistry, and the above example is the only report appearing before 2017 demonstrating this. As is often the case, fine-tuned selectivity control can be a matter of discovering fine-tuned ligand properties. Thus, Tran and Ellman recently reported a rhodium(I)-catalyzed C-2 alkylation of benzimidazoles **28** using acrylic systems **29** in the presence of the bidentate phosphine ligand dAr^Fpe

yielding exclusively branched alkylation products **30** in 12–96% yields (**Figure 11**) [22]. The use of ethyl methacrylate afforded products with a quaternary carbon at the C-2 site. The amide group can easily be converted to an aldehyde, thus making these products useful building blocks in medicinal chemistry.

2.3 Functionalization at C-4/C-7

Catalytic functionalizations at positions 1–3 are rather common in benzimidazoles. The activated nature of these positions makes direct functionalization chemistry feasible with a variety of catalytic systems as surveyed in Sections 2.1 and 2.2. Selective catalytic functionalization at positions 4–7 is significantly more challenging since these positions are less activated and also less chemically distinguishable from each other. Benzimidazoles that are pre-functionalized with some reactive functional group (mostly halogens), generated by de novo synthesis, are commonly used to achieve selectivity in these cases.

In order to obtain a key monomer for the construction of crystalline covalent organic frameworks (COFs), Xu et al. reported the double functionalization of 4,7-dibromobenzimidazole **31** using a Suzuki-Miyaura approach (**Figure 12**) [23]. Under rather standard cross-coupling conditions with 10 mol% of palladium(0) tetrakistriphenylphosphine and two equivalents of pinacol boronic ester **32**, double functionalization was achieved in 90% yield [23]. This is an example of the utility of benzimidazole functionalization in materials chemistry.

A great example of the utility of benzimidazole functionalization at C-4/C-7 in medicinal chemistry was reported by Auberson et al. in 2015 [24]. The 4-bromo-6-carbomethoxybenzimidazole **33** was treated with bisboronic ester **34** under catalytic action of Pd(dppf)Cl_2 (dppf = 1,1′-ferrocenediyl-bis(diphenylphosphine)), which afforded 92% yield of the boronic acid **35** (**Figure 13**). **35** was next employed as a coupling partner in a Suzuki-Miyaura coupling with heterocyclic bromide **36** to afford the complex product **37** in 77% yield [24]. This demonstrates an interesting strategy in which an accessible bromobenzimidazole can be transformed into a

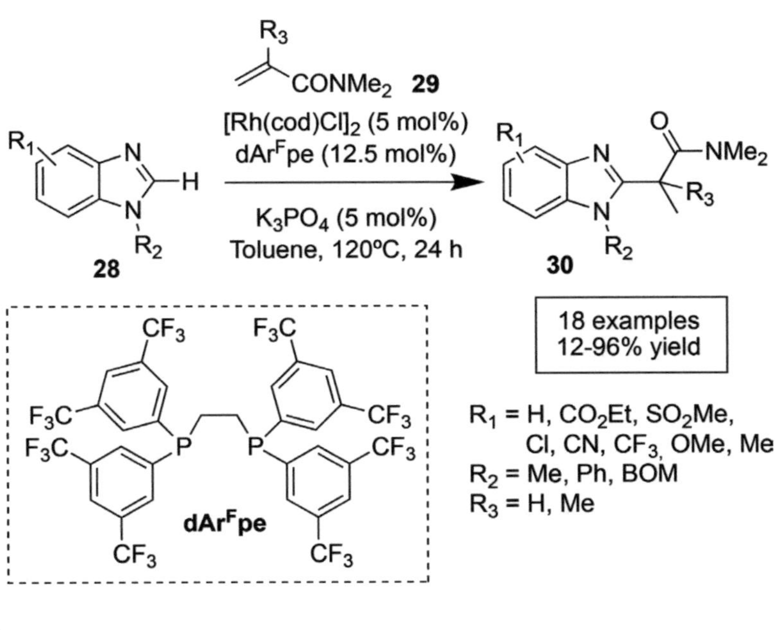

Figure 11.
Rh(I)-catalyzed branched alkylation at C-2.

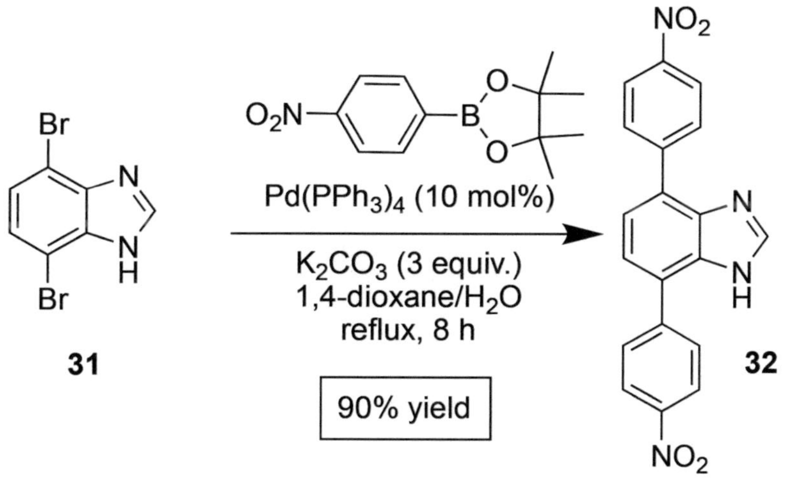

Figure 12.
Double Suzuki-Miyaura coupling to generate COF monomer.

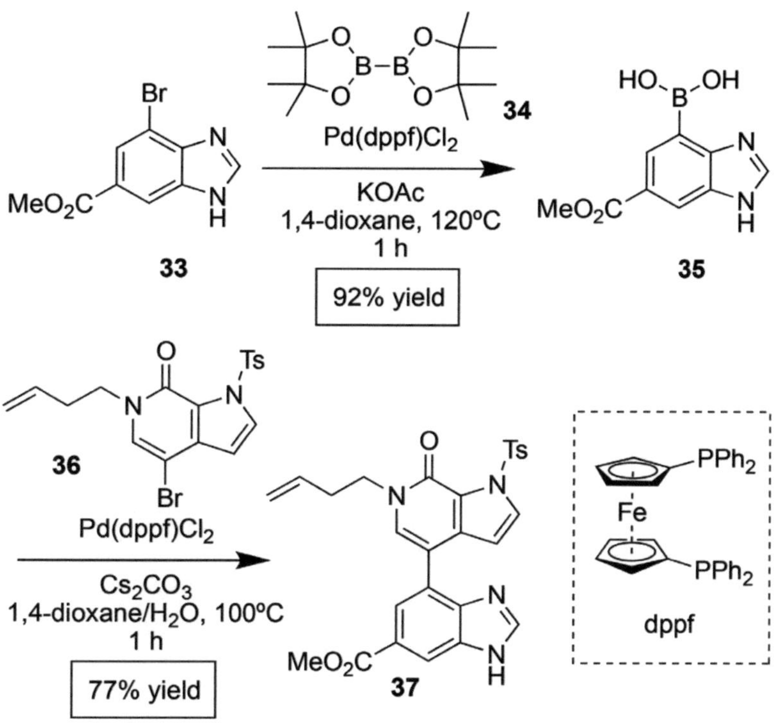

Figure 13.
Complex C-4 functionalization via borylation-cross-coupling sequence.

boronic acid for cross-coupling chemistry. Thus, it is overall a cross-coupling of two aryl bromides, really harnessing the full power of palladium catalysis.

Another powerful cross-coupling transformation of aryl halides is the Buchwald-Hartwig amination. The selective formation of C—N bonds clearly has numerous applications in medicinal chemistry. De la Fuente et al. have reported Buchwald-Hartwig functionalization with *N-tert*-butoxycarbonyl (Boc)-protected piperazine at position 4 of 4-chlorobenzimidazole **38** in the presence of Pd_2dba_3 and a phosphine ligand $P(tBu)_3$ (**Figure 14**) [25]. The [2-(trimethylsilyl)ethoxy] methyl acetal (SEM) protecting group at nitrogen was installed indiscriminately at the two nitrogen sites, so the starting material was effectively a 1:1 mixture of the 4- and 7-chlorobenzimidazole. However, the desired isomer was the 4-piperazinyl-benzimidazole **39** which could be isolated in a very good 43–49% yield (50% is theoretical maximum yield of one isomer) [25].

2.4 Functionalization at C-5 and C-6

Although most reported halogenated benzimidazoles are generated by de novo synthesis, in which the halogen is pre-functionalized on the starting materials used for assembling the benzimidazole scaffold, some examples exist of selective catalytic halogenations directly onto benzimidazole.

A practical and selective monobromination at C-5 has been reported by Das et al. in which sulfonic-acid functionalized silica acts as a heterogeneous acid catalyst system [26]. With only 13 wt% catalyst in the presence of an equivalent of N-bromosuccinimide (NBS), 77% yield of 5-bromobenzimidazole **40** was observed (**Figure 15**). Although the catalyst system is heterogeneous, its reactive Brønsted acidic sites are highly mobile on the catalyst surface and therefore achieve efficiency similar to that of homogeneous catalysts. Moreover, the catalyst could be recycled up to three times without loss of activity. The above study is interesting, particularly since no C-2 bromination was observed [26]. An early study by Smith et al. also used NBS as a brominating agent, but in the presence of pure silica gel acting as the heterogeneous catalyst, which yielded C-2 bromination only (67% yield), unless this position was occupied. They predominantly achieved C-2/C-5 dibromination when using two equivalents of NBS (60% yield) [27].

Cui et al. report a strategy for iodination at C-6 of **41** involving a pre-installed C-6 bromo substituent (**Figure 16**) [28]. By using a Pd(0)-catalyzed stannylation reaction, the C-6 tributyltin-substituted benzimidazole **42** can be generated albeit in low yields (14–29%). These products can further undergo oxidative iodination in the presence of molecular iodine in low to moderate yields (27–47% yields). The method was applied to radioiodination at C-6 with 125-I derived from $[^{125}I]NaI$ under oxidative conditions and afforded 43% radiochemical yield of the desired labeled product [28].

The strategy of bromination/Pd-catalyzed coupling has been employed for installment of various groups at positions C-5/C-6 in benzimidazoles. However, many are described only in the patent literature which often presents little information about conditions and chemical yields. Notably, this strategy has been employed for regioselective cyanation and carboxylation [29], alkylation (Negishi coupling)

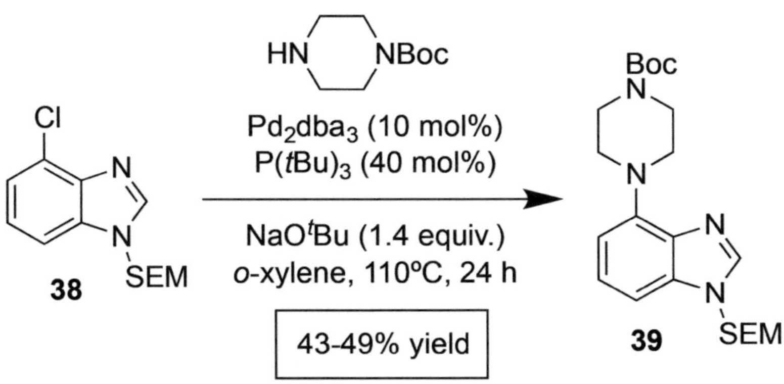

Figure 14.
Buchwald-Hartwig amination functionalization at C-4.

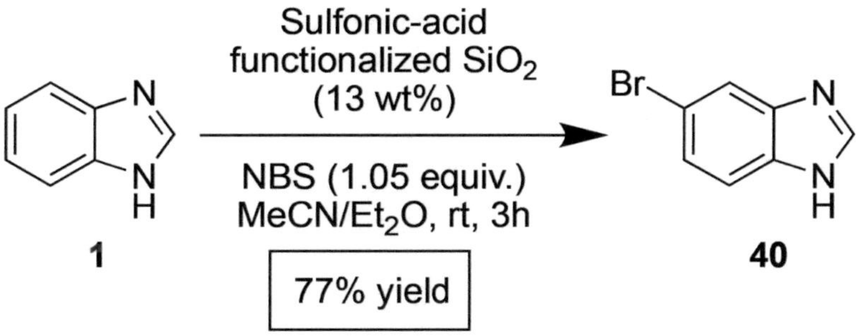

Figure 15.
Heterogeneously catalyzed C-5 bromination by functionalized silica.

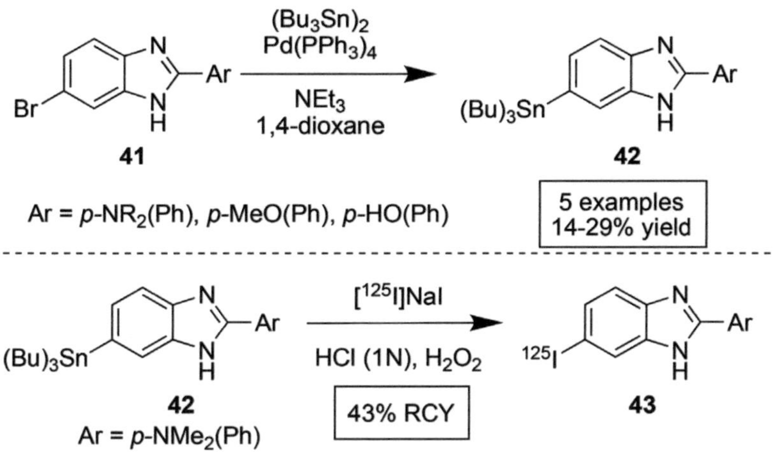

Figure 16.
Pd-catalyzed stannylation and subsequent radioiodination at C-5/C-6.

[30], alkynylation (Hiyama coupling) [31], and heteroarylation (Suzuki-Miyaura coupling) [32], thus demonstrating that a variety of combinations of halogens and cross-coupling chemistries are possible for synthesis of diverse benzimidazoles.

3. Conclusions

The need for new benzimidazoles with unique appendages and well-defined regioselectivity is undeniable from the viewpoints of medicinal and materials chemistry. The numerous applications, some of which are described herein, warrant further studies into the synthesis of novel analogues. This chapter has described the role of state-of-the-art catalytic chemistry in intermolecular functionalization of de novo assembled benzimidazole scaffolds. Particularly well developed are functionalizations at nitrogen and at position C-2 in the scaffold. *N*-Arylation and vinylations are effectively mediated by copper catalysts, whereas C-2 functionalization can be affected by a wider spectrum of palladium, rhodium, nickel, and cobalt catalysts involving direct C—H activation chemistry. A key success factor in enabling these chemistries has been the fine-tuning of ligand properties in the various catalytic systems. Functionalization of positions C-4 through C-7 is more challenging as these are less activated and less chemically distinguishable, so the use of pre-functionalization by de novo installment of halogens followed by contemporary cross-coupling chemistries is the most successful strategy to date. Even more powerful methods are anticipated to emerge in this area to make these positions also available for selective late-stage functionalization through direct C—H activation.

Acknowledgements

JHH wishes to acknowledge funding for this work through the Research Council of Norway (grant no. 275043 CasCat) and the Department of Chemistry at UiT The Arctic University of Norway. RF acknowledges the Northern Norway Regional Health Authority (grant no. SFP1196-14). The publication charges for this article have been funded by a grant from the publication fund of UiT The Arctic University of Norway.

Conflict of interest

The authors declare no conflicts of interest.

Author details

Jørn H. Hansen[1*] and Richard Fjellaksel[1,2]

1 UiT The Arctic University of Norway, Department of Chemistry, Organic Chemistry Group, Tromsø, Norway

2 UiT The Arctic University of Norway, Department of Pharmacy, Drug Transport and Delivery Group, Tromsø, Norway

*Address all correspondence to: jorn.h.hansen@uit.no

References

[1] Alamgir A, Black DSC, Kumar N. Synthesis, reactivity and biological activity of benzimidazoles. Topics in Heterocyclic Chemistry. 2007;**9**:87-118. DOI: 10.1007/7081_2007_088

[2] Kuwata S, Hahn FE. Complexes bearing protic *N*-heterocyclic carbene ligands. Chemical Reviews. 2018;**118**:9642-9677. DOI: 10.1021/acs.chemrev.8b00176

[3] Han Y, Cao H-T, Sun H-Z, Wu Y, Shan G-G, Su Z-M, et al. Effect of alkyl chain length on piezochromic luminescence of iridium(III)-based phosphors adopting 2-phenyl-1h-benzoimidazole type ligands. Journal of Materials Chemistry. 2014;**2**:7648-7655

[4] Prosenjit D, Yehoshoa B-D, Milstein D. Direct synthesis of benzimidazoles by dehydrogenative coupling of aromatic diamines and alcohols catalyzed by cobalt. ACS Catalysis. 2017;**7**:7456-7460. DOI: 10.1021/acscatal.7b02777

[5] Cernak T, Dykstra KD, Tyagarajan S, Vachal P, Krska SW. The medicinal chemist's toolbox for late stage functionalization of drug-like molecules. Chemical Society Reviews. 2016;**45**:546-576. DOI: 10.1039/C5CS00628G

[6] Correa A, Bolm C. Metal-catalyzed $C(sp^2)$–N bond formation. In: Taillefer M, Ma D, editors. Amination and Formation of sp^2 C-N Bonds: Topics in Organometallic Chemistry. Vol. 46. Berlin, Heidelberg: Springer; 2012

[7] Shieh W-C, Lozanov M, Repic O. Accelerated benzylation reaction utilizing dibenzyl carbonate as an alkylating reagent. Tetrahedron Letters. 2003;**44**:6943-6945. DOI: 10.1016/S0040-4039(03)01711-8

[8] Altman RA, Koval ED, Buchwald SL. Copper-catalyzed N-arylation of imidazoles and benzimidazoles. Journal of Organic Chemistry. 2007;**72**:6190-6199. DOI: 10.1021/jo070807a

[9] Shen G, Lv X, Qian W, Bao W. Cu_2O-catalyzed Ullmann-type reaction of vinyl bromides with imidazole and benzimidazole. Tetrahedron Letters. 2008;**49**:4556-4559. DOI: 10.1016/j.tetlet.2008.04.163

[10] Li Y, Jin J, Qian W, Bao W. An efficient and convenient $Cu(OAc)_2$/air mediated oxidative coupling of azoles via -H activation. Organic and Biomolecular Chemistry. 2010;**8**:326-330. DOI: 10.1039/b919396k

[11] Do H-Q, Daugulis O. Copper-catalyzed arylation of heterocycle C–H bonds. Journal of the American Chemical Society. 2007;**129**:12404-12405. DOI: 10.1021/ja075802+

[12] Bellina F, Calandri C, Cauteruccio S, Rossi R. Efficient and highly regioselective direct C-2 arylation of azoles, including free (NH)-imidazole, -benzimidazole and -indole, with aryl halides. Tetrahedron. 2007;**63**:1970. DOI: 10.1016/j.tet.2006.12.068

[13] Lyons TW, Sanford MS. Palladium-catalyzed ligand-directed C–H functionalization reactions. Chemical Reviews. 2010;**110**:1147-1169. DOI: 10.1021/cr900184e

[14] Biffis A, Centomo P, Del Zotto A, Zecca M. Pd metal catalysts for cross-couplings and related reactions in the 21st century: A critical review. Chemical Reviews. 2018;**118**:2249-2295. DOI: 10.1021/acs.chemrev.7b00443

[15] Gu Z-S, Chen W-X, Shao L-X. N-heterocyclic carbene-palladium(II)-1-methylimidazole complex-catalyzed direct C–H bond arylation of (Benz)imidazoles with aryl chlorides. Journal of Organic Chemistry. 2014;**79**:5806-5811. DOI: 10.1021/jo5010058

[16] Muto K, Hatakeyama T, Yamaguchi J, Itami K. C–H arylation and alkenylation of imidazoles by nickel catalysis: Solvent-accelerated imidazole C–H activation. Chemical Science. 2015;**6**:6792-6798. DOI: 10.1039/C5SC02942B

[17] Wang T-H, Ambre R, Wang Q, Lee W-C, Wang P-C, Liu Y, et al. Nickel-catalyzed heteroarenes cross coupling via tandem C–H/C–O activation. ACS Catalysis. 2018;**8**:11368-11376. DOI: 10.1021/acscatal.8b03436

[18] Nakao Y, Kanyiva KS, Oda S, Hiyama T. Hydroheteroarylation of alkynes under mild nickel catalysis. Journal of the American Chemical Society. 2006;**128**:8146-8147. DOI: 10.1021/ja0623459

[19] Ding Z, Yoshikai N. Mild and efficient C2-alkenylation of indoles with alkynes catalyzed by a cobalt complex. Angewandte Chemie International Edition. 2012;**51**:4698-4701. DOI: 10.1002/anie.201200019

[20] Tan KL, Bergman RG, Ellman JA. Intermolecular coupling of isomerizable alkenes to heterocycles via rhodium-catalyzed C–H bond activation. Journal of the American Chemical Society. 2002;**124**: 13964-13965. DOI: 10.1021/ja0281129

[21] Shih W-C, Chen W-C, Lai Y-C, Yu M-S, Ho J-J, Yap GPA, et al. The regioselective switch for amino-nhc mediated C–H activation of benzimidazole via Ni–Al synergistic catalysis. Organic Letters. 2012;**14**: 2046-2049. DOI: 10.1021/ol300570f

[22] Tran G, Confair D, Hesp KD, Mascitti C, Ellman JA. C2-selective branched alkylation of benzimidazoles by rhodium(I)-catalyzed C–H activation. Journal of Organic Chemistry. 2017;**82**:9243-9252. DOI: 10.1021/acs.joc.7b01723

[23] Xu H-S, Ding S-Y, An W-K, Wu H, Wang W. Constructing crystalline covalent organic frameworks from chiral building blocks. Journal of the American Chemical Society. 2016;**138**:11489-11492. DOI: 10.1021/jacs.6b07516

[24] Auberson YP, Troxler T, Zhang X, Yang CR, Feuerbach D, Liu Y-C, et al. From ergolines to indoles: Improved inhibitors of the human H3 receptor for the treatment of narcolepsy. ChemMedChem. 2015;**10**:266-275

[25] De la Fuente T, Martin-Fontecha M, Sallander J, Benhamu B, Campillo M, Medina RA, et al. Benzimidazole derivatives as new serotonin 5-HT6 receptor antagonists. Molecular mechanisms of receptor inactivation. Journal of Medicinal Chemistry. 2010;**53**:1357-1369

[26] Das B, Venkateswarlu K, Krishnaiah M, Holla H. An efficient, rapid and regioselective nuclear bromination of aromatics and heteroaromatics with NBS using sulfonic-acid-functionalized silica as a heterogeneous recyclable catalyst. Tetrahedron Letters. 2006;**47**:8693-8697. DOI: 10.1016/j.tetlet.2006.10.029

[27] Mistry AG, Smith K, Bye MR. A superior synthetic method for the bromination of indoles and benzimidazoles. Tetrahedron Letters. 1986;**27**:1051-1054

[28] Cui M, Ono M, Kimura H, Kawashima H, Liu BL, Saji H. Radioiodinated benzimidazole derivatives as single photon emission computed tomography probes for imaging of β-amyloid plaques in Alzheimer's disease. Nuclear Medicine and Biology. 2011;**38**:313-320

[29] Aliagas-Martin I, Crawford J, Lee W, Mathieu S, Rudolph J. Preparation of Benzoimidazolylmethyl pyrimidinediamine Derivatives and ANALOGS for Use as Serine/Threonine PAK1 Inhibitors. Patent No. WO 2013026914

[30] Kaur Bagal S, Brown AD, Kemp MI, Klute W, Marron BE, Miller DC, Skerratt SE, Suto MJ, West CW, Sanz LM. Benzimidazole and Imidazopyridine Derivatives as Sodium Channel Modulators. Patent No. WO2013114250A1

[31] Holmes I, Naylor A, Alber D, Powell JR, Major MR, Negoita-Giras M, Allen DR. Alkynyl Phosphine Gold Complexes for Treating Bacterial Infections. Patent No. WO2017093544A1

[32] Mortensen DS, Mederos MMD, Sapienza JJ, Albers RJ, Lee BG, Harris RL, Shevlin GI, Huang D, Schwarz KL, Packard GK, Parnes JS, Papa PW, Tehrani LR, Perrin-Ninkovic S, Riggs JR. Heteroaryl Compounds, Compositions Thereof, and their Use as Protein Kinase Inhibitors. Patent No. EP2090577B1

Chapter 3

X-Ray Crystal Structure Analysis of Selected Benzimidazole Derivatives

Aravazhi Amalan Thiruvalluvar, Gopalsamy Vasuki, Jayaraman Jayabharathi and Sivaraman Rosepriya

Abstract

This chapter describes the X-ray crystal structure analysis of selected benzimidazole derivatives, viz. BIP: 2-(1H-benzimidazol-2-yl)phenol, MBMPBI: 1-(4-methylbenzyl)-2-(4-methylphenyl)-1H-benzimidazole, DPBI: 1,2-diphenyl-1H-benzimidazole, PBIP: 2-(1-phenyl-1H-benzimidazol-2-yl)phenol, FPPBI: 2-(4-fluorophenyl)-1-phenyl-1H-benzimidazole and NPBIBHS: 2-(naphthalen-1-yl)-1-phenyl-1H-benzimidazole benzene hemisolvate. The BIP molecule is planar, and in the crystal, it is arranged in parallel planes, stabilised by π-π interactions and the hydrogen bonds. In MBMPBI, benzimidazole cores of the two independent (A and B) molecules are planar. Two C—H...N hydrogen bonds link B molecules only, forming centrosymmetric dimers with $R^2_2(8)$ ring motifs. In the DPBI molecule, the benzimidazole core is planar: one hydrogen-bond interaction (C—H...N) and C—H...π (three) interaction leading to the three-dimensional arrangement. In the PBIP molecule, the benzimidazole is nearly planar. The hydrogen bonds and a π-π stacking interaction are present in the crystal. In the FPPBI molecule, the benzimidazole unit is almost planar. The C—H...F hydrogen bonds and weak C—H...π interactions lead to a three-dimensional architecture in the crystal. In NPBIBHS, the naphthalene fragment lies out of the plane about the benzimidazole core unit. The C—H...N hydrogen bonds and C—H...π interactions lead to a three-dimensional architecture in the crystal.

Keywords: X-ray, single crystal, synthesis, structural analysis, inter and intramolecular hydrogen bonds, C—H...π and π...π interactions

1. Introduction

The X-ray diffraction technique is the most powerful technique of determining the relative atomic positions in a molecular structure. Furthermore, it is distinctively capable of providing precise evidence concerning bond lengths, bond angles, torsion angles and molecular dimensions. It is a well-known fact that hydrogen bonding is one of the crucial factors that contribute to the stability of a structure. Thus, it forms a part of the molecular conformation in that the symmetry and the subsequent packing of the molecules should yield the formation of as many hydrogen bonds as possible. This present chapter depicts the work carried out by the authors, on the crystal structure determination of selected biologically important new benzimidazole derivatives.

Literature survey shows that the benzimidazole is an aromatic ring system where an imidazole ring is fused to the 4 and 5 positions with a benzene ring. Benzimidazole derivatives in OLEDs are of current interest because of their thermal stability [1]. Benzimidazole derivatives are a part of vitamin B_{12} [2] and commercialised as anthelmintic and antihistaminic agents [3].

2. Synthetic approaches of benzimidazole compounds

Due to their possible biological and pharmacological activities, benzimidazoles synthesis has become a vital target in recent years [4]. Since our group is researching organic light emitting devices (OLEDs), we are concerned in using the MBMPBI [5] and DPBI [6] compounds as a ligand in the preparation of Ir(III) complexes and exploring further their electroluminescence (EL) properties. Furthermore, we are interested in using the PBIP [7], FPPBI [8] and NPBIBHS [9] compound as a ligand to study excited state intramolecular proton transfer (ESIPT) processes.

2.1 Synthesis of 2-(1H-benzimidazol-2-yl)phenol ($C_{13}H_{10}N_2O$): BIP

To 15 mmol of o-phenylenediamine in minimum 10 ml ethanol, a mixture of 15 mmol of o-hydroxybenzaldehyde and 60 mmol of ammonium acetate was added and refluxed at 90°C for 2 days. The reaction mixture was cooled and extracted with dichloromethane. The TLC monitored the completion of the reaction. The separated solid was purified by column chromatography (benzene: ethyl acetate (9:1)), after solvent evaporation, and the yield was 60% (**Figure 1**). Furthermore, a suitable single crystal is subjected to collect the X-ray diffraction data [4].

2.2 Synthesis of 1-(4-methylbenzyl)-2-(4-methylphenyl)-1H-benzimidazole ($C_{22}H_{20}N_2$): MBMPBI

To 15 mmol of o-phenylenediamine in minimum 10 ml ethanol, a mixture of 15 mmol of p-methylbenzaldehyde and 60 mmol of ammonium acetate was added and refluxed at 90°C (48 h). Purification of MBMPBI was made by following the procedure as that of BIP (column chromatography: benzene: ethyl acetate (9:1)), and the yield was 40% (**Figure 1**). Furthermore, a suitable single crystal is subjected to collect the X-ray diffraction data [5].

2.3 Synthesis of 1,2-diphenyl-1H-benzimidazole ($C_{19}H_{14}N_2$): DPBI

To 17 mmol of N-phenyl-o-phenylenediamine in minimum 10 ml ethanol, a mixture of 17 mmol of benzaldehyde and 60 mmol of ammonium acetate was added and refluxed at 90°C (4 h). Purification of DPBI was made by following the procedure as that of BIP (column chromatography: benzene: ethyl acetate (9:1)), and the yield was 50% (**Figure 1**). Furthermore, a suitable single crystal is subjected to collect the X-ray diffraction data [6].

2.4 Synthesis of 2-(1-phenyl-1H-benzimidazol-2-yl)phenol ($C_{19}H_{14}N_2O$): PBIP

To 17 mmol of N-phenyl-o-phenylenediamine in minimum 10 ml ethanol, a mixture of 17 mmol of o-hydroxybenzaldehyde and 60 mmol of ammonium acetate was added and refluxed at 90°C (4 h). Purification of PBIP was made by following

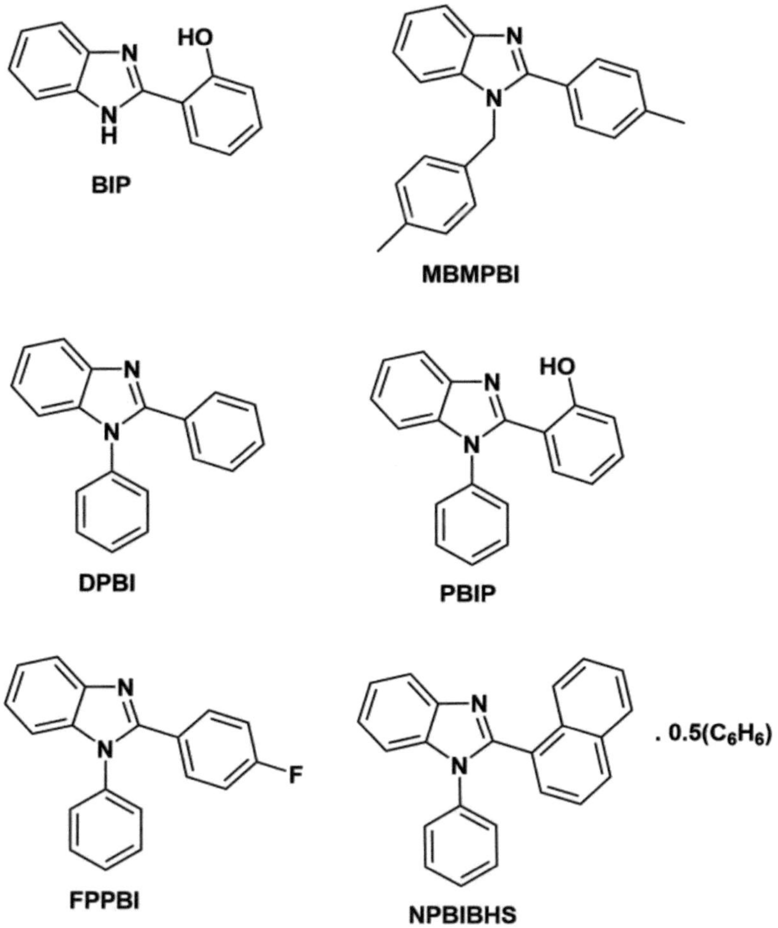

Figure 1.
Chemical structures of the studied compounds: BIP, MBMPBI, DPBI, PBIP, FPPBI and NPBIBHS.

the procedure as that of BIP (column chromatography-petroleum ether (60–80°C)), and the yield was 50% (**Figure 1**). Furthermore, a suitable single crystal is subjected to collect the X-ray diffraction data [7].

2.5 Synthesis of 2-(4-fluorophenyl)-1-phenyl-1H-benzimidazole ($C_{19}H_{13}FN_2$): FPPBI

To 17 mmol of N-phenyl-o-phenylenediamine in minimum 10 ml ethanol, a mixture of p-fluorobenzaldehyde (17 mmol) and 60 mmol of ammonium acetate was added and refluxed at 90°C (4 h). Purification of FPPBI was made by following the procedure as that of BIP (column chromatography-petroleum ether: ethyl acetate (9:1)), the yield was 50% (**Figure 1**). Furthermore, a suitable single crystal is subjected to collect the X-ray diffraction data [8].

2.6 Synthesis of 2-(naphthalen-1-yl)-1-phenyl-1H-benzimidazole benzene hemisolvate ($C_{23}H_{16}N_2$. $0.5C_6H_6$): NPBIBHS

To 17 mmol of N-phenyl-o-phenylenediamine in minimum 10 ml ethanol, a mixture of 17 mmol of 1-naphthaldehyde and 60 mmol of ammonium acetate was added and refluxed at 90°C (48 h). Purification of NPBIBHS was made by following the procedure as that of BIP (column chromatography-benzene as the eluent), and the yield was 50% (**Figure 1**). Furthermore, a suitable single crystal is subjected to collect the X-ray diffraction data [9].

3. Structural analysis of six benzimidazole compounds

3.1 Structural analysis of 2-(1H-benzimidazol-2-yl)phenol (BIP)

This section describes the determination of the crystal structure and molecular structure of BIP [4]. The direct method program SIR2011 [10] is used in solving the crystal structure. The SHELXL2013/4 [11] program was used to refine the structure.

This compound crystallises in the monoclinic system in the space group $P2_1/c$. Molecular formula: $C_{13}H_{10}N_2O$; molecular weight: 210.23; Z = 4; crystal data: a = 16.864(4) Å; b = 4.7431(8) Å; c = 12.952(2) Å; β = 102.34(2)°; V = 1012.1(3) $Å^3$; D_{cal} = 1.380 Mg m^{-3}; F_{000} = 440; final $R[F^2 > 2\sigma(F^2)] = 0.067$ and $wR(F^2) = 0.131$ for 1184 reflections observed with $I > 2\sigma(I)$.

From a difference Fourier map, H_1 attached to N_1 was located and freely refined with (N_1—H_1 = 0.91(2) Å). The outstanding H atoms were placed geometrically and permitted to ride on their parental atoms, with O—H = 0.82 and C—H = 0.93 Å for Csp^2 hydrogens; $U_{iso}(H) = kU_{eq}(C)$, where k = 1.5 for methyl and 1.2 for all other C-bonded H atoms.

This molecule is planar [maximum deviation = 0.016(2) Å]. The dihedral angle between the five-membered imidazole ring and the attached six-membered benzene ring is 0.37(13)°. An S(6) ring motif [12] is generated by the O—H...N hydrogen bond. The hydrogen bond involves the hydroxyl substituent (O_{26}) as the proton donor and the nitrogen (N_3) atom as the acceptor, which forms a six-membered ring. The N—H...O hydrogen bonds link the molecules, by making chains spreading in [001]. Four π-π assembling contacts concerning the five-membered ring, fused six-membered benzene ring and attached benzene ring system [The C_g-C_g distances increase from 3.6106(17) to 3.6668(17) Å].

The thermal displacement ellipsoid plot (**Figure 2**) at the 50% probability level was drawn using the program ORTEP-3 for Windows [13]. **Figure 3** presents the π-π interactions detected in the crystal structure, brought using the program PLATON [14]. The crystal structure packing view is shown in **Figure 4** [14].

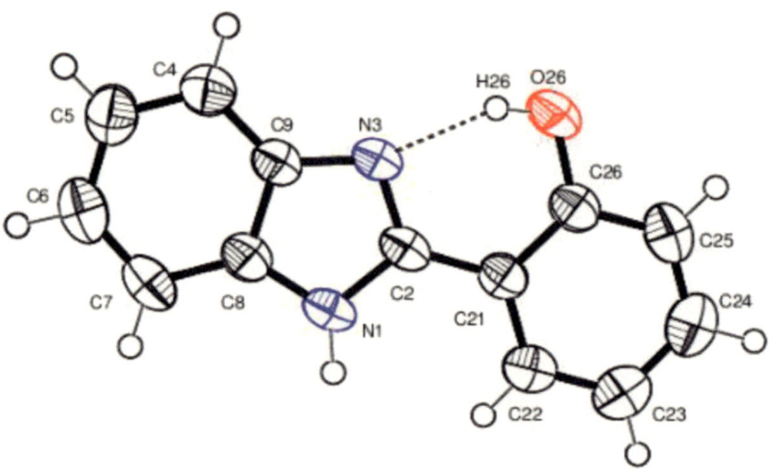

Figure 2.
The thermal displacement ellipsoid plot (at the 50% probability level).

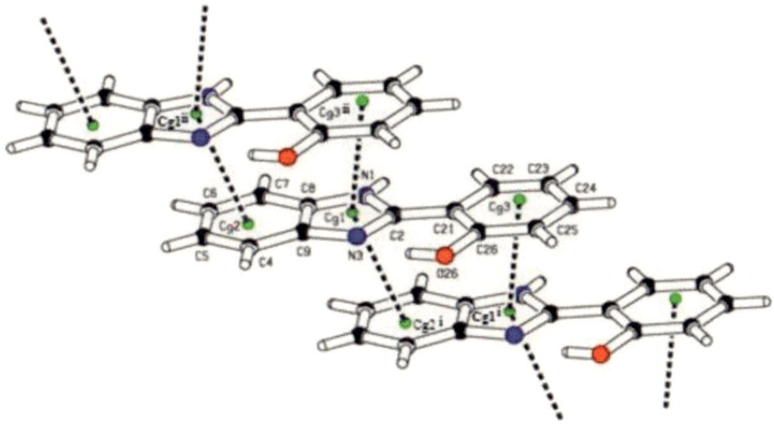

Figure 3.
The crystal structure, partially showing the formation of π-π interactions. Symmetry codes (i): x, −1 + y, z and (ii): x, 1 + y, z.

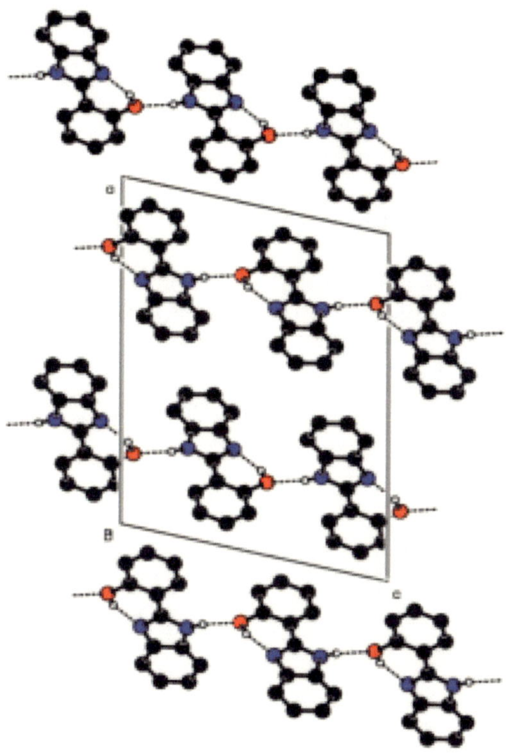

Figure 4.
The partial crystal packing with hydrogen bonds [14], viewed along the b axis.

3.2 Structural analysis of 1-(4-methylbenzyl)-2-(4-methylphenyl)-1H-benzimidazole (MBMPBI)

This section describes the determination of the crystal and molecular structure of MBMPBI [5]. The direct method program SIR2002 [15] is used in solving the crystal structure. The SHELXL97 [11] program was used to refine the structure.

This compound crystallises in the triclinic system in the space group $P\bar{1}$. Molecular formula: $C_{22}H_{20}N_2$; molecular weight: 312.40; Z = 4; crystal data: a = 9.6610(2) Å;

b = 10.2900(2) Å; c = 17.7271(3) Å; α = 84.437(2)°; β = 81.536(2)°; γ = 76.165(2)°; V = 1689.02(6) Å^3; D_{cal} = 1.229 Mg m^{-3}; F_{000} = 664; final $R[F^2 > 2\sigma(F^2)]$ = 0.039 and $wR(F^2)$ = 0.104 for 6452 observed reflections with $I > 2\sigma(I)$.

All the H atoms were placed geometrically and allowed to trip on their parental atoms, with C—H = 0.93 (Csp^2), 0.96 (methyl) and 0.97 Å (methylene) hydrogen atoms. $U_{iso}(H) = kU_{eq}(C)$, with k = 1.5 (—CH_3 H atoms) and 1.2 (for carbon-attached H atoms). The —CH_3 groups are disordered over two positions. So, they are refined as idealised disordered methyl groups with identical occupancy of the two locations.

Two crystallographically independent molecules A (first) and B (second) of this compound make the asymmetric unit. The planar [maximum deviations = 0.0161(8) Å for A (first) and 0.0276(8) Å for B (second)] benzimidazole least-squares plane and the benzene least-squares planes of the 4-methylbenzyl and 4-methylphenyl groups make dihedral angles of 76.64(3) and 46.87(4)° in A (first). The similar values in B (second) are 86.31(2) and 39.14(4)°. The two benzene rings make the dihedral angle of 73.73(3)° in A (first) and 80.69(4)° in B (second). The variation in the dihedral angles may be due to the H—H repulsions. The centrosymmetric dimers with $R^2_2(8)$ ring motifs [12] are formed by the two C_4B—H_4B...N_3B hydrogen bonds in B (second). The pattern contains a total of eight atoms in which two of them are donors, and two are acceptors, hence designated as $R^2_2(8)$. There are no corresponding interactions involving the A molecules.

The thermal displacement ellipsoid plot (for molecule A (first) only) (**Figure 5**) at the 30% probability level was drawn using the program ORTEP-3 for Windows [13]. The crystal structure packing view is shown in **Figure 6** [14].

3.3 Structural analysis of 1,2-diphenyl-1H-benzimidazole (DPBI)

This section describes the determination of the crystal structure and molecular structure of DPBI [6]. The direct method program SHELXS97 [11] is used in solving the crystal structure. The SHELXL97 [11] program was used to refine the structure.

This compound crystallises in the monoclinic system in the space group C2/c. Molecular formula: $C_{19}H_{14}N_2$; molecular weight: 270.32; Z = 8; crystal

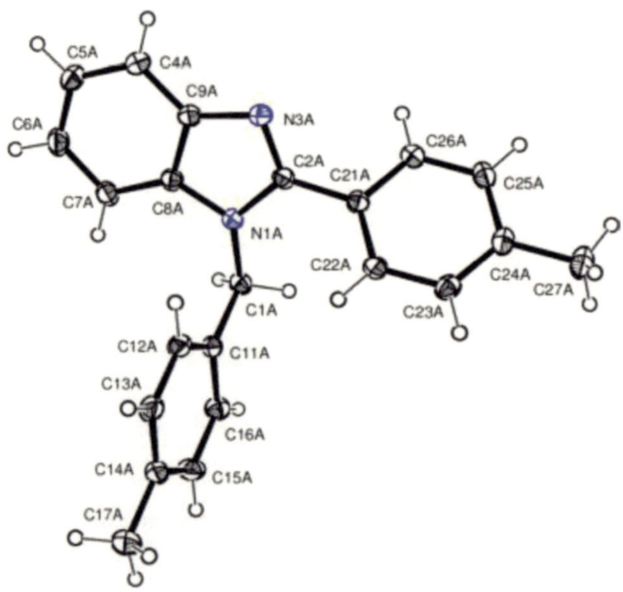

Figure 5.
The thermal displacement ellipsoid plot (at the 30% probability level).

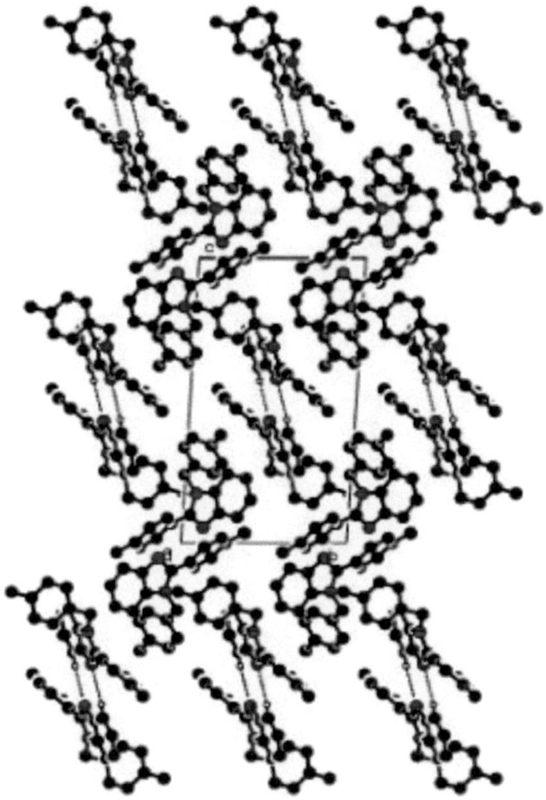

Figure 6.
The crystal packing with hydrogen bonds [14], viewed along the a axis.

data: a = 10.1878(3) Å; b = 16.6399(4) Å; c = 17.4959(5) Å; β = 106.205(3)°; V = 2848.13(14) $Å^3$; D_{cal} = 1.261 Mg m^{-3}; F_{000} = 1136; final $R[F^2 > 2\sigma(F^2)]$ = 0.052 and $wR(F^2)$ = 0.137 for 5803 reflections observed [I > 2σ(I)].

All the H atoms were placed geometrically and permitted to trip on their parental atoms, with C—H = 0.93 Å (Csp^2) and $U_{iso}(H) = kU_{eq}(C)$, where k = 1.5 for —CH_3 H atoms and 1.2 for all other H atoms.

The benzimidazole unit is planar [maximum deviation = 0.0102(6) Å]. The least-squares planes of the phenyl rings at N_1 and C_2 make angles of 55.80(2) and 40.67(3)° with the least-squares plane of the benzimidazole part. The least-squares planes of the phenyl rings at N_1 and C_2 make a dihedral angle of 62.37(3)°. One C—H...N hydrogen bond and three C—H...π interactions concerning the fused benzene ring and the five-membered imidazole rings are observed, forming a three-dimensional architecture in the crystal.

The thermal displacement ellipsoid plot (**Figure 7**) at the 50% probability level was drawn using the program ORTEP-3 for Windows [13]. **Figure 8** presents the C—H...π interactions in the crystal structure brought using the program PLATON [14]. The crystal structure packing view is shown in **Figure 9** [14].

3.4 Structural analysis of 2-(1-phenyl-1H-benzimidazol-2-yl)phenol (PBIP)

This section describes the determination of the crystal structure and molecular structure of PBIP [7]. The direct method program SHELXS86 [11] is used in solving the crystal structure. The SHELXL97 [11] program was used to refine the structure.

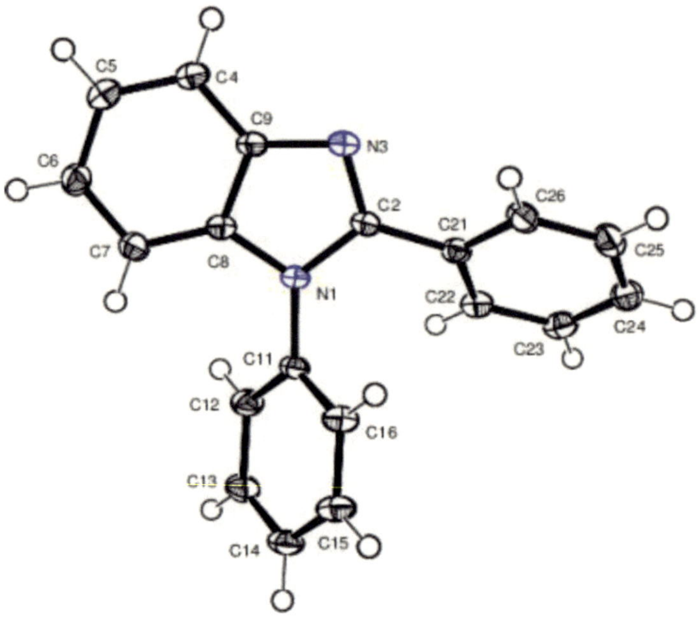

Figure 7.
The thermal displacement ellipsoid plot (at the 50% probability level).

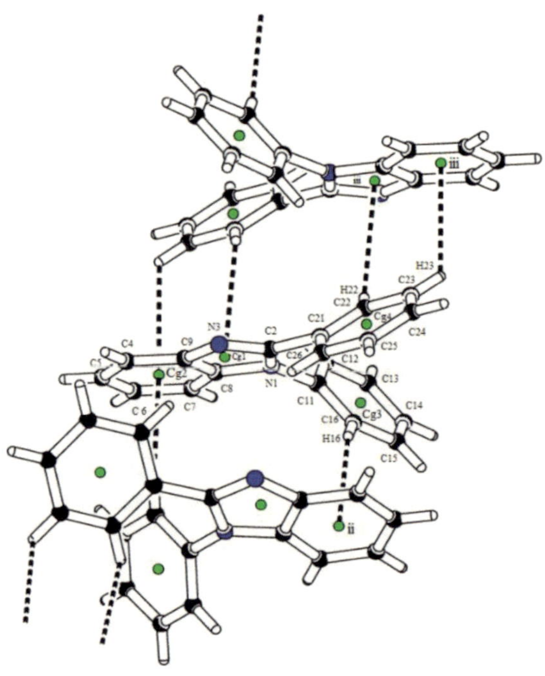

Figure 8.
The crystal structure, partially showing the formation of C—H...π interactions. Symmetry codes are (ii): −x, y, −z + 1/2 and (iii): −x, −y + 1, −z.

This compound crystallises in the triclinic system in the space group $P\bar{1}$ Molecular formula: $C_{19}H_{14}N_2O$; molecular weight: 286.32; Z = 2; crystal data: a = 8.1941(6) Å; b = 9.5983(14) Å; c = 10.3193(18) Å; α = 64.637(16)°; β = 80.356(10)°; γ = 83.610(9)°; V = 722.3(2) $Å^3$; D_{cal} = 1.316 Mg m^{-3}; F_{000} = 300; final $R[F^2 > 2\sigma(F^2)]$ = 0.059 and $wR(F^2)$ = 0.171 for 2420 observed reflections with $I > 2\sigma(I)$.

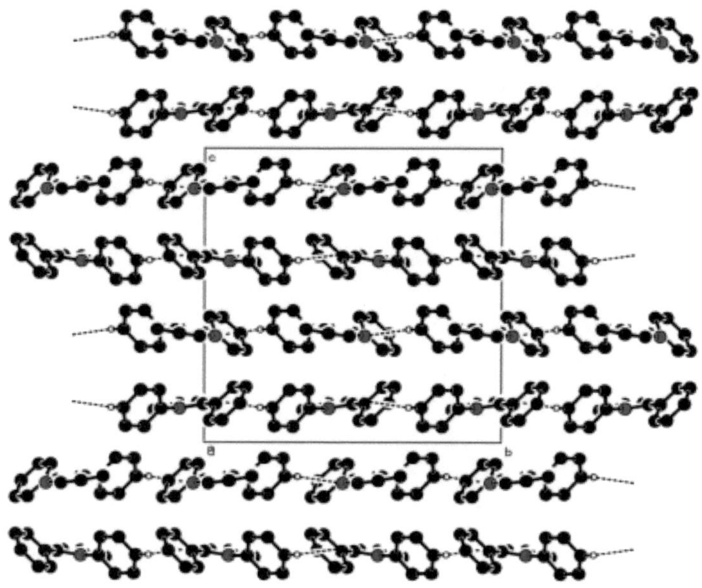

Figure 9.
The crystal packing with hydrogen bonds [14], viewed along the a axis.

A difference Fourier map was used to locate the H atom attached to O atom and refined freely with O_{26}—H_{26} = 0.97(3) Å. The outstanding H atoms were placed geometrically and permitted to trip on their parental atoms, with C—H = 0.95 Å, and with $U_{iso}(H) = 1.2U_{eq}$(parental atom).

The phenyl mean plane at N_1 and the benzene mean plane at C_2 makes angles of 68.98(6) and 20.38(7)°, respectively, with benzimidazole planar unit [maximum deviation = 0.0253(11) Å]. The phenyl and the adjacent benzene mean planes makes an angle of 64.30(7)°. An intramolecular S(6) ring motif [12] is generated by O—H...N hydrogen bond. The hydrogen bond involving has the hydroxyl substituent (O_{26}) as the proton donor and the nitrogen (N_3) atom as the acceptor, which forms a six-membered ring (N_3, C_2, C_{21}, C_{26}, O_{26} and H_{26}). The C—H...N and C—H...O hydrogen bonds links the molecules. There is a π-π assembling contact, with a centroid-centroid distance of 3.8428(12) Å.

The ORTEP-3 for Windows [13] was used to draw the thermal displacement ellipsoid plot (**Figure 10**) at the 50% probability level. **Figure 11** presents the π-π interactions observed in the crystal structure brought using the program PLATON [14]. The crystal structure packing view is shown in **Figure 12** [14].

The dashed lines indicate the intramolecular O—H...N hydrogen bond.

3.5 Structural analysis of 2-(4-fluorophenyl)-1-phenyl-1H-benzimidazole (FPPBI)

This section describes the determination of the crystal structure and molecular structure of FPPBI [8]. The crystal structure of FPPBI was solved by direct methods, using the program SIR2004 [16]. The crystal structure is refined by the program SHELXL97 [11].

This compound crystallises in the Monoclinic system in the space group $P2_1/n$. Molecular formula: $C_{19}H_{13}FN_2$; Molecular weight: 288.31; Z = 4; crystal data: a = 8.7527(4) Å; b = 10.1342(4) Å; c = 17.0211(6) Å; β = 104.187(4)°; V = 1463.75(11) $Å^3$; D_{cal} = 1.308 Mg m^{-3}; F_{000} = 600; final $R[F^2 > 2\sigma(F^2)] = 0.063$ and $wR(F^2) = 0.160$ for 5352 observed reflections with $(I > 2\sigma(I))$.

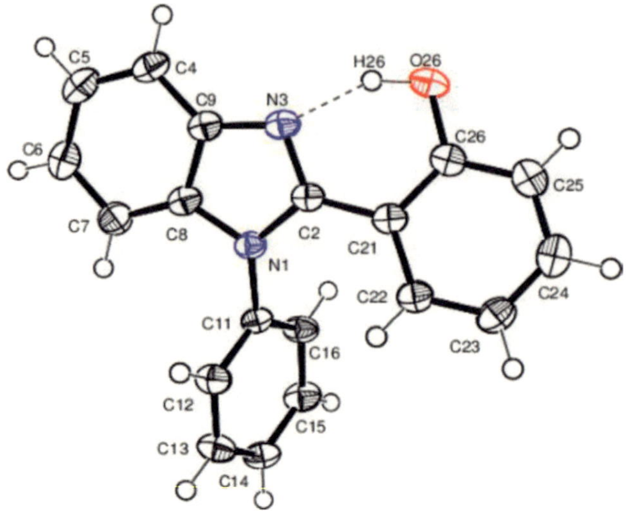

Figure 10.
The thermal displacement ellipsoid plot (at the 50% probability level).

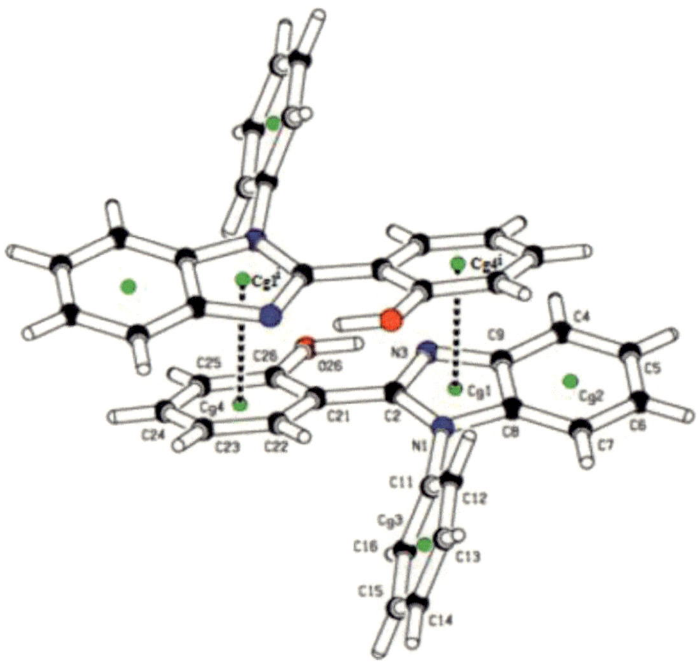

Figure 11.
The crystal structure, partially showing the formation of π-π stacking interactions. Symmetry code (i): 2 − x, −y, −z.

All the H atoms were placed geometrically and allowed to trip on their parental atoms, with C—H = 0.93 Å (Csp^2). $U_{iso}(H) = kU_{eq}(C)$, where k = 1.5 (CH_3H) and 1.2 (for all other carbon-attached H atoms).

The benzimidazole group is nearly planar [maximum deviation = 0.0342(9) Å]. The mean planes of the phenyl at N_1 and fluorobenzene at C_2 make dihedral angles of 58.94(3) and 51.43(3)°, respectively, with the benzimidazole least-squares plane. The phenyl and fluorobenzene mean planes make an angle of 60.17(6)°. Finally, three C—H...F hydrogen bonds and two weak C—H...π contacts connecting the fused benzene ring lead to a three-dimensional construction.

The ORTEP-3 for Windows [13] was used to draw the thermal displacement ellipsoid plot (**Figure 13**) at the 50% probability level. **Figure 14** presents the C—H...π interactions observed in the crystal structure, brought using the program PLATON [14]. The crystal structure packing view is shown in **Figure 15** [14].

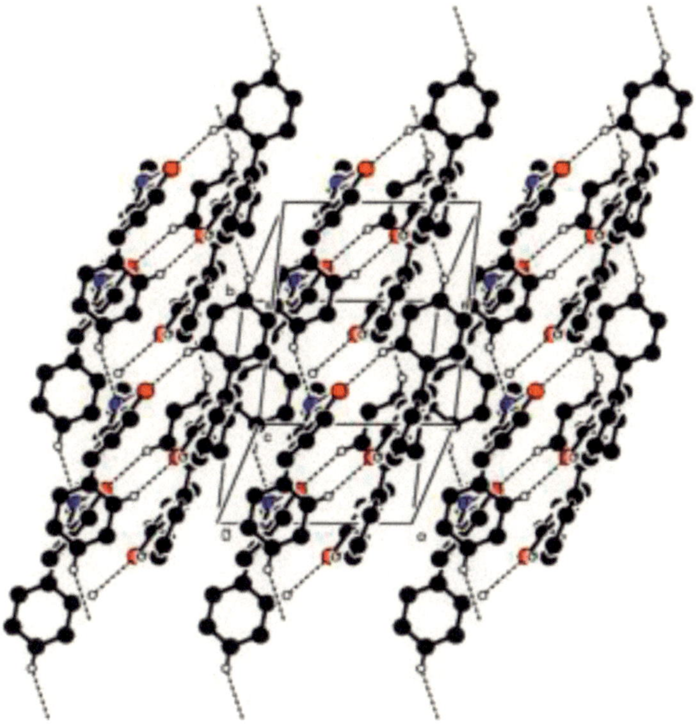

Figure 12.
The crystal packing with hydrogen bonds [14], viewed along the c axis.

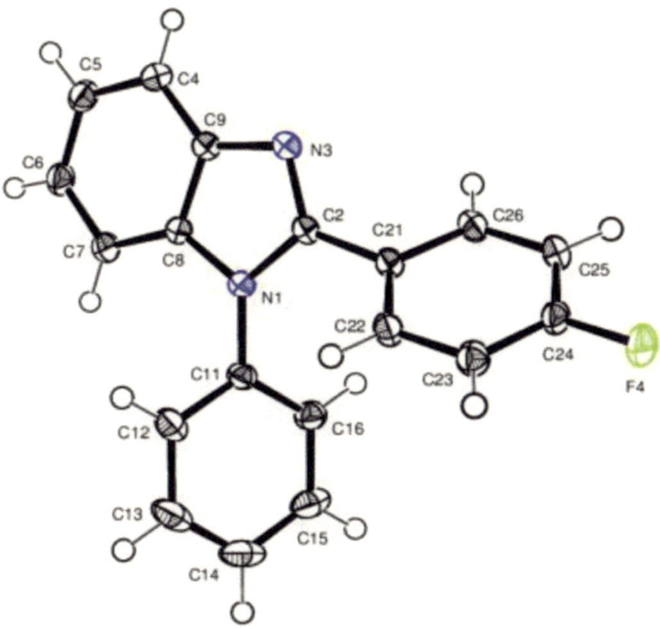

Figure 13.
The thermal displacement ellipsoid plot (at the 50% probability level).

3.6 Structural analysis of 2-(naphthalen-1-yl)-1-phenyl-1H-benzimidazole benzene hemisolvate (NPBIBHS)

This section describes the determination of the crystal structure and molecular structure of NPBIBHS [9]. The direct method program SHELXS2013 [11] was used to solve the crystal structure. SHELXL2013 [11] program is used to refine the crystal structure. This compound crystallises in the triclinic system in the space group P$\bar{1}$ Molecular formula: $C_{23}H_{16}N_2.0.5C_6H_6$; molecular weight: 359.43; Z = 2; crystal data: a = 8.5529(3) Å; b = 9.4517(3) Å; c = 11.8936(3) Å; α = 86.334(2)°;

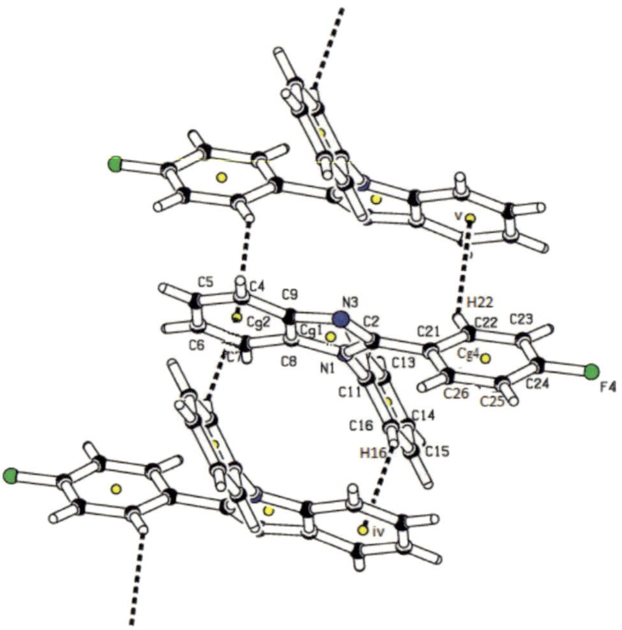

Figure 14.
The crystal structure, partially showing the formation of C—H...π interactions. Symmetry codes are (iv): −x, −y + 1, −z and (v): −x + 1, −y + 1, −z.

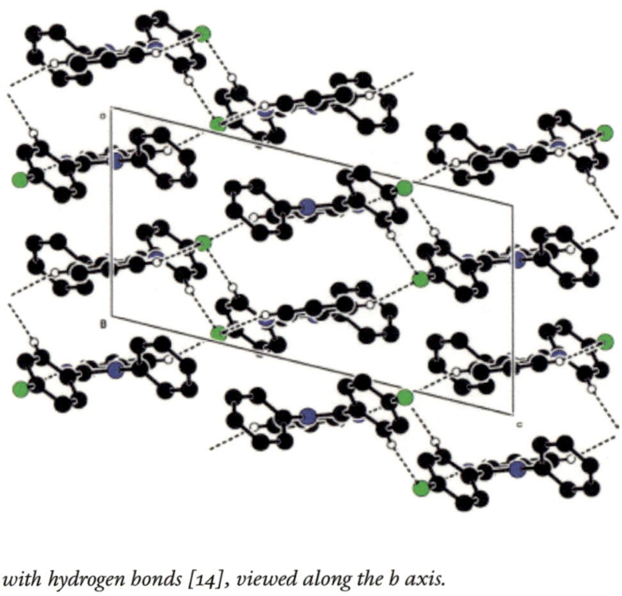

Figure 15.
The crystal packing with hydrogen bonds [14], viewed along the b axis.

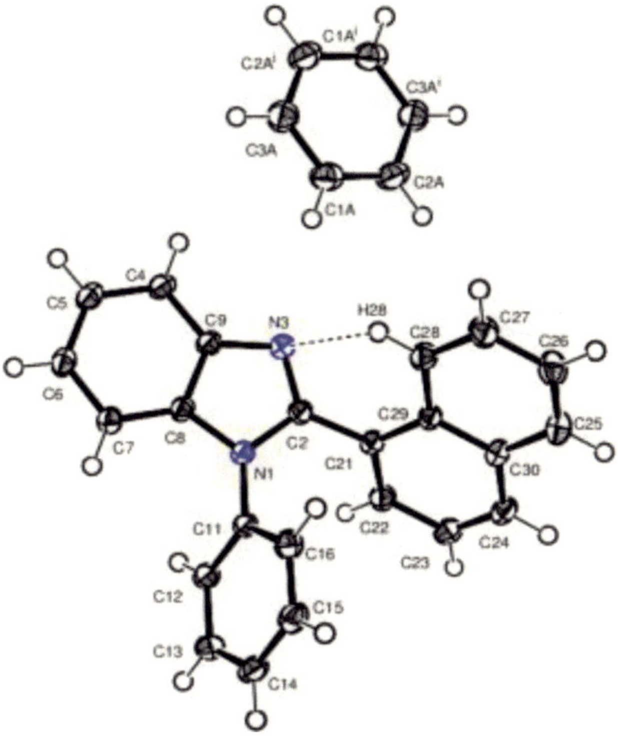

Figure 16.
The thermal displacement ellipsoid plot (at the 50% probability level).

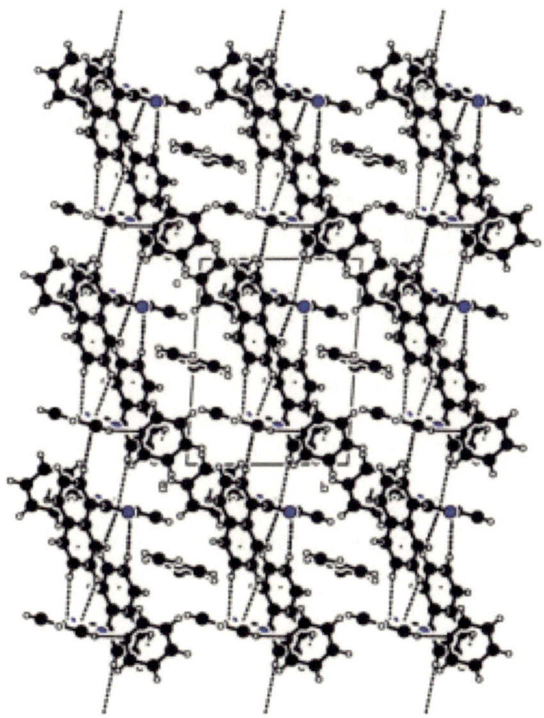

Figure 17.
The crystal structure, showing the formation of complex C—H…π interactions.

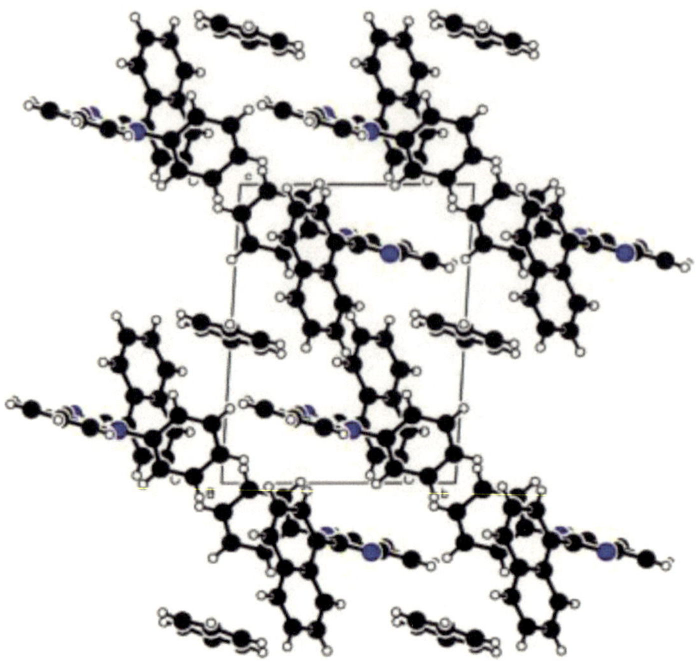

Figure 18.
The crystal packing with hydrogen bonds [14], viewed along the a axis.

β = 89.838(2)°; γ = 75.051(3)°; V = 926.94(5) Å^3; D_{cal} = 1.288 Mg m^{-3}; F_{000} = 378; final $R[F^2 > 2\sigma(F^2)]$ = 0.057 and $wR(F^2)$ = 0.160 for 9086 observed reflections with $I > 2\sigma(I)$.

All the H atoms were placed geometrically and permitted to trip on their parental atoms, with C—H = 0.95 Å (Csp^2) and $U_{iso}(H)$ = $kU_{eq}(C)$, where k = 1.5 (—CH_3 H) and 1.2 (for all other H).

The benzimidazole least-squares plane [maximum deviation = 0.0258(6) Å] and the naphthalene least-squares plane [maximum deviation = 0.0254(6) Å] make dihedral angle of 61.955(17)°. The least-squares planes of the imidazole ring and the phenyl ring make a dihedral angle of 61.73(4)°. An intramolecular S(6) ring motif [12] is generated by the C—H...N hydrogen bond. The hydrogen bond involving has the carbon atom (C_{28}) as the proton donor and the nitrogen atom (N_3) as the acceptor, which forms a six-membered ring. Seven weak C—H...π links concerning the attached ring system, the benzene solvent molecule, the imidazole and the phenyl rings are detected, to a three-dimensional architecture.

The thermal displacement ellipsoid plot (**Figure 16**) at the 50% probability level was drawn using the program ORTEP-3 for Windows [13]. **Figure 17** presents the C—H...π interactions observed in the crystal structure brought using the program PLATON [14]. The crystal structure packing view is shown in **Figure 18** [14].

4. Comparative study on the structural aspects of the six benzimidazole derivatives

Section 3 presents the X-ray crystal structure analyses of six closely related organic benzimidazole compounds. The MBMPBI compound is similar to 2-(1H-benzimidazol-2-yl)phenol (BIP) except for the presence of methylbenzyl at the first position of the benzimidazole unit, a methyl at the fourth position of the phenyl group and the absence of hydroxyl group. The DPBI compound is similar to 2-(1H-benzimidazol-2-yl)phenol (BIP) except for the presence of a phenyl group at the first position of the benzimidazole unit and the absence of an —OH group. The PBIP compound is similar to that of 2-(1H-benzimidazole-2-yl)phenol (BIP) except for the presence of a phenyl group at the first position of the benzimidazole unit and the absence of the —H atom. The FPPBI compound is like that of 1,2-diphenyl-1H-benzimidazole (DPBI) except for the presence of a fluorine atom at the fourth position of the phenyl group in the second location of the benzimidazole core.

All the six structures have the benzimidazole core essentially as the basic skeleton, with different groups (—H, —C_6H_4—OH, —CH_2—C_6H_4—CH_3, —C_6H_4—CH_3, —C_6H_5, —C_6H_5, —C6H5, —C_6H_4—OH, —C_6H_5, —C_6H_4-F, —C_6H_5, and —$C_{10}H_7$) as substituents. The structural determinations of the compounds have revealed several features, such as (1) the hydrogen bonds: O—H...N, N—H...O, C—H...N, C—H...O, C—H...F; (2) interactions C—H...π and (3) stacking interactions π-π.

A type of hydrogen bond operational among a soft acid CH and a soft base π-system is known as a C—H...π interaction. The most striking contacts are (1) the connections among the aliphatic C—H donors and the aromatic π-acceptors and (2) the connections among the aromatic C—H donors and aromatic π-acceptors. The non-covalent contacts that encompass the π systems in chemistry are the π-effects or the π-interactions.

Related crystal structures: 1293 articles match the search term 'Benzimidazole' on IUCr Journals Crystallography Journals Online (https://journals.iucr.org/) as on February 10, 2019. The search (IUCr Journals' paper reference codes are: bh2413: 2-(4-chlorophenyl)-1-phenyl-1H-benzimidazole, bi2334: 1-benzyl-2-phenyl-1H-benzimidazole, bv2218: 1-phenyl-2-[4-(trifluoromethyl)phenyl]-1H-benzimidazole, bx2457: 1-(4-bromobenzyl)-2-(4-bromophenyl)-1H-benzimidazole, ci2926: 1-benzyl-1H-benzimidazole, fy2081: 1-phenyl-2-p-tolyl-1H-benzimidazole, go2077: 2-(4-methoxyphenyl)-1-phenyl-1H-benzimidazole, hk2704: 2-p-tolyl-1-p-tolylmethyl-1H-benzimidazole, lh5659: 2-(3,4-difluorophenyl)-1H-benzimidazole and lh5706: 2-[4-(trifluoromethyl)phenyl]-1H-benzimidazole)) confirms that the geometry of the benzimidazole cores is similar in all the reported structures.

5. Conclusions

This chapter described the research work carried out by the authors on the crystal structure determination of some selected biologically important new benzimidazole derivatives by using X-rays. The detailed structural analyses on the bond lengths, bond angles, torsion angles and dihedral angles between the least-squares planes of these six benzimidazole derivatives indicate that in the compounds BIP, MBMPBI, DPBI, PBIP, FPPBI and NPBIBHS, the benzimidazole unit is essentially planar as expected and as revealed by the latest literature survey (https://journals.iucr.org/). The present X-ray study confirms that the benzimidazole skeleton has an imidazole planer five-membered heterocyclic ring fused with the benzene ring. The basic geometrical examination (bond lengths and bond angles) of the benzimidazole core in the BIP molecule are in good agreement with those observed in other closely related benzimidazole derivatives. All the substituents are in the expected positions around the benzimidazole units. The X-ray study confirms the molecular structure and atom connectivity of the above-studied compounds as shown in **Figures 2**, **5**, **7**, **10**, **13** and **16**. The O—H...N, N—H...O, C—H...N, C—H...O, C—H...F hydrogen bonds, C—H...π and the π-π interactions are effective in the stabilisation of the crystal structure. We are interested in studying the biological and photophysical properties of BIP, MBMPBI, DPBI, PBIP, FPPBI and NPBIBHS compounds. The benzimidazole derivatives are a sensitive fluorescent sensor for TiO_2 (P_{25}), Fe_2O_3, WO_3, Al_2O_3,

CuO, TiO_2 (H), ZnO, Cu-ZnO, Ag-ZnO, TiO_2 (R) and TiO_2 (A) nanoparticles. The benzimidazole-based iridium(III) complexes show green emission with maximum electroluminescent efficiencies at low voltage.

Acknowledgements

We acknowledge Dr. Anthony Linden and Professor R. J. Butcher for their help in collecting the single-crystal X-ray diffraction data.

Conflict of interest

There is no conflict of interest in writing this chapter.

Thanks

Aravazhi Amalan Thiruvalluvar thanks his wife Lilly for all her emotional support and help at various stages of this chapter writing and his son Uthaya Raj and daughter Manju Princy for their loving support.

List of abbreviations (a partial crystallographic information file (CIF) for the compound BIB only as an example)

_space_group_crystal_system	monoclinic
_space_group_name_H-M_alt	P 2_1/c
_chemical_formula_moiety	C13 H10 N2 O
_chemical_formula_weight	210.23
_cell_formula_units_Z	4
_cell_length_a	16.864(4)
_cell_length_b	4.7431(8)
_cell_length_c	12.952(2)
_cell_angle_alpha	90
_cell_angle_beta	102.34(2)
_cell_angle_gamma	90
_cell_volume	1012.1(3)
_exptl_crystal_density_diffrn	1.380
_exptl_crystal_F_000	440
_reflns_threshold_expression	I > 2(I)
_refine_ls_R_factor_gt	0.067
_refine_ls_wR_factor_gt	0.131
_reflns_number_gt	1184

Author details

Aravazhi Amalan Thiruvalluvar[1*], Gopalsamy Vasuki[1], Jayaraman Jayabharathi[2] and Sivaraman Rosepriya[3]

1 Department of Physics, Kunthavai Naacchiyaar Government Arts College for Women (Autonomous), Thanjavur, Tamil Nadu, India

2 Department of Chemistry, Annamalai University, Chidambaram, Tamil Nadu, India

3 Department of Physics, Rajah Serfoji Government College (Autonomous), Thanjavur, Tamil Nadu, India

*Address all correspondence to: thiruvalluvar.a@gmail.com

References

[1] Cross EM, White KM, Moshrefzadeh RS, Francis CV. Azobenzimidazole compounds and polymers for nonlinear optics. Macromolecules. 1995;**28**(7):2526-2532. DOI: 10.1021/ma00111a055

[2] Brown KL. Chemistry and enzymology of vitamin B_{12}. Chemical Reviews. 2005;**105**(6):2075-2150. DOI: 10.1021/cr030720z

[3] Spasov AA, Yozhitsa IN, Bugaeva LI, Anisimova VA. Benzimidazole derivatives: Spectrum of pharmacological activity and toxicological properties (a review). Pharmaceutical Chemistry Journal. 1999;**33**(5):232-243. DOI: 10.1007/BF02510042

[4] Prakash SM, Thiruvalluvar A, Rosepriya S, Srinivasan N. 2-(1H-Benzimidazol-2-yl)phenol. Acta Crystallographica Section E: Crystallographic Communications. 2014;**E70**:o184. DOI: 10.1107/S1600536814001366

[5] Rosepriya S, Thiruvalluvar A, Jayamoorthy K, Jayabharathi J, Linden A. 1-(4-Methylbenzyl)-2-(4-methylphenyl)-1H-benzimidazole. Acta Crystallographica Section E: Crystallographic Communications. 2011;**E67**:o3519. DOI: 10.1107/S160053681105077X

[6] Rosepriya S, Thiruvalluvar A, Jayamoorthy K, Jayabharathi J, Öztürk Yildirim S, Butcher RJ. 1,2-Diphenyl-1H-benzimidazole. Acta Crystallographica Section E: Crystallographic Communications. 2012;**E68**:o3283. DOI: 10.1107/S1600536812044960

[7] Thiruvalluvar A, Rosepriya S, Jayamoorthy K, Jayabharathi J, Öztürk Yildirim S, Butcher RJ. 2-(1-Phenyl-1H-benzimidazol-2-yl)phenol. Acta Crystallographica Section E: Crystallographic Communications. 2013;**E69**:o62. DOI: 10.1107/S1600536812049859

[8] Jayamoorthy K, Rosepriya S, Thiruvalluvar A, Jayabharathi J, Butcher RJ. 2-(4-Fluorophenyl)-1-phenyl-1H-benzimidazole. Acta Crystallographica Section E: Crystallographic Communications. 2012;**E68**:o2708. DOI: 10.1107/S1600536812035155

[9] Srinivasan N, Thiruvalluvar A, Rosepriya S, Prakash SM, Butcher RJ. 2-(Naphthalen-1-yl)-1-phenyl-1H-benzimidazole benzene hemisolvate. Acta Crystallographica Section E: Crystallographic Communications. 2014;**E70**:o55-o56. DOI: 10.1107/S160053681303331X

[10] Burla MC, Caliandro R, Camalli M, Carrozzini B, Cascarano GL, Giacovazzo C, et al. SIR2011: A new package for crystal structure determination and refinement. Journal of Applied Crystallography. 2012;**45**:357-361. DOI: 10.1107/S0021889812001124

[11] Sheldrick GM. A short history of SHELX. Acta Crystallographica Section A: Foundations and Advances. 2008;**A64**:112-122. DOI: 10.1107/S0108767307043930

[12] Bernstein J, Davis RE, Shimoni L, Chang N-L. Patterns in hydrogen bonding: Functionality and graph set analysis in crystals. Angewandte Chemie International Edition in English. 1995;**34**:1555-1573. DOI: 10.1002/anie.199515551

[13] Farrugia LJ. WinGX and ORTEP for windows: An update. Journal of Applied Crystallography. 2012;**45**:849-854. DOI: 10.1107/S0021889812029111

[14] Spek AL. Structure validation in chemical crystallography. Acta

Crystallographica Section D: Structural Biology. 2009;**D65**:148-155. DOI: 10.1107/S090744490804362X

[15] Burla MC, Camalli M, Carrozzini B, Cascarano GL, Giacovazzo C, Polidori G, et al. SIR2002: The program. Journal of Applied Crystallography. 2003;**36**:1103. DOI: 10.1107/S0021889803012585

[16] Burla MC, Caliandro R, Camalli M, Carrozzini B, Cascarano GL, De Caro L, et al. SIR2004: An improved tool for crystal structure determination and refinement. Journal of Applied Crystallography. 2005;**38**:381-388. DOI: 10.1107/S002188980403225X

Chapter 4

Synthesis and Pharmacological Profile of Benzimidazoles

Kantharaju Kamanna

Abstract

Benzimidazoles are a class of heterocyclic, aromatic compounds which share a fundamental structural characteristic of six-membered benzene fused to five-membered imidazole moiety. Molecules having benzimidazole motifs showed promising application in biological and clinical studies. Nowadays it is a moiety of choice which possesses many pharmacological properties extensively explored with a potent inhibitor of various enzymes involved in a wide range of therapeutic uses which are antidiabetic, anticancer, antimicrobial, antiparasitic, analgesics, antiviral, and antihistamine, as well as used in cardiovascular disease, neurology, endocrinology, ophthalmology, and more. The increased interest for benzimidazole compounds has been due to their excellent properties, like increased stability, bioavailability, and significant biological activity. This book chapter mainly discussed recent synthetic methods developed for the benzimidazole derivatives and their pharmacological properties exemplified on several derivatives.

Keywords: benzimidazole, heterocycle, medicinal chemistry, structure activity relationship, biological activity

1. Introduction

The biological application of benzimidazole nucleus is discovered way back 1944, when Woolley speculated that benzimidazoles resemble purine-like structure and elicit some biological application [1]. Hence benzimidazole structure found isosters of naturally occurring nucleotides, which allows them to contact easily with the biopolymers of the living system. Later, Brink discovered 5,6-dimethylbenzimidazole as a degradation product of vitamin B_{12} and subsequently found some of its analogs having vitamin B_{12}-like activity [2, 3]. These initial study reports emerged to explore various decorated benzimidazole motif discoveries by the medicinal chemist. Over the few decades of active research, benzimidazole has evolved as an important heterocyclic nucleus due to its wide range of pharmacological applications. Hence, it's worth to understand the basic chemistry and structure of such a wonderful molecule. Benzimidazole is formed by the fusion of benzene and imidazole moiety, and numbering system according to the IUPAC is depicted in **Figure 1**. Historically, the first benzimidazole was prepared in 1872 by Hoebrecker, who obtained 2,5 (or 2,6)-dimethylbenzimidazole by the reduction of 2-nitro-4-methylacetanilide [4]. Benzimidazoles which contain a hydrogen atom attached to nitrogen in the 1-position readily tautomerize, and this may be depicted in **Figure 1**. This basic "6 + 5" heterocyclic structure is shared by another class of chemical compounds existing in nature shown in **Figure 2**.

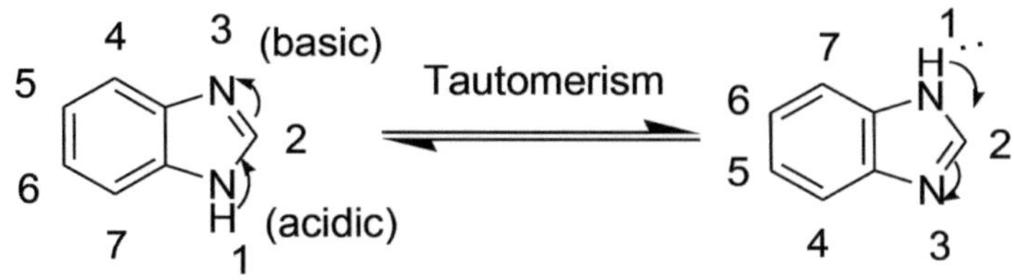

Figure 1.
Tautomeric forms of benzimidazole.

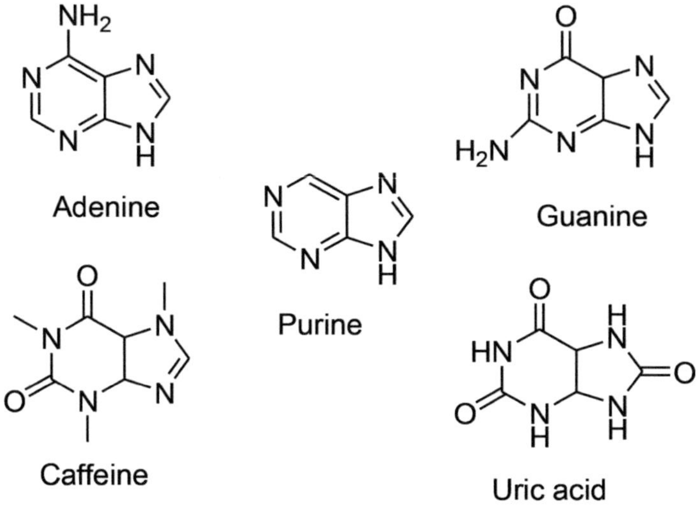

Figure 2.
Common biomolecules with a "6 + 5" heterocyclic structure.

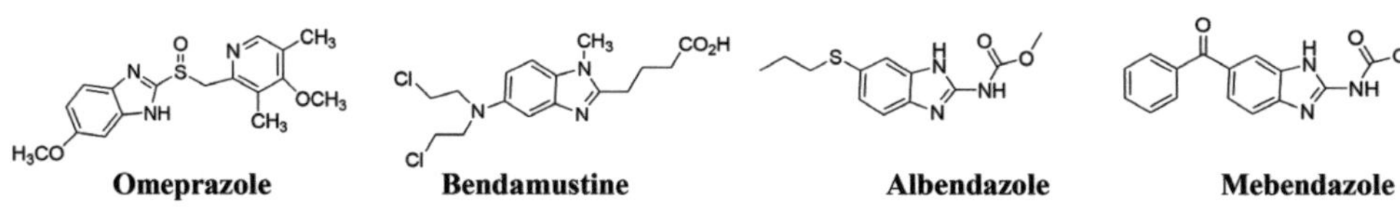

Figure 3.
Benzimidazole-based drugs.

Among the members of this group of molecules are well-known building blocks for biopolymers, such as adenine and guanine, two of the five nucleic acid bases, uric acid, and caffeine. From this basic structural similarity, it is not too surprising that benzimidazole nucleus has emerged biologically as an important pharmacophore with a privileged structure in medicinal chemistry. Nowadays it is a moiety of choice which possesses many pharmacological properties. The most prominent benzimidazole compound in nature is *N*-ribosyl-dimethylbenzimidazole, which serves as an axial ligand for cobalt in vitamin B_{12}. The pharmacological application of benzimidazole analogs found potent inhibitors of various enzymes involved and therapeutic uses including as antidiabetic, anticancer, antimicrobial, antiparasitic, analgesics, antiviral, antihistamine, and also neurological, endocrinological, and ophthalmological drugs [5–13].

The use of benzimidazole started many years back in 1990 onward, a vast number of benzimidazole analogs synthesis were reported, which resulted in increased stability, bioavailability, and significant biological activity. Some of the well-known active drugs with benzimidazole ring are mentioned in **Figure 3**, omeprazole, bendamustine, albendazole, and mebendazole. This chapter is mainly focused on the chemistry of the benzimidazoles and on the recently reported synthesis and mechanisms, structural aspects, and pharmacological applications with biological and clinical studies.

2. Overview of benzimidazole synthesis

Experimentally the simple method for the synthesis of benzimidazole derivatives begin with benzene containing nitrogen functions at *ortho*-position to each other *o*-phenylenediamine (OPD) is well documented. In this section several synthetic methodologies are grouped according to the starting material which is used for the benzimidazole motif synthesis reviewed.

2.1 Synthesis of benzimidazoles by the reaction of substituted aldehyde with OPD

The reaction of OPD with aromatic/aliphatic aldehyde under suitable condition for the synthesis of 2-substituted benzimidazoles is well-known. Since the reaction involved oxidation, it required oxidative condition. The oxidation reaction may be carried out in the presence of air or more conveniently by oxidizing agent such as cupric acetate first introduced by Weidenhagen [14–16]. This method reported the reaction of OPD with aldehyde in the presence of water or alcoholic solution in the presence of cupric acetate. The formed cuprous salt of benzimidazole is decomposed with hydrogen sulfide which gave free benzimidazole after filtration. This method isolated excellent yields of 2-substituted benzimidazoles of alkyl, aryl, and heterocyclic substituted moiety. Further Wright's group reported the synthesis of *N*-alkylated benzimidazoles using *N*-alkylated-*o*-phenylenediamine with aldehydes gave good yields of 1-substituted benzimidazole. The mechanism found the initial formation of a Schiff intermediate by the reaction of aldehyde with one of the amines of OPD, followed by cyclization to form the product (**Figure 4**). The researcher observed that, the reaction between OPD and aldehyde in the absence of a specific oxidizing agent results to either 2-substituted benzimidazoles or aldimines (**Figure 5**) product formation in some cases aldimines major and some cases 2-substituted benzimidazoles are the major or both form exists in equal amounts. Rao and Smith et al. independently reviewed the reaction between OPD and Aldehydes (**Figure 6**) as a simple and efficient method to synthesize benzimidazoles [17, 18]. Numerous methods

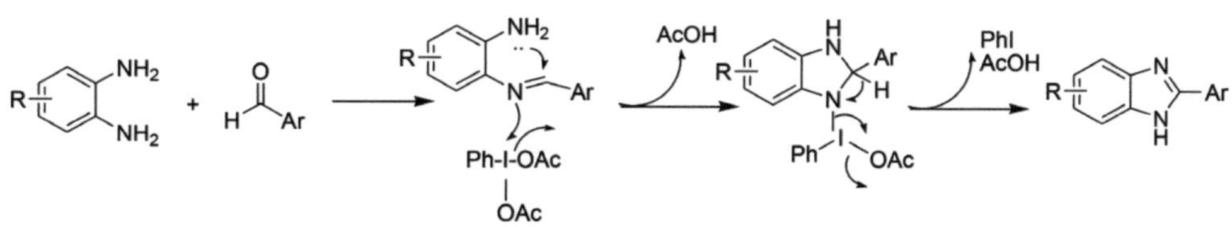

Figure 4.
Mechanism of formation of benzimidazole catalyzed by oxidizing agent [PhI (OAc)$_2$].

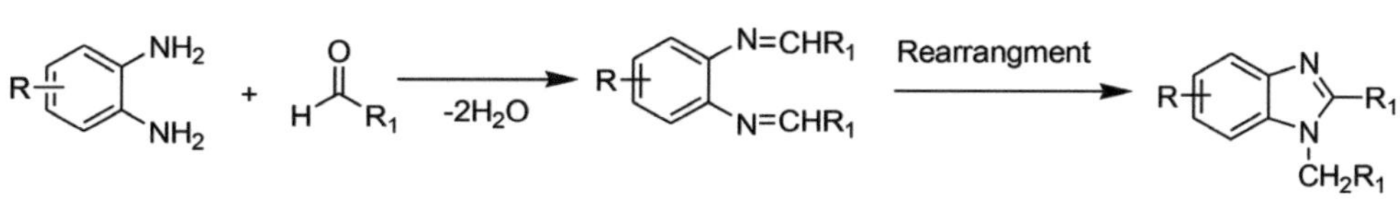

Figure 5.
Synthesis of benzimidazoles via aldiminic intermediates in the absence of catalyst.

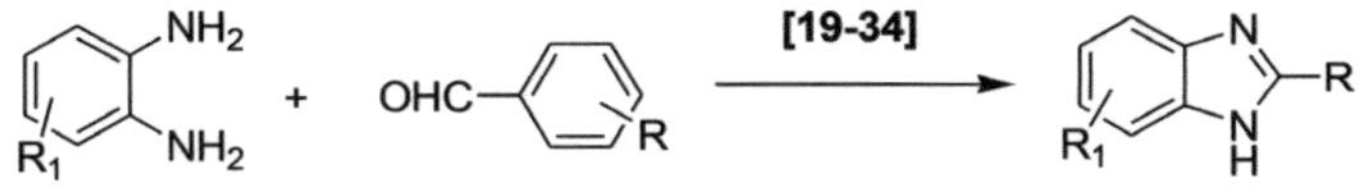

Figure 6.
Synthesis of benzimidazole derivatives from OPD and aldehydes.

are reported for the condensation of substituted OPD with aryl/alkyl/heterocyclic aldehydes catalyzed by different oxidizing agents or metal triflate such as Sc $(OTf)_3$ or Yb $(OTf)_3$ [19], sulfamic acid [20], H_2O_2-HCl [21], $FeBr_3$ [22], PhI $(OAc)_2$ [23], $LaCl_3$ [24], H_5IO_6-SiO_2 [25], Ce $(NO_3)_3.6H_2O$ [26], $NaHSO_4$-SiO_2 [27], mercuric oxide [28], chloranil [29], manganese dioxide [30], and I_2/TBHP [31] and more methods [32–34]. This method isolated excellent yields of 2-substituted benzimidazoles with alkyl, aryl, and heterocyclic substituted moiety (**Figure 6**).

2.2 Synthesis of benzimidazoles by the reaction of aryl/alkyl/heterocyclic acid chloride with OPD

Other synthetic routes involved carboxylic acid with an OPD-required harsh condition in the presence of a strong acid at elevated temperatures with poor yield reported for the benzimidazole. Alternatively, a two-step synthesis is reported, wherein the OPD is treated with one equivalent of an acid chloride derivative and the resulting mono-acylated product is subjected to cyclodehydration under various conditions such as heating in aqueous acids/solvents or by greener methods such as glycerol [35], ionic liquid [Hbim] BF_4 [36], agro-waste extract WEPBA [37], hetero-polyacid [38], $BF_3.Et_2O$ [39], zeolite [40], KF-Al_2O_3 [41], and more (**Figure 7**).

2.3 Synthesis of benzimidazoles by the reaction of substituted alcohol or amines with *o*-nitroarylamines

Researcher demonstrated alternative substrate *o*-nitroarylamine reaction with substituted alcohol or amines by using various reducing/redox agents ($FeCl_3$) for the synthesis of benzimidazole in a single step. This procedure has got commercial importance due to reasonable yield isolation (**Figure 8**) [42–44].

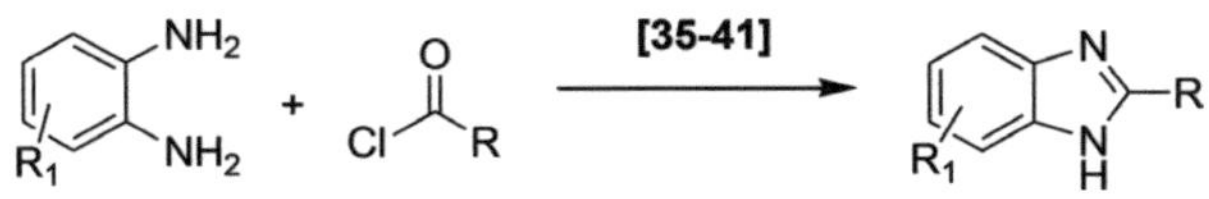

Figure 7.
Synthesis of benzimidazole starting OPD with acyl chloride derivatives.

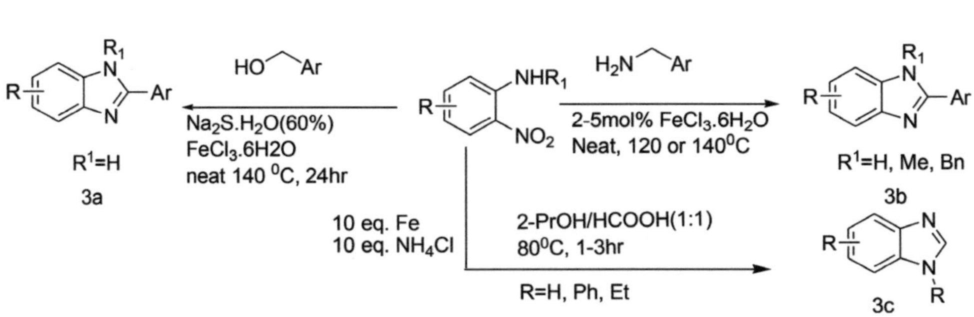

Figure 8.
Synthesis of benzimidazoles starting o-nitroarylamines.

2.4 Synthesis of benzimidazoles by the reaction of aldehyde or EAA with *o*-substituted arylamines

One-pot three-component reaction of 2-haloanilines, aldehydes, and NaN_3 is also reported for the synthesis of benzimidazole [45]. The reaction catalyzed CuCl (5 mol%), and 5 mol% of TMEDA was reacted in DMSO at 120°C which gave the product good yields (**4a**). The reaction showed tolerance toward aliphatic, heterocyclic aldehydes, and functional groups such as ester, nitro, and chloro on aromatic afforded the desired products in moderate yields. Bahrami et al. reported useful synthetic methodology for the synthesis of benzimidazoles using catalytic redox cycling based on (Ce(IV)/Ce(III))/H_2O_2 redox-mediated oxidation of the Schiff intermediate derived from differently substituted aromatic 1,2-phenylendiamines/2-thiol with a variety of aromatic aldehydes which resulted in isolation of the product in good yield (**4b**) [46]. Further, Bao et al. found Brønsted acid-catalyzed (TsOH) cyclization reactions of 2-amino anilines with ethylacetoaceate (EAA) under oxidant-, metal-, and radiation-free conditions (**4c**). In this method various 2-substituted benzimidazoles are obtained with different groups such as methyl, chloro, nitro, and methoxy linked on benzene rings which were tolerated (**Figure 9**) [47].

2.5 Synthesis of benzimidazoles by *C-H* amination of *N*-substituted amidines

Researcher demonstrated oxidative C-H amination of *N*''-aryl-N'-tosyl/N'-methylsulfonylamidines and *N,N'-bis*(aryl)amidines using iodobenzene as a catalyst to obtain 1,2-disubstituted benzimidazoles in the presence of *m*-CPBA which gave target product moderate to high yields (**5a**) [48]. Alternatively, other research group reported intramolecular *N*-arylations of amidines mediated by KOH in DMSO at 120°C (**5b**). The method allows diversely substituted products in moderate to very good yields (**Figure 10**) [49].

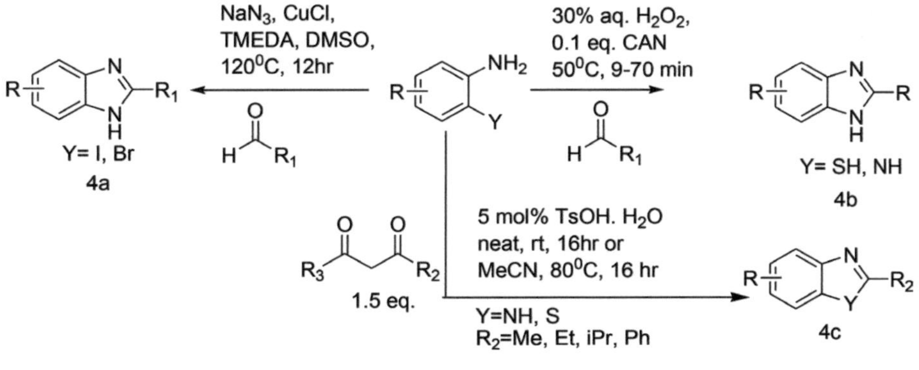

Figure 9.
One-pot three-component reaction for the synthesis of benzimidazole.

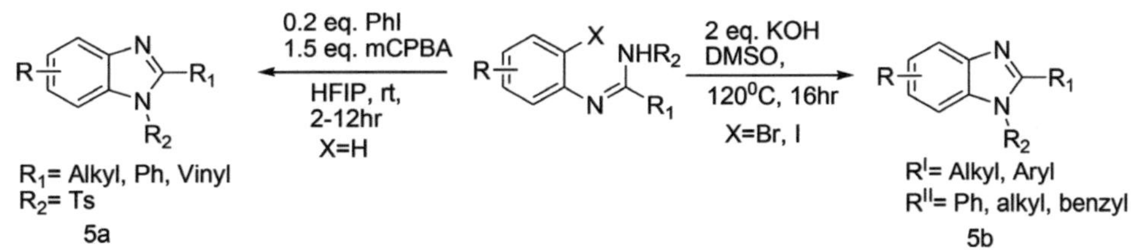

Figure 10.
Synthesis of benzimidazoles using N-substituted amidines.

2.6 Functionalization of benzimidazole to 2-substituted (hetero)aryl benzimidazole

Shao et al. recently reported the synthesis of benzimidazoles via direct C–H bond arylation in the presence of a NHC-Pd(II)-Im complex. The method is tolerable to various activated and deactivated (hetero)aryl chlorides to get 2-(hetero) aryl benzimidazoles in high yields. It is a facile and an alternative methodology for the direct C–H bond arylation of (*benz*)imidazoles (**Figure 11**) [50].

2.7 Synthesis by the reaction of *N*-substituted formamides with OPD derivatives

Bhanage et al. demonstrated efficient and convenient one-pot protocol synthesis of a benzimidazole derivative using various OPD derivatives and *N*-substituted formamides (C1 sources) in a zinc acetate-catalyzed cyclization in the presence of poly(methylhydrosiloxane) to afford corresponding products in good yields (**Figure 12**) [51].

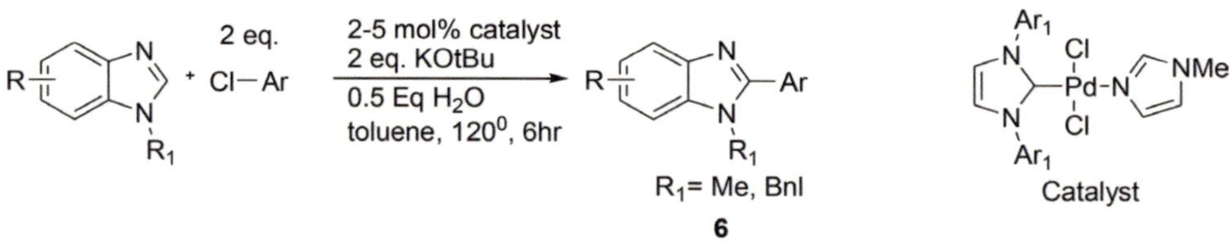

Figure 11.
Synthesis of 2-substituted (hetero)aryl benzimidazoles.

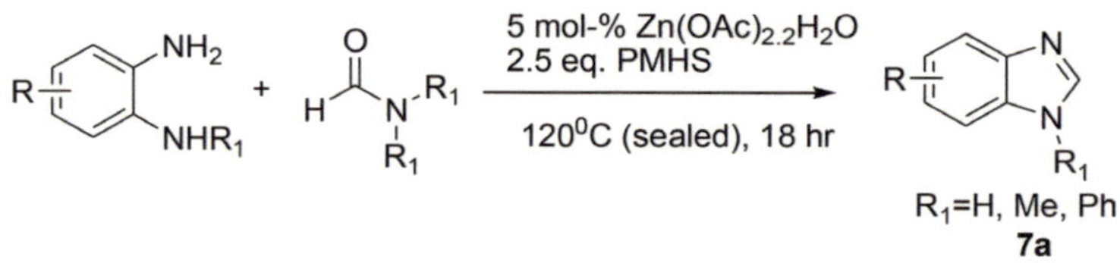

Figure 12.
N-Substituted formamides as C1 sources for the synthesis of benzimidazole.

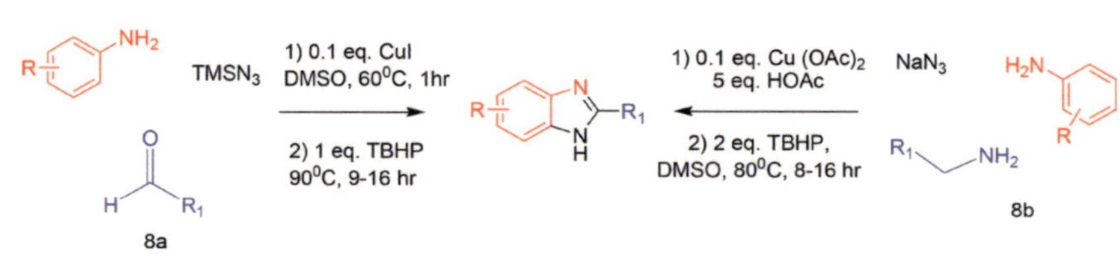

Figure 13.
One-pot three-component reaction.

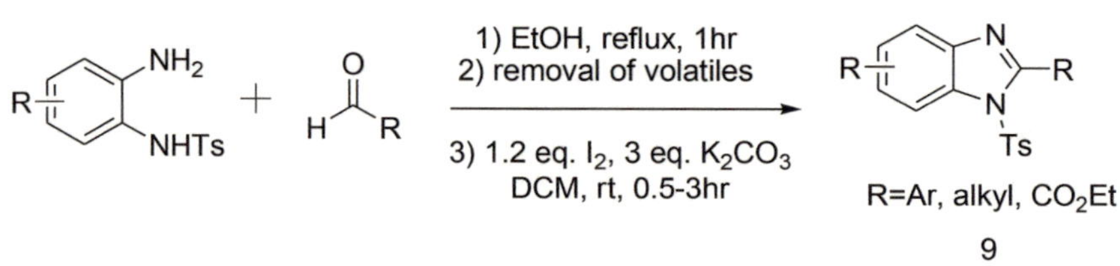

Figure 14.
Synthesis of 1,2-disubstituted benzimidazole.

2.8 Synthesis by one-pot three-component reaction

Punniyamurthy's group reported copper-catalyzed one-pot, three-component reaction of *N*-aryl imines, in which imine acts as a directing group by chelating to the metal center, which affords a potential route for the transformation of the commercial aryl amines, aldehydes, and azides into valuable benzimidazole with vast substrate scope and diversity (**8a**). Further, the same group is reported in copper(II)-catalyzed oxidative cross-coupling of anilines, primary alkyl amines, and sodium azide in the presence of TBHP at moderate temperature (**8b**). This one-pot protocol involves a domino C-H functionalization, transamination, *ortho*-selective amination, and cyclization sequence. The method is found tolerable to broad functional group and can be extended to the coupling of benzyl alcohols (**Figure 13**) [52, 53].

2.9 Synthesis of 1,2-disubstituted benzimidazole

Chang et al. demonstrated intramolecular C—H amidation using molecular iodine under basic conditions. The imine substrates required were readily prepared by condensation of aldehydes with OPD derivatives. The reaction is carried out in the absence of metal-free cyclization, works well with crude imines, and allows synthesis of series of *N*-substituted benzimidazoles. This method is tolerable to a variety of aromatic, aliphatic, and cinnamic aldehydes to produce diverse 1,2-disubstituted benzimidazoles (**Figure 14**) [54].

3. Pharmacological profile of benzimidazole derivatives

Benzimidazole moiety came in scenic after discovery of it as an integral part of the structure of the vitamin B_{12} in the 1950s. In the early 1960s, it was developed as plant fungicides and later as veterinary anthelminthic. Further, a variety of veterinary anthelmintics were developed and marketed, including parbendazole, fenbendazole, oxfendazole, and cambendazole. In 1962 the first benzimidazole developed and licensed for human use was thiabendazole, and present more derivatives of benzimidazole that have been clinically approved are albendazole, mebendazole, and flubendazole as anthelmintic; omeprazole, lansoprazole, and pantoprazole as proton pump inhibitors; astemizole as antihistaminic; enviradine as antiviral; and candesartan cilexetil and telmisartan as antihypertensives. In literature various substituted derivatives of benzimidazole demonstrated various therapeutic agents such as anticancer, antiproliferative, antimicrobials, antivirals, antiparasites, anthelmintic activity, anticonvulsant, antioxidants, anti-inflammatory, antihypertensive, immunomodulators, proton pump inhibitors, anticoagulants, hormone modulators, and CNS stimulants as well as antidepressants, antidiabetics, anti-HIV, lipid level modulators, etc. and have made an important scaffold for the development of new therapeutic agents (**Figure 15**) [10, 12, 13, 55–79].

3.1 Anticancer activity

Yang et al. optimized the solubility problem of lead benzimidazole (**1**) through introducing *N*-methylpiperazine groups at the 2-position showing preliminary in vitro anticancer activities [80]. Raghavan et al. demonstrated synthesis and evaluation of 1-(4-methoxyphenethyl)-1H-benzimidazole-5-carboxylic acid derivatives (**2**). Caused maximum cell death in leukemic cells with a micromolar concentration [81]. Omar et al. demonstrated synthesis and docking studies of new series benzimidazole-pyrrole and tetracycline conjugates (**3**) tested against lung cancer cell line A549 and breast cancer cell line MCF-7 and found these molecules exhibited

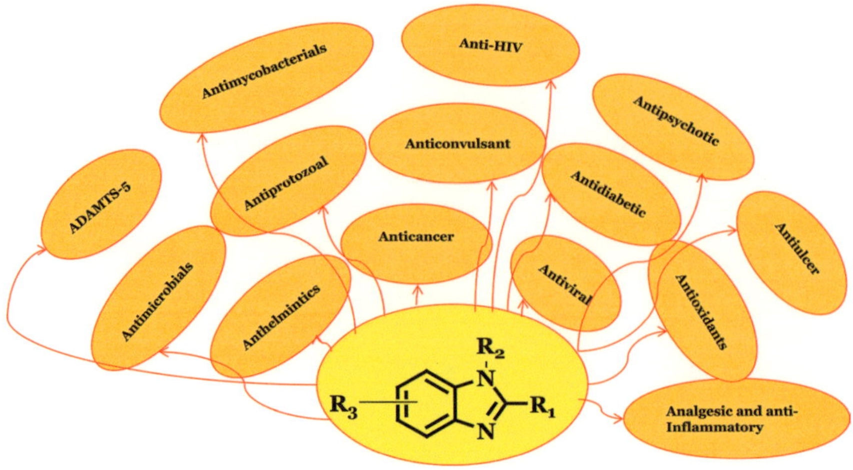

Figure 15.
Pharmacological profile of benzimidazole nucleus.

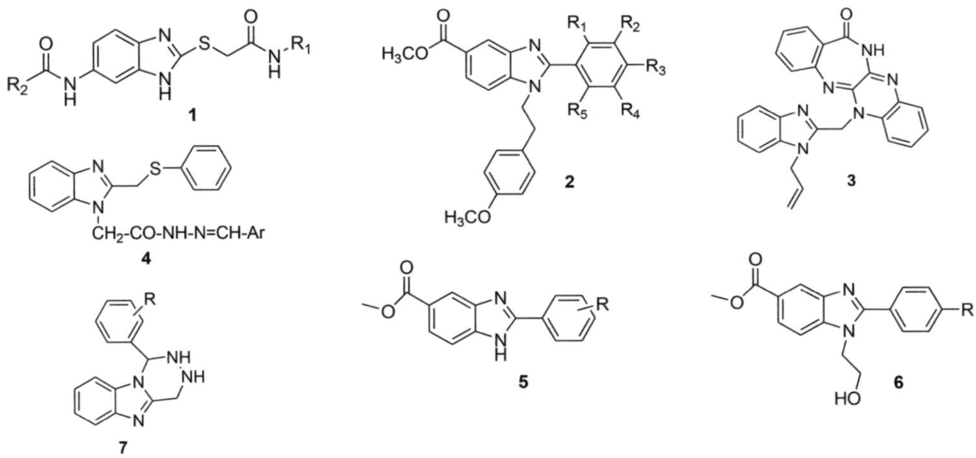

Figure 16.
Benzimidazole derivatives with anticancer effect.

remarkable higher activity than standard [82]. Karthikeyan et al. discovered derivatives of benzimidazoles 2-(phenyl)-3*H*-benzo[d]imidazole-5-carboxylic acids (**4**) and its methyl esters for anti-breast cancer agents [83]. Yoon et al. demonstrated novel benzimidazole derivatives (**5**) in sirtuin inhibitors (SIRT1 and SIRT2) with antitumor activities [84]. El-Nassan's group showed novel 1,2,3,4-tetrahydro[1,2,4] triazino[4,5-*a*]benzimidazole (**6**) analogs of aryl and heteroaryl groups showing antitumor activity in human breast adenocarcinoma cell line (MCF-7) [85]. Singh and Tandon demonstrated 2-aryl-substituted 2-*bis*-1*H*-benzimidazoles (**7**) evaluated as a topoisomerase-I inhibitor, and more benzimidazole derivatives are in line for the development of drug candidates (**Figure 16**) [86, 87].

3.2 Antiviral activity

Benzimidazole and its derivatives showed antiviral activity via contact with different virus particles such as human cytomegalovirus (HCMV), human herpes simplex virus (HSV-1), human immunodeficiency virus (HIV), and hepatitis-B and hepatitis-C virus (HBV and HCV). Luo et al. demonstrated the hepatitis-B virus inhibition by

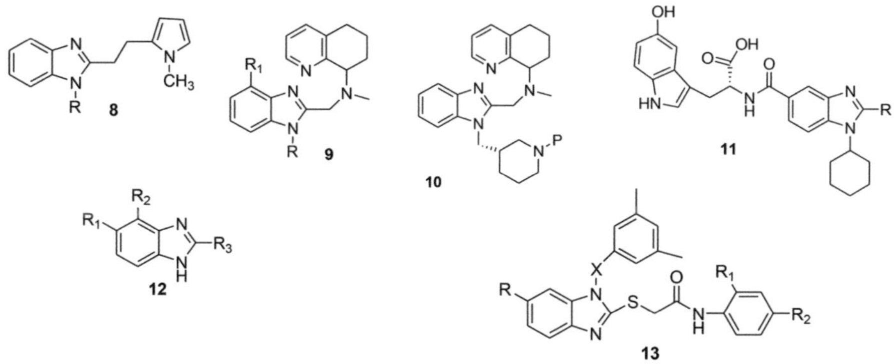

Figure 17.
Benzimidazole derivatives with antiviral activity.

novel benzimidazole derivatives (**8**) in HepG2.2.15 cell line [88]. Gudmundsson et al. discovered alkyl and cyclic alkyl amine substituted *N*-(1*H*-benzimidazol-2-ylmethyl)-5,6,7,8-tetrahydro-8-quinolinamines (**9**) and screened them for anti-HIV-1 activity as CXCR4 antagonists [89]. Miller et al. demonstrated stereochemically defined *N*-substituted benzimidazoles containing cyclic alkyl amine side chains (**10**), and its SAR analogs showed $CXCR_4$ antagonist activity as anti-HIV agents [90]. Beaulieu et al. demonstrated few benzimidazole-based allosteric inhibitors (**11**) that bind to thumb pocket I of the HCV NS5B polymerase inhibition to HCV NS5B [91]. In another work, Wubulikasimu et al. evaluated a series of benzimidazoles bearing a heterocyclic ring as oxadiazole, thiadiazole, and triazole (**12**) for their inhibition against *Coxsackieviruses* B_3 and B_6 in Vero cells [92]. Monforte et al. reported N-1-aryl-benzimidazole 2-substituted analogs (**13**) inhibit HIV-1 nonnucleoside reverse transcriptase inhibitors (NNRTIs) [93]. Some of these analogs inhibited the replication of HIV at nanomolar concentration with low cytotoxicity (**Figure 17**) [94].

4. Conclusions

The modern drug discovery more emphasizes on benzimidazole nucleus containing pharmacophore extensively applied in the biological and clinical studies. In this book chapter reviewed, recent optimized synthetic methods reported by various research groups for the synthesis of benzimidazole derivatives are exemplified. Further, the therapeutic use of benzimidazole in important areas such as antidiabetic, anticancer, antimicrobial, antiparasitic, analgesics, antiviral, antihistamine, and more is discussed. In spite of the active, exhaustive, and target-based research on the development of many drug-like molecule development, the number of molecules that made its way to the market and clinic is not measurable. It can be probably due to lack of a comprehensive compilation of various research reports in each activity capable of giving an insight into the SAR of the compounds. The biological profiles of these new generations of benzimidazole would represent a fruitful matrix for further development of better medicinal agents.

Acknowledgements

I would like to thank my PhD students for contributing to this book chapter in experimental and literature collection. The authors thank the University Grants Commission and DST-FIST, New Delhi, India, VGST-GoK, for the financial support to establish research laboratory and instrumental facility. Author is also grateful to the host university, RCUB, for financial and infrastructure support.

Conflict of interest

The authors confirm that this book chapter content has no conflicts of interest.

Abbreviations

2-PrOH	2-propanol
AcOH	acetic acid
ABTS	2,2′-azino-bis(3-ethylbenzothiazoline-6-sulfonic acid)
CAN	ceric ammonium nitrate
CH_3CN	acetonitrile
$Cu(OAc)_2$	cupric acetate
CuCl	copper chloride
CuI	copper iodide
DCM	dichloromethane
DMF	*N,N*-dimethylformamide
DMSO	dimethyl sulfoxide
DPPH	2,2-diphenyl-1-picrylhydrazyl
EtOH	ethanol
EAA	ethyl acetoacetate
H_2O_2	hydrogen peroxide
H_5IO_6-SiO_2	silica-supported periodic acid
HCl	hydrochloric acid
HCOOH	formic acid
HFIP	1,1,1,3,3,3-hexafluoro-2-propanol
I_2	iodine
KOH	potassium hydroxide
KOtBu	potassium tertiary butoxide
m-CPBA	*m*-chloroperoxybenzoic acid
NaN_3	sodium azide
NH_4Cl	ammonium chloride
OPD	*o*-phenylenediamine
$PhI(OAc)_2$	benzene (diacetoxyiodo)
PhI	iodobenzene
PMHS	poly(methylhydrosiloxane)
r.t	room temperature
SAR	structure activity relationship
TBHP	*tert*-butyl hydroperoxide
TMEDA	*N,N,N',N'*-tetramethylethylenediamine
$TMSN_3$	trimethylsilyl azide
TsOH	*p*-toluene sulfonic acid
WEPBA	water extract of papaya bark ash
$Zn(OAc)_2$	zinc acetate

Author details

Kantharaju Kamanna
Department of Chemistry, Peptide and Medicinal Chemistry Research Laboratory, Rani Channamma University, Belagavi, Karnataka, India

*Address all correspondence to: kk@rcub.ac.in

References

[1] Woolley DW. Some biological effects produced by benzimidazole and their reversal by purines. The Journal of Biological Chemistry. 1944;**152**:225-232

[2] Brink NG, Flokers K. Vitamin-B_{12}. Vi. 5,6-Dimethylbenzimidazole, a degradation product of vitamin-B_{12}. Journal of the American Chemical Society. 1949;**71**:2951. DOI: 10.1021/ja01176a532

[3] Epstein SS. Effect of some benzimidazoles on a vitamin B12-requiring alga. Nature. 1960;**188**: 143-144. DOI: 10.1038/188143a0

[4] Wright JB. Chemistry of benzimidazoles. Chemical Reviews. 1951;**48**:397-541. DOI: 10.1021/cr60151a002

[5] McKellar QA, Scott EW. The benzimidazole anthelmintic agents—A review. Journal of Veterinary Pharmacology and Therapeutics. 1990;**13**:223-247. DOI: 10.1111/j.1365-2885.1990.tb00773.x

[6] Spasov AA, Yozhitsa IN, Bugaeva LI, Anisimova VA. Benzimidazole derivatives: Spectrum of pharmacological activity and toxicological properties (a review). Pharmaceutical Chemistry Journal. 1999;**33**:232-243. DOI: 10.1007/BF02510042

[7] Rossignol JF, Maisonneuve H. Benzimidazoles in the treatment of trichuriasis: A review. Annals of Tropical Medicine and Parasitology. 1984;**78**:135-144. DOI: 10.1080/00034983.1984.11811787

[8] Patil A, Ganguly S, Surana S. A systematic review of benzimidazole derivatives as an antiulcer agent. Rasayan Journal of Chemistry. 2008;**1**:447-460

[9] Boiani M, Gonzalez M. Imidazole and benzimidazole derivatives as chemotherapeutic agents. Mini Reviews in Medicinal Chemistry. 2005;**5**:409-424. DOI: 10.2174/1389557053544047

[10] Narasimhan B, Sharma D, Kumar P. Benzimidazole: A medicinally important heterocyclic moiety. Medicinal Chemistry Research. 2012;**21**:269-283. DOI: 10.1007/s00044-010-9533-9

[11] Sivakumar R, Pradeepchandran R, Jayaveera KN, Kumarnallasivan P, Vijaianand PR, Venkatnarayanan R. Benzimidazole: An attractive pharmacophore in medicinal chemistry. International Journal of Pharmaceutical Research. 2011;**3**:19-31

[12] Geeta Y, Swastika G. Structure activity relationship (SAR) study of benzimidazole scaffold for different biological activities: A mini-review. European Journal of Medicinal Chemistry. 2015;**97**:419-443. DOI: 10.1016/j.ejmech.2014.11.053

[13] Yogita B, Om S. The therapeutic journey of benzimidazoles: A review. Bioorganic & Medicinal Chemistry. 2012;**20**:6208-6236. DOI: 10.1016/j.bmc.2012.09.013

[14] Hofmann K. Imidazole and its Derivatives Part-1. New York: Wiley Interscience; 1953

[15] Preston PN. Synthesis, reactions, and spectroscopic properties of benzimidazoles. Chemical Reviews. 1974;**74**(3):279-314. DOI: 10.1021/cr60289a001

[16] John BW. The chemistry of the benzimidazoles. Chemical Reviews. 1951;**48**:398-541. DOI: 10.1021/cr60151a002

[17] James GS, Isaac H. Organic redox reactions during the interaction of

o-phenylenediamine with benzaldehyde. Tetrahedron Letters. 1971;**38**:351-3544

[18] Veeranagaiah V, Rao NVS, Ratnam CV. Studies in the formation of heterocyclic rings containing nitrogen. Proceedings of the Indian Academy of Science, Section A. 1974;**79**:230-235. DOI: doi.org/10.1007/BF03051324

[19] Liyan F, Wen C, Lulu K. Highly chemoselective synthesis of benzimidazoles in $Sc(OTf)_3$-catalyzed system. Heterocycles. 2015;**91**:2306. DOI: 10.3987/COM-15-1332

[20] Heravi MM, Derikvand F, Ranjbar L. Sulfamic acid-catalyzed, three-component, one-pot synthesis of [1,2,4]triazolo/benzimidazolo quinazolinone derivatives. Synthetic Communications. 2010;**40**:677-685. DOI: 10.1080/00397910903009489

[21] Bahrami K, Khodaei MM, Kavianinia I. H_2O_2/HCl as a new and efficient system for synthesis of 2-substituted benzimidazoles. Journal of Chemical Research. 2006;**12**:783-784. DOI: 10.3184/030823406780199730

[22] Ma H, Han X, Wang Y, Wang J. A simple and efficient method for synthesis of benzimidazoles using $FeBr_3$ or $Fe(NO_3)_3 \cdot 9H_2O$ as catalyst. ChemInform. 2007;**38**:49. DOI: 10.1002/chin.200749146

[23] Du L-H, Wang Y-G. A rapid and efficient synthesis of benzimidazoles using hypervalent iodine as oxidant. Synthesis. 2007;**5**:675-678. DOI: 10.1055/s-2007-965922

[24] Venkateswarlu Y, Kumar SR, Leelavathi P. Facile and efficient one-pot synthesis of benzimidazoles using lanthanum chloride. Organic and Medicinal Chemistry Letters. 2013;**3**:7. DOI: 10.1186/2191-2858-3-7

[25] Sontakke VA, Ghosh S, Lawande PP, Chopade BA, Shinde VS. A simple, efficient synthesis of 2-aryl benzimidazoles using silica supported periodic acid catalyst and evaluation of anticancer activity. ISRN Organic Chemistry. 2013:1-7. DOI: 10.1155/2013/453682

[26] Martins GM, Puccinelli T, Gariani RA, Xavier FR, Silveira CC, Mendes SR. Facile and efficient aerobic one-pot synthesis of benzimidazoles using $Ce(NO3)3 \cdot 6H_2O$ as promoter. Tetrahedron Letters. 2017;**58**:1969-1972. DOI: 10.1016/j.tetlet.2017.04.020

[27] Kumar KR, Satyanarayana PVV, Reddy BS. $NaHSO_4$-SiO_2 promoted synthesis of benzimidazole derivatives. Archives of Applied Science Research. 2012;**4**:1517-1521

[28] Rombi M, Dick PR. Chemical Abstracts. 1972;**80**:108526s. FR2178385

[29] Harnish H. Chemical Abstracts. 1975;**83**:79240y. DE2346316

[30] Bhatnagar KS, George MV. Oxidation with metal oxides-II: Oxidation of chalcone phenylhydrazones, pyrazolines, o-aminobenzylidine anils and o-hydroxy benzylidine with MnO_2. Tetrahedron. 1968;**24**:1293-1298

[31] Moumita S, Asish RD. I_2/TBHP promoted oxidative C-N bond formation at room temperature: Divergent access of 2-substituted benzimidazoles involving ring distortion. Tetrahedron Letters. 2018;**59**:2520-2525. DOI: 10.1016/j.tetlet.2018.05.028

[32] Bhenki C, Karhale S, Helavi V. 5-Sulfosalicylic acid as an efficient organocatalyst for environmentally benign synthesis of 2-substituted benzimidazoles. Iranian Journal of Catalysis. 2016;**6**:409-413

[33] Gan Z, Tian Q, Shang S, Luo W, Dai Z, Wang H, et al. Imidazolium

chloride-catalyzed synthesis of benzimidazoles and 2-substituted benzimidazoles from o-phenylenediamines and DMF derivatives. Tetrahedron. 2018;**74**:7450-7456. DOI: 10.1016/j.tet.2018.11.014

[34] Chakraborty A, Debnath S, Ghosh T, Maiti DK, Majumdar S. An efficient strategy for N-alkylation of benzimidazoles/imidazoles in SDS-aqueous basic medium and N-alkylation induced ring opening of benzimidazoles. Tetrahedron. 2018;**74**:5932-5941. DOI: 10.1016/j.tet.2018.08.029

[35] Bachhav HM, Bhagat SB, Telvekar VN. Efficient protocol for the synthesis of quinoxaline, benzoxazole and benzimidazole derivatives using glycerol as green solvent. Tetrahedron Letters. 2011;**52**:5697-5701. DOI: 10.1016/j.tetlet.2011.08.105

[36] Nadaf RN, Siddiqui SA, Daniel T, Lahoti RJ, Srinivasan K. Room temperature ionic liquid promoted regioselective synthesis of 2-aryl benzimidazoles, benzoxazoles and benzothiazoles under ambient conditions. Journal of Molecular Catalysis A: Chemical. 2014;**214**:155-160. DOI: 10.1016/j.molcata.2003.10.064

[37] Kantharaju K, Hiremath PB. One-pot, green approach synthesis of 2-aryl substituted benzimidazole derivatives catalyzed by water extract of papaya bark ash. Asian Journal of Chemistry. 2018;**30**:1634-1638. DOI: 10.14233/ajchem.2018.21296

[38] Heravi MM, Sadjadi S, Oskooie HA, Shoar RH, Bamoharram FF. Heteropolyacids as heterogeneous and recyclable catalysts for the synthesis of benzimidazoles. Catalysis Communications. 2008;**9**:504-507. DOI: 10.1016/j.catcom.2007.03.011

[39] Tandon VK, Kumar M. BF_3-Et_2O promoted one-pot expeditious and convenient synthesis of 2-substituted benzimidazoles and 3,1,5-benzoxadiazepines. Tetrahedron Letters. 2004;**45**:4185-4187. DOI: 10.1016/j.tetlet.2004.03.117

[40] Heravi MM, Tajbakhsh M, Ahmadi NA, Mohajerani B. Zeolites. Efficient and eco-friendly catalysts for the synthesis of benzimidazoles. Monatshefte für Chemie. 2006;**137**:175-179. DOI: 10.1007/s00706-005-0407-7

[41] Khalili SB, Sardarian AR. KF/Al_2O_3: An efficient solid heterogeneous base catalyst in one-pot synthesis of benzimidazoles and bis-benzimidazoles at room temperature. Monatshefte fuer Chemie. 2012;**143**:841-846. DOI: 10.1007/s00706-011-0647-7

[42] Nguyen TB, Ermolenko L, Al-Mourabit A. Sodium sulfide: A sustainable solution for unbalanced redox condensation reaction between o-nitroanilines and alcohols catalyzed by an iron-sulfur system. Synthesis. 2015;**47**:1741-1748. DOI: 10.1055/s-0034-1380134

[43] Nguyen TB, Bescont JL, Ermolenko L, Mourabit AA. Cobalt- and iron-catalyzed redox condensation of o-substituted nitrobenzenes with alkylamines: A step- and redox-economical synthesis of diazaheterocycles. Organic Letters. 2013;**15**:6218-6221. DOI: 10.1021/ol403064z

[44] Hanan EJ, Chan BK, Estrada AA, Shore DG, Lyssikatos JP. Mild and general one-pot reduction and cyclization of aromatic and heteroaromatic 2-nitroamines to bicyclic 2H-imidazoles. Synlett. 2010;**18**: 2759-2764. DOI: 10.1002/chin.201111140

[45] Kim Y, Kumar MR, Park N, Heo Y, Lee S. Copper-catalyzed, one-pot three-component synthesis of benzimidazoles by condensation and C-N bond formation. The Journal of Organic Chemistry. 2011;**76**:9577-9583. DOI: 10.1021/jo2019416

[46] Bahrami K, Khodaei MM, Naali F. Mild and highly efficient method for the synthesis of 2-arylbenzimidazoles and 2-arylbenzothiazoles. The Journal of Organic Chemistry. 2008;**73**:6835-6837. DOI: 10.1021/jo8010232

[47] Mayo MS, Yu X, Zhou X, Feng X, Yamamoto Y, Bao M. Convenient synthesis of benzothiazoles through Brønsted acid catalyzed cyclization of 2-amino thiophenols/anilines with β-diketones. Organic Letters. 2014;**16**:764-767. DOI: 10.1021/ol403475v

[48] Alla SK, Kumar RK, Sadhu P, Punniyamurthy T. Iodobenzene catalyzed C-H amination of N-substituted amidines using m-chloroperbenzoic acid. Organic Letters. 2013;**15**:1334-1337. DOI: 10.1021/ol400274f

[49] Baars H, Beyer A, Kohlhepp SV, Bolm C. Transition-metal-free synthesis of benzimidazoles mediated by KOH/DMSO. Organic Letters. 2014;**16**:536-539. DOI: 10.1021/ol403414v

[50] Gu Z-S, Chen W-X, Shao L-X. N-Heterocyclic carbene-palladium(II)-1-methylimidazole complex-catalyzed direct C-H bond arylation of (benz) imidazoles with aryl chlorides. The Journal of Organic Chemistry. 2014;**79**:5806-5811. DOI: 10.1021/jo5010058

[51] Nale DB, Bhanage BM. N-Substituted formamides as C1-sources for the synthesis of benzimidazole and benzothiazole derivatives by using zinc catalysts. Synlett. 2015;**26**:2831-2834. DOI: 10.1055/s-0035-1560319

[52] Mahesh D, Sadhu P, Punniyamurthy T. Copper(I)-catalyzed regioselective amination of N-aryl imines using TMSN3 and TBHP: A route to substituted benzimidazoles. The Journal of Organic Chemistry. 2015;**80**:1644-1650. DOI: 10.1021/jo502574u

[53] Mahesh D, Sadhu P, Punniyamurthy T. Copper(II)-catalyzed oxidative cross-coupling of anilines, primary alkyl amines and sodium azide using TBHP: A route to 2-substituted benzimidazoles. The Journal of Organic Chemistry. 2016;**81**:3227-3234. DOI: 10.1021/acs.joc.6b00186

[54] Hu Z, Zhao T, Wang M, Wu J, Yu W, Chang J. I_2-mediated intramolecular C-H amidation for the synthesis of N-substituted benzimidazoles. The Journal of Organic Chemistry. 2017;**82**:3152-3158. DOI: 10.1021/acs.joc.7b00142

[55] Medzhitov R. Inflammation 2010: New adventures of an old flame. Cell. 2010;**140**:771-776. DOI: 10.1016/j.cell.2010.03.006

[56] Gulcan HO, Mavideniz A, Sahin MF, Orhan IE. Benzimidazole-derived compounds designed for different targets of Alzheimer's disease. Current Medicinal Chemistry. 2019. DOI: 10.2174/0929867326666190124123208. E-pub Abstract Ahead of Print

[57] Dilip D, Tai-Lin C, Yi-wen L, Tsai-Yi Y, Ming-Hsi W, Tung-Hu T, et al. Novel N-mustard-benzimidazoles/benzothiazoles hybrids, synthesis and anticancer evaluation. Anti-Cancer Agents in Medicinal Chemistry. 2017;**17**:1741-1755. DOI: 10.2174/1871520617666170522120200

[58] Ritchu S, Sandeep J, Sandeep A, Deepika S, Neelam J. Synthesis, characterization and molecular Docking studies of novel N-(benzimidazol-1-ylmethyl)-4-chlorobenzamide analogues for potential anti-inflammatory and antimicrobial activity. Anti-Inflammatory & Anti-Allergy Agents in Medicinal Chemistry. 2018;**17**:16-31. DOI: 10.2174/1871523017666180426125141

[59] Gurmeet S, Yogita B, Gulshan B, Kumar R, Design G. Synthesis and PASS

assisted evaluation of novel 2-substituted benzimidazole derivatives as potent anthelimintics. Medicinal Chemistry. 2014;**10**:418-425. DOI: 10.2174/15734064 1004140421115518

[60] Prasanna AD, Saleel AL. Design and synthesis of mannich bases as benzimidazole derivatives as analgesic agents. Anti-Inflammatory & Anti-Allergy Agents in Medicinal Chemistry. 2015;**14**:35-46. DOI: 10.2174/1871523014 666150312164625

[61] Ibrahim HS, Albakri ME, Mahmoud WR, Allam HA, Reda AM, Abdel-Aziz HA. Synthesis and biological evaluation of some novel thiobenzimidazole derivatives as anti-renal cancer agents through inhibition of c-MET kinase. Bioorganic Chemistry. 2019;**85**:337-348. DOI: 10.1016/j.bioorg.2019.01.006

[62] Wu Z, Bao XL, Zhu WB, Wang YH, Phuong Anh NT, Wu XF, et al. Design, synthesis, and biological evaluation of 6-benzoxazole benzimidazole derivatives with antihypertension activities. ACS Medicinal Chemistry Letters. 2018;**10**:40-43. DOI: 10.1021/acsmedchemlett.8b00335

[63] Salehi N, Mirjalili BBF, Nadri H, Abdolahi Z, Forootanfar H, Samzadeh-Kermani A, et al. Synthesis and biological evaluation of new N-benzylpyridinium-based benzoheterocycles as potential anti-Alzheimer's agents. Bioorganic Chemistry. 2018;**83**:559-568. DOI: 10.1016/j.bioorg.2018.11.010

[64] Bharadwaj SS, Poojary B, Nandish SKM, Kengaiah J, Kirana MP, Shankar MK, et al. Efficient synthesis and in silico studies of the benzimidazole hybrid scaffold with the quinolinyloxadiazole skeleton with potential α-glucosidase inhibitory, anticoagulant, and antiplatelet activities for type-II diabetes mellitus management and treating thrombotic disorders. ACS Omega. 2018;**3**:12562-12574. DOI: 10.1021/acsomega.8b01476

[65] Mostafa AS, Gomaa RM, Elmorsy MA. Design and synthesis of 2-phenyl benzimidazole derivatives as VEGFR-2 inhibitors with anti-breast cancer activity. Chemical Biology & Drug Design. 2019;**93**:454-463. DOI: 10.1111/cbdd.13433

[66] Vlaminck J, Cools P, Albonico M, Ame S, Ayana M, Bethony J, et al. Comprehensive evaluation of stool-based diagnostic methods and benzimidazole resistance markers to assess drug efficacy and detect the emergence of anthelmintic resistance: A Starworms study protocol. PLoS Neglected Tropical Diseases. 2018;**12**:e0006912. DOI: 10.1371/journal.pntd.0006912

[67] Al-Blewi FF, Almehmadi MA, Aouad MR, Bardaweel SK, Sahu PK, Messali M, et al. Design, synthesis, ADME prediction and pharmacological evaluation of novel benzimidazole-1,2,3-triazole-sulfonamide hybrids as antimicrobial and antiproliferative agents. Chemistry Central Journal. 2018;**12**:110. DOI: 10.1186/s13065-018-0479-1

[68] Asati V, Ghode P, Bajaj S, Jain SK, Bharti SK. 3D-QSAR and molecular docking studies on oxadiazole substituted benzimidazole derivatives: Validation of experimental inhibitory potencies towards COX-2. Current Computer-Aided Drug Design. 2018. DOI: 10.2174/1 573409914666181003153249

[69] Bistrović A, Krstulović L, Stolić I, Drenjančević D, Talapko J, Taylor MC, et al. Synthesis anti-bacterial and anti-protozoal activities of amidinobenzimidazole derivatives and their interactions with DNA and RNA. Journal of Enzyme Inhibition and Medicinal Chemistry. 2018;**33**:1323-1334. DOI: 10.1080/14756366.2018.1484733

[70] Hussain A, AlAjmi MF, Rehman MT, Khan AA, Shaikh PA, Khan RA. Evaluation of transition metal complexes of benzimidazole-derived scaffold as promising anticancer chemotherapeutics. Molecules. 2018;**23**:pii: E1232. DOI: 10.3390/molecules23051232

[71] Gangrade A, Pathak V, Augelli-Szafran CE, Wei HX, Oliver P, Suto M, et al. Preferential inhibition of Wnt/β-catenin signaling by novel benzimidazole compounds in triple-negative breast cancer. International Journal of Molecular Sciences. 2018;**19**:pii: E1524. DOI: 10.3390/ijms19051524

[72] Al Ajmi MF, Hussain A, Rehman MT, Khan AA, Shaikh PA, Khan RA. Design, synthesis, and biological evaluation of benzimidazole-derived biocompatible copper(II) and zinc(II) complexes as anticancer chemotherapeutics. International Journal of Molecular Sciences. 2018;**19**:pii: E1492. DOI: 10.3390/ijms19051492

[73] Akhtar MJ, Khan AA, Ali Z, Dewangan RP, Rafi M, Hassan MQ, et al. Synthesis of stable benzimidazole derivatives bearing pyrazole as anticancer and EGFR receptor inhibitors. Bioorganic Chemistry. 2018;**78**:158-169. DOI: 10.1016/j.bioorg.2018.03.002

[74] Koronkiewicz M, Chilmonczyk Z, Kazimerczuk Z, Orzeszko A. Deoxynucleosides with benzimidazoles as aglycone moiety are potent anticancer agents. European Journal of Pharmacology. 2018;**820**:146-155. DOI: 10.1016/j.ejphar.2017.12.018

[75] Ajani OO, Tolu-Bolaji OO, Olorunshola SJ, Zhao Y, Aderohunmu DV. Structure-based design of functionalized 2-substituted and 1,2-disubstituted benzimidazole derivatives and their in vitro antibacterial efficacy. Journal of Advanced Research. 2017;**8**:703-712. DOI: 10.1016/j.jare.2017.09.003

[76] Liu HB, Gao WW, Tangadanchu VKR, Zhou CH, Geng RX. Novel aminopyrimidinyl benzimidazoles as potentially antimicrobial agents: Design, synthesis and biological evaluation. European Journal of Medicinal Chemistry. 2018;**143**:66-84. DOI: 10.1016/j.ejmech.2017.11.027

[77] Bistrović A, Krstulović L, Harej A, Grbčić P, Sedić M, Koštrun S, et al. Design, synthesis and biological evaluation of novel benzimidazole amidines as potent multi-target inhibitors for the treatment of non-small cell lung cancer. European Journal of Medicinal Chemistry. 2018;**143**:1616-1634. DOI: 10.1016/j.ejmech.2017.10.061

[78] Farahat AA, Ismail MA, Kumar A, Wenzler T, Brun R, Paul A, et al. Indole and benzimidazole bichalcophenes: Synthesis, DNA binding and antiparasitic activity. European Journal of Medicinal Chemistry. 2018;**143**:1590-1596. DOI: 10.1016/j.ejmech.2017.10.056

[79] Abdelgawad MA, Bakr RB, Omar HA. Design, synthesis and biological evaluation of some novel benzothiazole/benzoxazole and/or benzimidazole derivatives incorporating a pyrazole scaffold as antiproliferative agents. Bioorganic Chemistry. 2017;**74**:82-90. DOI: 10.1016/j.bioorg.2017.07.007

[80] Xiang P, Zhou T, Wang L, Sun CY, Hu J, Zhao YL, et al. Novel benzothiazole, benzimidazole and benzoxazole derivatives as potential antitumor agents: Synthesis and preliminary in vitro biological evaluation. Molecules. 2012;**17**:873-883. DOI: 10.3390/molecules17010873

[81] Gowda NR, Kavitha CV, Chiruvella KK, Joy O, Rangappa KS, Raghavan SC. Synthesis and biological evaluation of novel

1-(4-methoxyphenethyl)-1H-benzimidazole-5-carboxylic acid derivatives and their precursors as antileukemic agents. Bioorganic & Medicinal Chemistry Letters. 2009;**19**:4594-4600. DOI: 10.1016/j.bmcl.2009.06.103

[82] Omar MA, Shaker YM, Galal SA, Ali MM, Kerwin SM, Li J, et al. Synthesis and docking studies of novel antitumor benzimidazoles. Bioorganic & Medicinal Chemistry. 2012;**20**: 6989-6901. DOI: 10.1016/j.bmc.2012.10.010

[83] Liu T, Sun C, Xing X, Jing L, Tan R, Luo Y, et al. Synthesis and evaluation of 2-[2-(phenylthiomethyl)-1H-benzo [d] imidazol-1-yl] acetohydrazide derivatives as antitumor agents. Bioorganic & Medicinal Chemistry Letters. 2012;**22**:3122-3125. DOI: 10.1016/j.bmcl.2012.03.061

[84] Karthikeyan C, Solomon VR, Lee H, Trivedi P. Synthesis and biological evaluation of 2-(phenyl)-3H-benzo[d] imidazole-5-carboxylic acids and its methyl esters as potent anti-breast cancer agents. Arabian Journal of Chemistry. 2017;**10**:S1788-S1794. DOI: 10.1016/j.arabjc.2013.07.003

[85] Yoon YK, Ali MA, Wei AC, Choon TS, Osman H, Parang K, et al. Synthesis and evaluation of novel benzimidazole derivatives as sirtuin inhibitors with antitumor activities. Bioorganic & Medicinal Chemistry. 2014;**22**:703-710. DOI: 10.1016/j.bmc.2013.12.029

[86] El-Nassan HB. Synthesis antitumor activity and SAR study of novel [1 2 4] triazino [4 5-a] benzimidazole derivatives. European Journal of Medicinal Chemistry. 2012;**53**:22-27. DOI: 10.1016/j.ejmech.2012.03.028

[87] Singh M, Tandon V. Synthesis and biological activity of novel inhibitors of topoisomerase I: 2-Arylsubstituted 2-bis-1H-benzimidazoles. European Journal of Medicinal Chemistry. 2011;**46**:659-669. DOI: 10.1016/j.ejmech.2010.11.046

[88] Luo Y, Yao JP, Yang L, Feng CL, Tang W, Wang GF, et al. Design and synthesis of novel benzimidazole derivatives as inhibitors of hepatitis B virus. Bioorganic & Medicinal Chemistry. 2010;**18**:5048-5055. DOI: 10.1016/j.bmc.2010.05.076

[89] Gudmundsson KS, Sebahar PR, Richardson LD, Miller JF, Turner EM, Catalano JG, et al. Amine substituted N-(1H-benzimidazol-2ylmethyl)-5678-tetrahydro-8-quinolinamines as CXCR4 antagonists with potent activity against HIV-1. Bioorganic & Medicinal Chemistry Letters. 2009;**19**:5048-5052. DOI: 10.1016/j.bmcl.2009.07.037

[90] Miller JE, Turner EM, Gudmundsson AS, Jenkinson S, Spaltenstein A, Thomsan N, et al. Novel N-substituted benzimidazole CXCR4 antagonists as potential anti-HIV agents. Bioorganic & Medicinal Chemistry Letters. 2010;**20**:2125-2128. DOI: 10.1016/j.bmcl.2010.02.053

[91] Beaulieu PL, Dansereau N, Duan J, Garneau M, Gillard J, McKercher G, et al. Benzimidazole thumb pocket I finger-loop inhibitors of HCV NS5B polymerase: Improved drug-like properties through C-2 SAR in three sub-series. Bioorganic & Medicinal Chemistry Letters. 2010;**20**:1825-1829. DOI: 10.1016/j.bmcl.2010.02.003

[92] Tonelli M, Simone M, Tasso B, Novelli F, Boido V, Sparatore F, et al. Antiviral activity of benzimidazole derivatives II antiviral activity of 2-phenylbenzimidazole derivatives. Bioorganic & Medicinal Chemistry. 2010;**18**:2937-2953. DOI: 10.1016/j.bmc.2010.02.037

[93] Wubulikasimu R, Yang Y, Xue F, Luo X, Shao D, Li Y, et al. Synthesis and biological evaluation of novel

benzimidazole derivatives bearing a heterocyclic ring at 4/5 position. Bulletin of the Korean Chemical Society. 2013;**34**:2297-2304. DOI: 10.5012/bkcs.2013.34.8.2297

[94] Monforte A-M, Ferro S, Luca LD, Surdo GL, Morreale F, Pannecouque C, et al. Design and synthesis of N1-aryl-benzimidazoles 2-substituted as novel HIV-1 nonnucleoside reverse transcriptase inhibitors. Bioorganic & Medicinal Chemistry. 2014;**22**(4):1459-1467. DOI: 10.1016/j.bmc.2013.12.045

Section 3

The Therapeutic Potential of Benzimidazoles

Chapter 5

Antidiabetogenic Features of Benzimidazoles

Alexander A. Spasov, Pavel M. Vassiliev, Vera A. Anisimova and Olga N. Zhukovskaya

Abstract

Literature data on the insulinogenic effect of 2-aminobenzimidazole prompted us to investigate its novel derivatives, particularly those containing an additional fused cycle in C1,2-α position, including imidazole, dihydroimidazole, or tetrahydropyrimidine ring. Consensus analysis of the hypoglycemic effect of these compounds performed with IT Microcosm and PASS system revealed that activity is mostly characteristic for N^9-2,3-dihydroimidazo[1,2-a]benzimidazole derivatives. Substructural analysis of hypoglycemic activity identified substituents that determine the greatest pharmacological effect. According to the in silico assessment of the ADME properties, RU-254 was nominated as a lead compound due to the most optimal calculated and experimental activity and pharmacokinetic parameters. Preclinical studies have shown that identified compound has a pronounced insulinogenic effect and hypoglycemic effect, both in intact animals and in animals with experimental diabetes mellitus. RU-254 also reduces the level of glycated hemoglobin upon chronic administration, slightly decreases the activity of DPP-4, and increases the average number of Langerhans islets in the pancreas. Pharmaceutical drug formulation of RU-254 was developed and investigated for pharmacokinetic, pharmacodynamic, and toxicological properties. The dosage form of the drug under the name limiglidol (compound RU-254, diabenol) was evaluated in the full cycle of clinical studies that confirmed the safety, tolerability, and prominent antidiabetic properties of the drug.

Keywords: in silico, IT Microcosm, consensus prediction, antidiabetic effect, aminobenzimidazoles, cyclic benzimidazoles, pharmacodynamics, pharmacokinetics, toxicology, diabenol

1. Introduction

The history of drug discovery for the treatment of diabetes mellitus was and still is strongly determined by achievements in the field of fundamental medicine. Initially, the role of the pancreas and islets of Langerhans in the development of this pathology was proved; later, the structure of insulin, insulin receptor, and glucose transporters was deciphered; the role of the liver glycogenolysis and gluconeogenic enzymes, contributing to increased glucose output and hyperglycemia, was established; molecular mechanisms for the development of insulin resistance, the importance of the incretin system and Na^+/glucose transporters in the kidneys, and intestinal α-glucosidase were revealed, which led to the introduction of novel antidiabetic drugs into clinic [1–4].

The basis of insulin resistance at the cellular level primarily resides in the disruption of insulin signaling pathway at the level of the insulin receptor and insulin receptor substrate (IRS) proteins. The underlying mechanism of this phenomenon is impaired phosphorylation of serine amino acid residues, catalyzed by a number of intracellular protein kinases. The muscles, liver, and adipose tissue are the primary target organs of concern for the development of insulin resistance [5]. It was established that the severity of insulin resistance correlates, first of all, with intracellular lipid accumulation [6]. It is intracellular lipids that hamper signal transmission from the insulin receptor and cause a decrease in insulin-dependent glucose uptake. The pivotal role of AMP-dependent protein kinase (AMPK), which is an energy "sensor" of the cell, is also established, since AMPK through TORCI, the first mTOR-based protein complex, serves as a metabolic switch between catabolic and anabolic processes of the cell. Metformin is a biguanide derivative, which is the first-line drug for the treatment of type 2 diabetes. In 2001, it was shown that the molecular mechanism of its action is at least in part mediated by AMPK [7]. It is believed that indirect activation of AMPK by metformin-induced Ser172 phosphorylation determines its pleiotropic effects [8].

At the same time, it is important to note that course of type 2 diabetes mellitus characterized by several consecutive phases. It begins with primary insulin resistance and compensatory hyperinsulinemia with the subsequent development of β-cell dysfunction, thus creating the need for administration of insulin secretagogues or insulin formulations at the late stages of the disease [9–11].

Given that the previous works described the insulinogenic effect of the antihelminthic drug mebendazole [12], which can be considered as a new scaffold (2-aminobenzimidazole or cyclic guanidine) that exhibits an insulinogenic effect, we performed an experimental study of the novel cyclic guanidine derivatives, designed by introduction of additional fused cycle (imidazole, dihydroimidazole, and tetrahydropyrimidine).

2. Results

The synthesis of novel 2-aminobenzimidazole (AmBI) [13, 14] derivatives and fused benzimidazole derivatives was carried out, including N^9-imidazo[1,2-*a*]benzimidazoles (N^9-ImBI) [15–17], N^1-imidazo[1,2-*a*]benzimidazoles (N^1-ImBI) [18, 19], N^9-2,3-dihydroimidazo[1,2-*a*]benzimidazoles (N^9-DhImBI) [20, 21], N^1-2,3-dihydroimidazo[1,2-*a*]benzimidazoles (N^1-DhImBI) [18, 19], and 2,3,4,10-tetrahydropyrimido[1,2-*a*]benzimidazoles (PrmBI) [22–24].

In order to identify the most promising antidiabetic substances using IT Microcosm [25, 26] and PASS computer systems [27, 28], a step-by-step detailed in silico analysis of the hypoglycemic properties of the new compounds was carried out. Programs DruLiTo [29] and QikProp [30] were employed to assess key ADME properties and characteristics.

Hypoglycemic effect of the newly obtained derivatives was initially studied in rats upon intraperitoneal administration at a dose of 50 mg/kg. Blood sampling was carried out 4 hours after treatment with test compounds. Blood glucose concentration was determined with the glucose oxidase method using a commercial Glucose FKD kit [31]. The ratio of glucose concentrations in the blood plasma of the experimental and control group animals served as an indicator of hypoglycemic activity [32].

It was found that among condensed benzimidazole derivatives, a number of substances exceeded hypoglycemic activity of metformin, which served as a reference drug.

2.1 In silico study

IT Microcosm [25, 26] and PASS [27, 28] computer systems were employed to determine the most promising chemical class of compounds. A training set of known hypoglycemic drugs and a library of tested benzimidazole derivatives were subjected to a consensus prediction of the level of hypoglycemic activity. The average informativity coefficient K_{Pr} was calculated and used as a metric for comparison of AmBI, N^9-ImBI, N^1-ImBI, N^9-DhImBI, N^1-DhImBI, and PrmBI derivatives. K_{Pr} value ranges from 0 for inactive compounds to 5 for highly active compounds.

According to value of K_{Pr}, the potential of benzimidazole derivatives classes as sources of substances with hypoglycemic activity decreases in the following order: N^9-DhImBI (K_{Pr} = 4.50) > PrmBI (K_{Pr} = 4.25) > AmBI (K_{Pr} = 2.50) > N^1-ImBI (K_{Pr} = 2.00) > N^1-DhImBI (K_{Pr} = 1.25) > N^9-ImBI (K_{Pr} = 0.25) [33].

Thus, it was shown that N^9-2,3-dihydroimidazo[1,2-*a*]benzimidazole and 2,3,4,10-tetrahydropyrimido[1,2-*a*]benzimidazole derivatives have the most promising blood glucose lowering activity. That is, these tricyclic structures containing embedded guanidine group turned out to be more active than 2-aminobenzimidazole derivatives.

Subsequently, employing substructural analysis [34] and analysis via median [35] and supremal [36] estimates, the class of N^9-2,3-dihydroimidazo[1,2-*a*]benzimidazoles was selected as the most promising for the development of hypoglycemic compounds (**Figures 1** and **2**). It was shown that this scaffold is more preferable than 2,3,4,10-tetrahydropyrimido[1,2-a]benzimidazole.

Substructural analysis [34] of the level of hypoglycemic activity among the N^9-2,3-dihydroimidazo[1,2-*a*]benzimidazole derivatives allowed us to reveal a chemical feature (substituent) that largely determines high hypoglycemic activity—diethylaminoethyl substituent at the N^9 atom of the N^9-DhImBI scaffold.

According to the frequency analysis of physicochemical parameters [37] of experimentally studied derivatives of N^9-DhImBI scaffold, a significant feature of high hypoglycemic activity was revealed—a charge on the internal imidazole cycle of the condensed system, namely, $Q(Imid1)_{cs} \geq -0.109$, which is a characteristic for compound RU-254 (diabenol).

Taken together, the results of a complex consensus in silico analysis of the hypoglycemic activity of six classes of benzimidazole derivatives revealed 9-diethylaminoethyl-2,3-dihydroimidazo[1,2-*a*]benzimidazole dihydrochloride (RU-254, diabenol) as the most promising highly active compound.

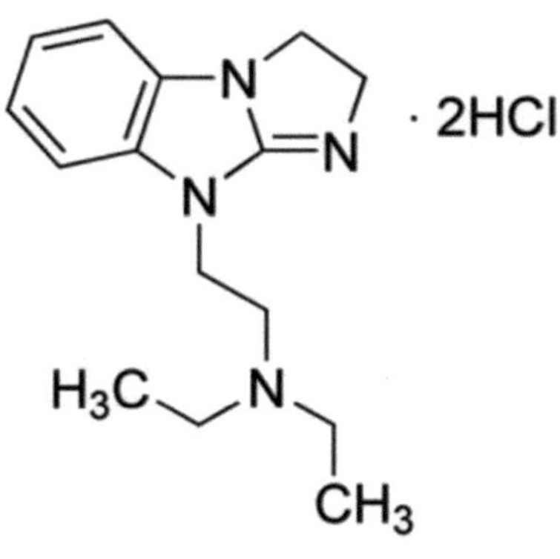

To assess the feasibility of a further study of the pharmacological properties of compound RU-254, we calculated parameters of drug-likeliness and ADME properties (absorption, distribution, metabolism, excretion) for RU-254 and reference antidiabetic drugs metformin and glibenclamide.

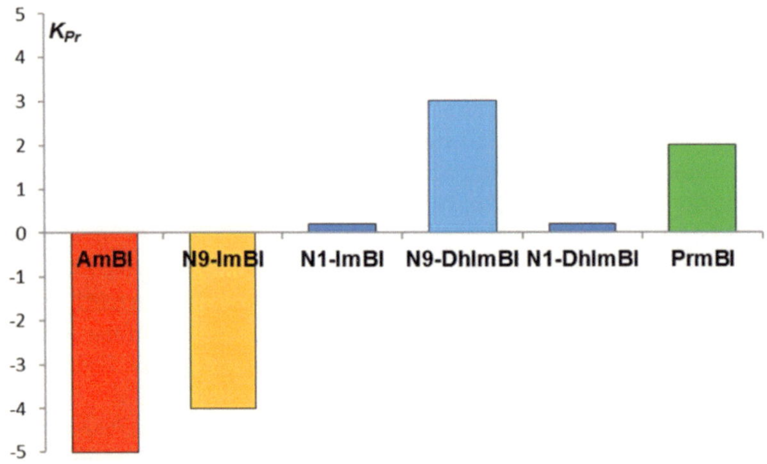

Figure 1.
Informativity coefficients describing the influence of basic benzimidazole structure on high hypoglycemic activity level (according to the substructural analysis).

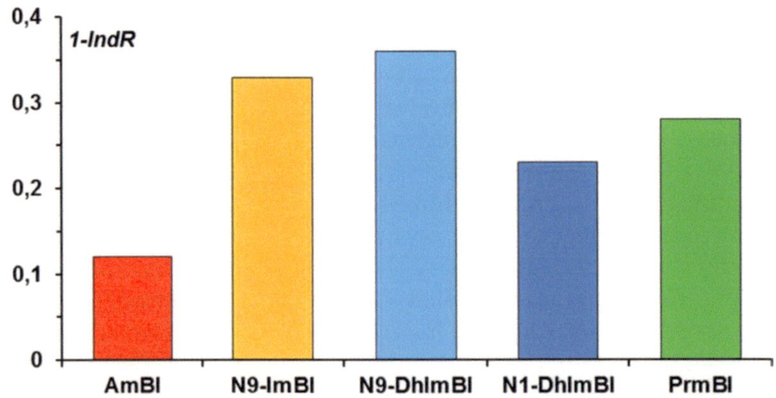

Figure 2.
Supremal evaluations of the effect of basic benzimidazole structure on high hypoglycemic activity level.

Using the DruLiTo program [29], it was found that diabenol satisfies the boundary conditions of all eight drug-likeliness filters, while metformin and glibenclamide correspond only for two of them.

Water solubility, serum albumin binding parameters, cellular permeability, and absorbability through the gastrointestinal tract for the three aforementioned substances were calculated with QikProp program [30]. A comparative analysis of the obtained characteristics showed that water solubility and the degree of binding to serum albumin of diabenol are higher than that of glibenclamide and lower than that of metformin. Indicators of bioavailability and absorbability through the gastrointestinal tract in diabenol are higher than that of glibenclamide and metformin. Thus, in terms of the total pharmacokinetic characteristics calculated in the QikProp program, diabenol is superior to metformin and glibenclamide. It should be noted that the calculated values of pharmacokinetic parameters of all three compounds are in the ranges that are recognized as appropriate for drug molecules.

Summarizing the results of the evaluation of ADME properties obtained using two computational approaches, it can be argued that diabenol in regard of its calculated drug-like and pharmacokinetic characteristics is not inferior to metformin and glibenclamide and is a very promising substance for performing advanced preclinical studies.

2.2 Synthesis

Synthesis of 9-diethylaminoethyl-2,3-dihydroimidazo[1,2-*a*]benzimidazole dihydrochloride (RU-254, diabenol) is readily realized through condensation of 2-amino-1-diethylaminoethylbenzimidazole with an excess of dibromoethane and subsequent transformation of the resulting of 9-diethylaminoethyl-2,3-dihydroimidazo[1,2-*a*]benzimidazole dihydrobromide to the base and the desired dihydrochloride [21].

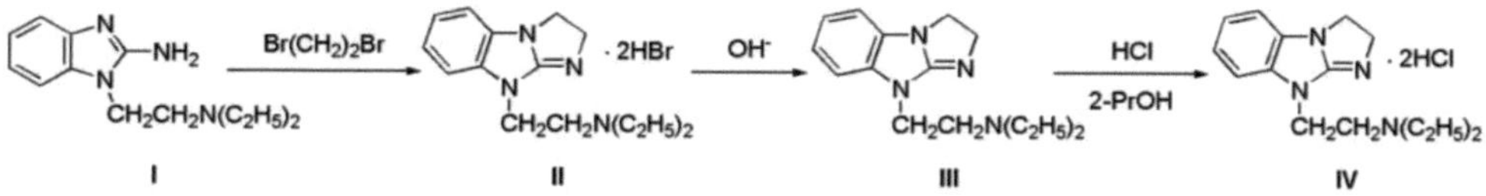

Example. A stirred suspension of 69.6 g (0.3 mol) of 2-amino-1-diethylamino-ethylbenzimidazole (I) in 104 ml (1.2 mol) of dibromoethane is gently heated in a glycerin bath. At 60–70°C, the initial amine dissolves completely, and at 100–105°C, an exothermic cyclization reaction occurs (the bath is set aside at the beginning), accompanied by strong boiling up of the reaction mass while temperature rises to 140°C and a heavy colorless precipitate begins to form. After 5–7 minutes, the reaction virtually ends, and, in order to complete it, the mixture is heated for an additional 20 minutes at 140–145°C. After that, 80 ml of DMF are added to the thick mass with vigorous stirring, and the mixture is heated for another 10–15 minutes. Cooling to 20–25°C, filtering the precipitate, and washing with DMF (3 × 20 ml) and acetone (3 × 25 ml) afford 106 g of dihydrobromide (II) in 84% yield. The latter is dissolved in 230 ml of water and boiled for 10 minutes with 3–5 g of activated carbon. Carbon is filtered off, and the filtrate after cooling is brought to pH 10 with 40% sodium hydroxide solution. The light yellow oil (III) which separates on the surface is extracted with toluene. The toluene extracts are washed with water and dried with anhydrous potassium carbonate. The desiccant is filtered off and washed with toluene. Combined toluene fractions are acidified by gradual addition of a saturated solution of hydrogen chloride in 2-propanol to pH 1. The heavy colorless precipitate of dihydrochloride (IV) is filtered off after 4–5 hours at 20–25°C, washed with acetone, and dried at 100–110°C for 2–3 hours to a constant weight. Yield is 75 g (90.9%) from dihydrobromide (III) and 75% from the initial amine (II). The crude product can be recrystallized from 2-propanol to pharmacopeia grade purity.

2.3 Preclinical studies

Diabenol had a pronounced hypoglycemic effect and antihyperglycemic activity in carbohydrate tolerance tests performed on white outbred rats and rabbits. The compound studied showed a marked decrease of glycemia in animals with impaired glucose tolerance (in rats with severe streptozotocin diabetes and insulin resistance syndrome), in rats with alloxan diabetes, and in rabbits with acute insulin deficiency, induced with administration of anti-insulin guinea pig serum. In experiments on pancreatomic dogs, diabenol did not reduce blood glucose but enhanced the hypoglycemic effect of exogenously administered insulin [38–40], thus confirming insulin-mediated mechanism of action.

Detailed study of antidiabetic action revealed not only pancreatotropic but also extrapancreatic components of diabenol action. Its pancreatotropic effect is determined by the enhancement of phase 1 insulin secretion, especially in glucose-stimulated conditions (**Figures 3** and **4**).

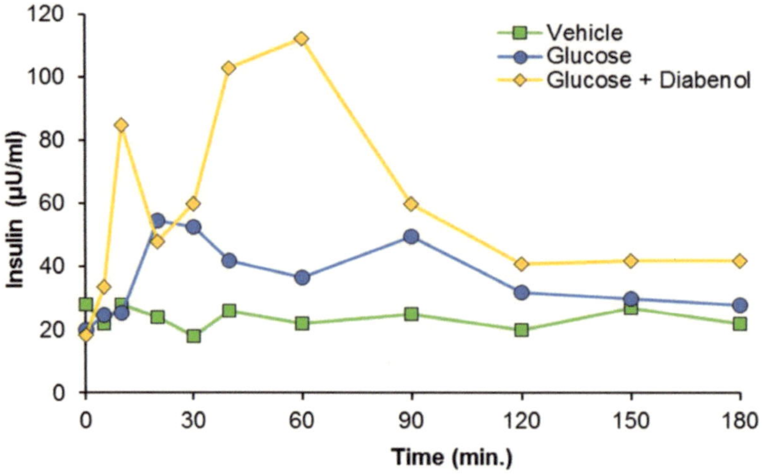

Figure 3.
Effect of diabenol (10 mg/kg, intravenously) in blood insulin levels during glucose tolerance test (1 g/kg) in cats.

Diabenol increases the insulin-dependent glucose uptake in muscles of rat diaphragm. Under conditions of alloxan-induced diabetes in rats, diabenol restored liver glycogen content and glycolysis rate and inhibited glycogenolysis in insulin-dependent organs and tissues (liver, striated muscles) while having no significant effect on these parameters in kidneys, which are insulin-independent organs [38].

It could be assumed that increased insulinotropic effect of diabenol is associated with a possible incretinomimetic effect. Studies [41] showed the ability of diabenol to inhibit the incretin-degrading DPP-4 enzyme, leading to a modulation of the insulin response. In our studies, diabenol also inhibited DPP-4, but in substantially higher concentrations (IC_{50} 1.35–2.05 mM), which cannot be achieved in the animal's body. Along with that, a 28-day administration of diabenol to rats with streptozotocin-induced diabetes was found to slightly and statistically insignificantly decrease the plasma activity of DPP-4 [42], which could be attributed to the action of its metabolites.

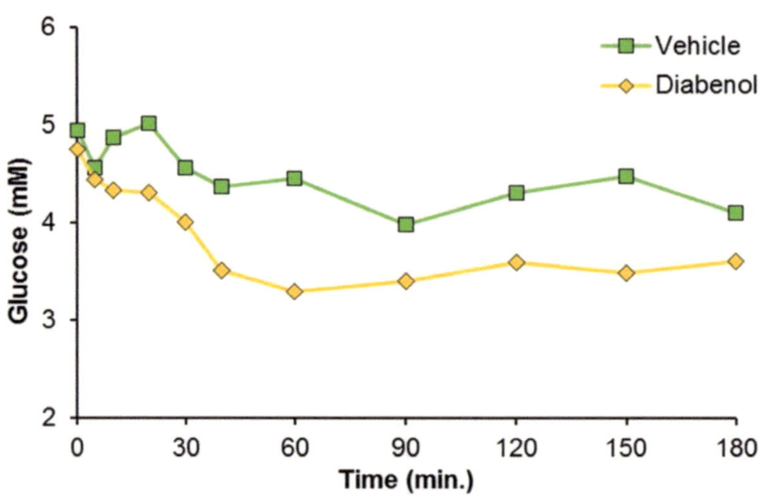

Figure 4.
Effect of diabenol (10 mg/kg, intravenously) on the basal portal vein blood glucose levels in cats [38].

Long-term administration of diabenol to streptozotocin-nicotinamide-induced diabetic rats allowed us to obtain interesting and valuable results. Oral administration of diabenol in a dose of 25 mg/kg for 4 weeks reduced blood glucose levels and volume of consumed liquid by more than 2 times, and level of glycated hemoglobin by 2.2%, and increased the content of C-peptide [43]. Moreover, diabenol administration resulted in a significant increase in the average number of islets of Langerhans in the splenic region of the pancreas and a significant increase in the area of the pancreatic β-cells (**Table 1**) [44].

The studied compound did not affect the apoptosis index (fraction of caspase-3 positive cells) and the proliferation index (PCNA-positive and Ki-67-positive cells) of the endocrinocytes of Langerhans islets. That is, diabenol had a cytoprotective effect on the cells. These data confirm the possibility of increasing the synthetic activity of β-cells under the influence of diabenol [44].

Given the complex nature of type 2 diabetes mellitus and aiming to increase the effectiveness of antidiabetic therapy in clinical practice, combination drugs (fixed combinations) are actively used to simultaneously target several key pathogenesis factors of the underlying disease or its complications [7, 9]. The optimal ratios of diabenol with metformin (1:4) and glibenclamide (5:1) were determined in experiments on rats with streptozotocin-nicotinamide-induced diabetes. Administration of these fixed combinations proved to be effective in terms of key metabolic markers, including blood glucose level, dynamics of glycated hemoglobin reduction, C-peptide level, and recovery of pancreatic β-cells, and has a positive impact on carbohydrate metabolism—liver glycogen content and glycogenolysis [45].

A very important aspect of diabetes pathogenesis is the activation of lipid peroxidation, which facilitates development of β-cell dysfunction and peripheral insulin resistance [11]. In order to address this issue in clinical practice, combination therapy regimens for diabetes have begun to include an antioxidant, for example, lipoic acid [46]. At the first stage of our study, the direct effect of some antidiabetic agents on free radical processes was studied in vitro. It was established that diabenol is a scavenger of superoxide anion, hydroxyl, and peroxyl radicals; rosiglitazone is active only against the superoxide anion, gliclazide has an antiradical effect in experiments with DPPH, and metformin and glibenclamide were unable to interfere with these processes. At the same time, the established direct antioxidant properties of some studied drugs are difficult to be expected in vivo, since they require relatively high concentrations to exert antiradical activity in vitro [47, 48].

Experimental groups	Islet area (μm^2)	Volume fraction of islets (%)	Relative number of β-cells (%)	Volume fraction of β-cells (%)	Nuclear area of β-cells (μm^2)
Intact control	15,448.2 ± 9819.4	11.0 ± 1.2	63.8 ± 7.2	74.2 ± 5.6	26.4 ± 3.7
Streptozotocin-nicotinamide-induced diabetes	12,801.5 ± 11,252.3	5.1 ± 2.3*	47.1 ± 3.5*	55.3 ± 6.1*	30.5 ± 6.2
Streptozotocin-nicotinamide induced diabetes +25 mg/kg diabenol	9559.6 ± 11,513.8	7.5 ± 1.5	54.3 ± 9.5	63.1 ± 4.6	25.4 ± 6.4

**Statistically significant compared with the intact control group.*

Table 1.
Morphometric parameters of pancreatic islets in splenic region of the pancreas of streptozotocin-nicotinamide-induced diabetic rats after administration of diabenol for 21 days (M ± m) [44].

A further study [48] determined the optimal ratios for a combination of diabenol and lipoic acid (2.8: 1 and 5.6: 1). Its activity was studied in a streptozotocin-nicotinamide-induced diabetic rat model. It was established that this combination possesses a more pronounced antidiabetic effect than monotherapy with diabenol. The more important finding of this study is a significantly reduced content of lipid peroxidation products in the liver, pancreas, and kidneys. In the pancreas under streptozotocin intoxication conditions, β-cells were significantly preserved by the combined treatment with diabenol and lipoic acid.

It is known that diabetes is associated with the increased thrombogenic potential of the blood and impaired rheology properties [49]. This effect is attributed not only to hyperosmolarity of the blood due to hyperglycemia but also to an increased aggregation of platelets and red blood cells. Among the antidiabetic agents used in clinical practice, only gliclazide has a direct inhibitory effect on platelets [8]. For other drugs, a similar effect is observed only with prolonged therapy. Given the fact that diabetes increases the frequency of thrombosis events, we studied the effect of diabenol on aggregation properties of platelets and red blood cells and its influence on microcirculation in experimental diabetes.

It was established that diabenol, both in vitro and in the conditions of the whole organism, has an antiplatelet activity. Probably, the effect on functional activity of platelets is determined by the influence of diabenol on balance of prostacyclin and thromboxane A2 systems. Diabenol showed an antithrombogenic effect on the model of thrombosis of the carotid induced by electric current and in systemic adrenaline-collagen thrombosis, exceeding the activity of gliclazide [50–52].

Diabenol reduced the aggregability and increased the erythrocyte deformability in normal conditions and, more profoundly, in experimental diabetes. Using fluorescent probes, it has been shown that diabenol was able to increase electronegativity and reduces the microviscosity of the red blood cells membrane, which results in the increase in their deformability [53–56].

Amelioration of thrombogenic potential and blood viscosity gives diabenol ability to enhance the survival of skin graft (a model of the diabetic foot) in both intact and alloxan-induced diabetic animals [57].

Toxicological study of the diabenol pharmaceutical substance and the dosage form (tablets containing 0.2 g of the active ingredient) involved acute and chronic toxicity, examination of cumulative properties, immunotoxicity, effects on carcinogenesis, and transplacental action. The therapeutic dose of the drug had no adverse effects. Subtoxic doses of diabenol lead to a sharp decrease in pancreatic β-cell secretion along with platelet hemorrhages.

Upon long-term administration of diabenol to low-cancer NMRI mice, transgenic HER-2/neu mice, and LIO rats with drinking water, no toxic or carcinogenic effect was observed. An interesting fact has been demonstrated, that is, in NMRI mice, diabenol delayed the development of age-related disorders of the extra function and increased the life span of animals. The drug inhibited the occurrence of spontaneous tumors, reduced the incidence of malignant lymphomas, and inhibited the onset and development of colon cancer induced with 1,2-dimethylhydrazine in rats. The authors of this study conclude that diabenol has an anticarcinogenic and geroprotective effect. Hence, both diabenol and metformin contain a guanidine group in their structure and exert experimentally proven antitumor effect and geroprotective activity [58–60].

The data on the pharmacokinetic study of diabenol established the values of the drug half-life and average retention time, which suggest that the substance undergoes significantly rapid elimination. The drug penetrates well into organs and tissues, especially in those with a high degree of vascularization. An important role in the processes of elimination of a compound is played by processes of its metabolism [61].

2.4 Clinical studies

Clinical studies of diabenol were conducted on 180 patients with type 2 diabetes. The drug was administered orally in solid dosage form (0.2 g tablets) 2 times a day. The study design was a randomized controlled comparative study of efficacy, tolerability, and safety. Glidiab (gliclazide) served as a reference drug. The course of diabenol administration ameliorated both fasting and postprandial hyperglycemia, reduced glycated hemoglobin level by 1.1% at the end of the third month, and increased postprandial insulin levels. Diabenol reduced platelet aggregation, increased erythrocyte deformability, and reduced their aggregability, thus normalizing coagulation hemostasis [62, 63].

Clinical and laboratory parameters of patients allowed to conclude that diabenol administered in a dose of 0.4 g per day for 3 months had no adverse or toxic effects [63].

3. Conclusions

Based on the studies performed, it can be stated that aminobenzimidazole is a universal privileged substructure that can be used as a source for development of novel antidiabetic agents. Introduction of structural modifications by means of transition to tricyclic structures led us to the identification of N^9-2,3-dihydroimidazo[1,2-*a*] benzimidazole as an optimal scaffold, and in this chemical class of compounds, agents with high glucose lowering activity were identified.

As a result, RU-254, or diabenol, was developed as the most active compound. Both in the experimental and clinical settings, it restores insulin secretion and ameliorates peripheral tissue glucose uptake. Another important effect of diabenol that has been established experimentally and confirmed in clinical studies is a reduction of thrombogenic potential and viscosity of blood. It has been demonstrated that diabenol, much like the biguanide derivative metformin, exerts an anticarcinogenic and geroprotective effect in rodents.

Thus, a new cyclic aminobenzimidazole derivative, diabenol, containing a guanidine group, combines pharmacological effects characteristic for both biguanide derivatives (reduction of hyperglycemia and liver glycogenolysis, improved glucose tolerance, anticarcinogenic and geroprotective effects) and for sulfonylurea derivative gliclazide (restoring insulin secretion, antiplatelet, and antiradical activity). It is quite possible that all of the identified effects are not a manifestation of the multi-target action but are pleiotropic effects of diabenol.

Author details

Alexander A. Spasov[1], Pavel M. Vassiliev[1]*, Vera A. Anisimova[2] and Olga N. Zhukovskaya[2]

1 Volgograd State Medical University, Volgograd, Russia

2 Research Institute for Physical and Organic Chemistry, South Federal University, Rostov-on-Don, Russia

*Address all correspondence to: pvassiliev@mail.ru

References

[1] Blaslov K, Naranda FS, Kruljac I, Renar IP. Treatment approach to type 2 diabetes: Past, present and future. World Journal of Diabetes. 2018;**9**:209-219. DOI: 10.4239/wjd.v9.i12.209

[2] Dedov II, Shestakova MV. Diabetes Mellitus: Diagnosis, Treatment, Prevention. Moscow: Medical Information Agency; 2011. p. 808

[3] Spasov AA, Chepurnova MV. Scientific approaches to the combination therapy of type 2 diabetes mellitus. Bulletin of Volgograd State Medical University. 2011;**1**:8-12

[4] Spasov AA, Petrov VI, Chepliaeva NI, Lenskaya KV. Fundamental bases of search of medicines for therapy of a diabetes mellitus type 2. Vestnik Rossiĭskoĭ Akademii Meditsinskikh Nauk. 2013;**2**:43-49. DOI: 10.15690/vramn.v68i2.548

[5] Tkachuk VA, Vorotnikov AV. Molecular mechanisms of insulin resistance. Diabetes Mellitus. 2014;**2**: 29-40. DOI: 10.14341/DM2014229-40

[6] Snel M, Jonker JT, Schoones J, Lamb H, de Roos A, Pijl H, et al. Ectopic fat and insulin resistance: Pathophysiology and effect of diet and lifestyle interventions. International Journal of Endocrinology. 2012;**2012**:1-18. DOI: 10.1155/2012/983814

[7] Steinberg GR, Kemp BE. AMPK in health and disease. Physiological Reviews. 2009;**89**:1025-1078. DOI: 10.1152/physrev.00011.2008

[8] Siluk D, Kraliszan R. Antiaggregatory activity of hypoglycaemic sulphonylureas. Diabetologia. 2002;**7**:1034-1037. DOI: 10.1007/s00125-002-0855-0

[9] Ametov AS. Type 2 Diabetes. Problems and Solutions. Moscow: GEOTAR-Media; 2015. p. 280, p. 352

[10] Ametov AS, Kondratyeva LV, Prudnikova MA. Metformin—More than the gold standard. In: Ametov AS, editor. Type 2 Diabetes. Problems and Solutions. Moscow: GEOTAR-Media; 2015. pp. 202-256

[11] Balabolkin MI, Klebanova EM, Kreminskaya VM. Treatment of Diabetes Mellitus and Its Complications. Moscow: Medicine; 2005. pp. 288-304

[12] Caprio S, Ray TK, Boden G, et al. Improvement of metabolic control in diabetic patients during mebendazole administration: Preliminary studies. Diabetologia. 1984;**27**:52-55

[13] Anisimova VA, Koshchienko YV, Pyatin BM et al. The Method of Obtaining Derivatives of 2-Amino-1-Aminoalkylbenzimidazoles. USSR Authorship certificate №. 1149592. Bulletin of Inventions. 1984

[14] Anisimova VA, Spasov AA, Kosolapov VA, et al. Synthesis and pharmacological activity of 3-(N,N-disubstituted)acetamide-1-R-2-aminobenzimidazolium chlorides. Khimiko-Farmatsevticheskii Zhurnal. 2012;**46**:6-10. DOI: 10.30906/0023-1134-2012-46-9-6-10

[15] Anisimova VA, Simonov AM. 3-Acyl-substituted imidazo[1,2-a] benzimidazoles. Khimiya Geterotsiklicheskikh Soedinenii. 1976;**1**:121-125

[16] Anisimova VA, Spasov AA, Kucheryavenko AF, et al. Synthesis and pharmacological activity of 2-(hetaryl) imidazo[1,2-a]benzimidazoles. Khimiko-Farmatsevticheskii Zhurnal. 2002;**36**:12-17

[17] Simonov AM, Anisimova VA, Borisova TA. Preparation of imidazo[1,2-a]benzimidazole derivatives from 1-alkyl-,

1-aralkyl-2-iminobenzimidazoline-3-acetic acids and their esters. Khimiya Geterotsiklicheskikh Soedinenii. 1973;**1**:111-114

[18] Anisimova VA, Zhukovskaya ON, Kuzmenko TA. Synthesis of benzimidazole derivatives and condensed systems based on it. In: Spasov AA, Petrov VI, Minkin VI, editors. Antidiabetic Potential of Benzimidazoles. Volgograd: VolgSMU; 2016. pp. 56-106

[19] Spasov AA, Dudchenko GP, Voronkova MP. Antidiabetic properties of the diabenol compound. In: Spasov AA, Petrov VI, Minkin VI, editors. Antidiabetic Potential of Benzimidazoles. Volgograd: VolgSMU; 2016. pp. 151-166

[20] Anisimova VA, Levchenko MV, Koshchienko YV, Pozharskiy AF. The method of obtaining 9-substituted 2,3-dihydroimidazo[1,2-a] benzimidazole or their salts. USSR Authorship certificate №. 952847. Bulletin of Inventions. 1982;**31**:126

[21] Anisimova VA, Levchenko MV, Kovalev GV. Synthesis and pharmacological activity of some 2,3-dihydroimidazo[1,2-a] benzimidazoles and intermediate products of their synthesis. Khimiko-Farmatsevticheskii Zhurnal. 1987;**21**:313-319

[22] Anisimova VA, Spasov AA, Tolpygin IE, Ogasov AA, Stepanov AV. Synthesis and pharmacological activity of N-aryloxyethyl-substituted 9H-2,3-dihydroimidazoyl and 10H-2,3,4,10-tetrahydropyrimido[1,2-a] benzimidazoles. Khimiko-Farmatsevticheskii Zhurnal. 2006;**40**:23-26

[23] Anisimova VA, Tolpygin IE, Spasov AA. Synthesis and pharmacological activity of 10-alkylaminoethyl-2,3,4,10-tetrahydropyrimido[1,2-a] benzimidazoles. Khimiko-Farmatsevticheskii Zhurnal. 2012;**46**:3-8

[24] Anisimova VA, Spasov AA, Kosolapov VA. Synthesis and pharmacological activity of 2,3-dihydroimidazo- and 2,3,4,10-tetrahydropyrimido[1,2-a] benzimidazolyl-N-acetic acids. Khimiko-Farmatsevticheskii Zhurnal. 2012;**46**:15-20

[25] Vassiliev PM, Kochetkov AN. IT Microcosm. State Registration Certificate for Software Program 2011618547 (Russian); 2011

[26] Vassiliev PM, Spasov AA, Kosolapov VA, Kucheryavenko AF, Gurova NA, Anisimova VA. Consensus drug design using IT microcosm. In: Gorb L, Kuz'min V, Muratov E, editors. Application of Computational Techniques in Pharmacy and Medicine. Dordrecht Netherlands: Springer Science + Business Media; 2014. pp. 369-431

[27] Filimonov DA, Poroikov VV, Gloriozova TA, Lagunin AA. PASS: State Registration Certificate for Software Program 2006613275 (Russian); 2006

[28] Filimonov DA, Poroikov VV. Prediction of the spectrum of biological activity of organic compounds. Rossiiskii Khimicheskii Zhurnal. 2006;**50**:66-75

[29] Drug Likeness Tool (DruLITo) [Internet]. 2017. Available from: http://www.niper.gov.in/pi_dev_tools/DruLiToWeb/DruLiTo_index.html [Accessed: Jan 15, 2019]

[30] QikProp: Rapid ADME Predictions of Drug Candidates [Internet]. 2017. Available from: https://www.schrodinger.com/QikProp [Accessed: Jan 15, 2019]

[31] Severin SE, Solovyeva GA, editors. Workshop on Biochemistry: A Textbook. 2nd ed. Moscow: MSU; 1989. p. 509

[32] Larsen SD, Connell MA, Cudahy MM, Evans BR, May PD, Meglasson MD, et al. Synthesis and biological activity of analogues of the antidiabetic/antiobesity agent 3-guanidinopropionic acid: Discovery of a novel aminoguanidinoacetic acid antidiabetic agent. Journal of Medicinal Chemistry. 2001;**44**:1217-1230. DOI: 10.1021/jm000095f

[33] Vassiliev PM, Spasov AA, Lenskaya KV, Poroikov VV, Filimonov DA, Anisimova VA. Planning of in silico screening and experimental study of hypoglycemic cyclic guanidine derivatives. Kuban Scientific Medicinal Bulletin. 2014;**6**(148):11-15

[34] Vassiliev PM, Spasov AA, Lenskaya KV, Anisimova VA. Substructural analysis of the hypoglycemic activity of cyclic guanidine derivatives. Bulletin of Volgograd State Medical University. 2014;**3**(51):28-30

[35] Lenskaya KV, Vassiliev PM, Spasov AA, Anisimova VA. Analysis of the prospects of chemical classes of cyclic guanidine derivatives with the method of median estimates. Vestn. Novosti Meditsinskoĭ Tekhniki. 2015;**3**:2-8. DOI: 10.12737/13202

[36] Lenskaya KV, Vassiliev PM, Spasov AA, Anisimova VA. Analysis of the prospects of chemical classes of cyclic guanidines with the method of supremal assessment. Bulletin of Volgograd State Medical University. 2015;**2**(54):98-100

[37] Lenskaya KV, Cheplyaeva NI, Vassiliev PM, Spasov AA, Anisimova VA. Frequency analysis of the dependence of hypoglycemic activity on the physicochemical parameters of cyclic guanidine derivatives. In: Proceedings of the XXth Anniversary Russian National Congress "Man and Drug". Moscow; 2013. p. 372

[38] Spasov AA, Dudchenko GP, Turchaeva AF, Kovalev SG. Development of hypoglycemic drugs with antiplatelet properties based on condensed benzimidazole derivatives. Bulletin of Volgograd State Medical University. 1995;**1**:33-36

[39] Spasov AA, Gavrilova EE. Diabenol—A new antidiabetic agent with hemobiological properties. Bulletin of Volgograd State Medical University. 1997;**3**:47-51

[40] Spasov AA, Yozhitsa IN, Bugaeva LI, Anisimova VA. Benzimidazole derivatives: Spectrum of pharmacological activity and toxicological properties (a review). Pharmaceutical Chemistry Journal. 1999;**33**:232-243. DOI: 10.1007/BF02510042

[41] Zolotov NN, Kreminskaya VM. Dipeptidyl peptidase IV inhibiting agent and pharmaceutical composition based on it. Patent RU2485952. 2013

[42] Spasov AA, Cheplyaeva NI, Lenskaya KV, Snigur GL. The effect of a limiglidol on DPP-4 and morphological features of pancreatic islets in streptozotocin diabetes model. Eksperimental'naia i Klinicheskaia Farmakologiia. 2015;**78**:8-12

[43] Spasov AA, Voronkova MP, Snigur GL. An experimental model of type 2 diabetes mellitus. Biomedicine. 2011;**3**:12-18

[44] Cheplyaeva NI. Pharmacological properties of the combination of hypoclycemic substance diabenol and α-lipoic acid [thesis]. Volgograd: Volgograd State Medical University; 2012

[45] Chepurnova MV. Pharmacological properties of combined hypoglycemic substances on the basis of the drug diabenol [thesis]. Volgograd: Volgograd State Medical University; 2011

[46] Evans JL, Goldfine ID. Alpha-lipoic acid: A multifunctional antioxidant that improves insulin sensitivity

in patient with type 2 diabetes. Diabetes Technology & Therapeutics. 2000;**2**:401-413

[47] Spasov AA, Kosolapov VA, Cheplyaeva NI. Antioxidant activity of oral hypoglycemic agents. Problems of Endocrinology. 2011;**4**:21-24

[48] Spasov AA, Kosolapov VA, Cheplyaeva NI. Comparative characteristics of antioxidant properties of hypoglycemic agents diabenol and gliclazide. Eksperimental'naia i Klinicheskaia Farmakologiia. 2011;**74**:14-16

[49] Morel O, Kossler L, Ohimann P. Diabetes and the platelets: Toward new therapentic paradigms for diabetic atherothrombosis. Atheroscherosis. 2010;**2**:367-376

[50] Kucheryavenko AF, Spasov AA, Petrov VI, Anisimova VA. Antiaggregant activity of a new benzimidazole derivative. Bulletin of Experimental Biology and Medicine. 2014;**156**:796-798. DOI: 10.1007/s10517-014-2453-9

[51] Kucheryavenko AF, Spasov AA, Smirnov AV. Antithrombotic activity of a new hypoglycemic compound limiglidole in mouse model of cell thrombosis. Bulletin of Experimental Biology and Medicine. 2015;**159**:41-43. DOI: 10.1007/s10517-015-2885-x

[52] Spasov AA, Kucheryavenko AF, Chepurnova MV, Lenskaya KV. Antithrombotic activity of hypoglycemic agents. Regional Blood Circulation and Microcirculation. 2011;**10**:95-98

[53] Degtyarev AN, Kucheryavenko AF, Spasov AA, Ostrovskiy OV. Benzimidazole derivatives as the basis for the creation of drugs that affect the rheological properties of blood. In: Proceedings of Actual Problems of Experimental and Clinical Pharmacology; May 1999; Saint-Petersburg. Saint-Petersburg: Politekhnika; 1999. p. 120

[54] Degtyarev AN, Ostrovskiy OV, Shakhova NI, Anisimova VA. Evaluation of the corrective effect of new compounds, benzimidazole derivatives, with the "increased viscosity syndrome". In: Proceedings of Actual Problems of Experimental and Clinical Pharmacology; May 1999; Saint-Petersburg. Saint-Petersburg: Politekhnika; 1999. p. 59

[55] Spasov AA, Petrov VI, Anisimova VA. New hypoglycemic agent with hemorheological properties. In: Proceedings of IVth Russian Diabetological Congress; 19-22 May 2008; Moscow. Moscow; 2008. p. 336

[56] Kucheryavenko AF, Spasov AA, Naumenko LV. The influence of new hypoglycemic agent limiglidol on the parameters of hemostasis in experimental diabetes. Problems of Endocrinology. 2015;**61**:51-56. DOI: 10.14341/probl201561151-56

[57] Vasilyeva SV, Galenko-Yaroshevskiy VP, Khropova TN, Tegay AV, Uvarov AV. The effect of imidazobenzimidazole derivatives RU-185 and RU-254 on the viability of the skin graft. Bulletin of Experimental Biology and Medicine. 2002;**2**:90

[58] Popovich IG, Zabezhinskiĭ MA, Anikin IV, Tyndyk ML, Spasov AA, Anisimov VN. Inhibition of 1,2-dimethylhydrazine-induced carcinogenesis in rat gut by the antidiabetic drug Diabenol. Voprosy Onkologii. 2004;**50**:562-566

[59] Popovich IG, Zabezhinskiĭ MA, Egormin PA, Anikin IV, Tyndyk ML, Semenchenko AV, et al. Effect of antidiabetic drug diabenol on parameters of biological age, life span and tumor development in NMRI and HER-2/neu mice. Advances in Gerontology. 2004;**15**:80-90

[60] Popovich IG, Zabezhinski MA, Egormin PA, Tyndyk ML, Anikin IV, Spasov AA, et al. Insulin in aging and cancer: Antidiabetic drug diabenol as geroprotector and anticarcinogen. The International Journal of Biochemistry & Cell Biology. 2005;**37**:1117-1129. DOI: 10.1016/j.biocel.2004.08.002

[61] Spasov AA, Dudchenko GP, Smirnova LA, Gavrilova ES. Pharmacodynamic and pharmacokinetic properties of the compound RU-254. Bulletin of Volgograd State Medical University. 1999;**5**:26-34

[62] Dedov II, Balabolkin MI, Spasov AA, Petrov VI. New domestic hypoglycemic agent with hemorheological properties—diabenol (clinical studies). In: Proceedings of IVth Russian Diabetological Congress; 19-22 May 2008; Moscow. Moscow; 2008. p. 268

[63] Petrov VI, Spasov AA, Nedogoda SV, Voronkova MP. Clinical efficacy of the drug diabenol in type 2 diabetes mellitus. In: Spasov AA, Petrov VI, Minkin VI, editors. Antidiabetic Potential of Benzimidazoles. Volgograd: VolgSMU; 2016. pp. 451-464

Chapter 6

Development of Benzimidazole Compounds for Cancer Therapy

Puranik Purushottamachar,
Senthilmurugan Ramalingam and Vincent C.O. Njar

Abstract

A fact that is largely unknown in the lay press and even the scientific community is that today cancer kills more people worldwide than tuberculosis (TB), malaria, and human immunodeficiency virus (HIV) combined. Benzimidazole is a heterocyclic aromatic organic compound considered to be a useful pharmacophore in a variety of impactful drugs. The purpose of this review is to highlight the benzimidazole-containing agents that are currently in clinical use or in clinical development as anticancer drugs. It is hoped that this review would function as comprehensive working reference of research accomplishment in the field of discovery and development of benzimidazole-based anticancer drugs.

Keywords: benzimidazole derivatives, privileged pharmacophore, anticancer drugs/agents

1. Introduction

Benzimidazole (**1**) (**Figure 1**) is used as the major scaffold or as a moiety on other scaffolds for the development of a variety of drugs [1–4]. The wide range of pharmacological activities of benzimidazole-containing agents are attributed to the unique fused benzene and imidazole rings, which can interact in a noncovalent manner with a range of biological targets due to the presence of an electron-rich aromatic system and the two hetero-nitrogen atoms [5, 6]. Because of the ability of benzimidazole derivative to interact with a variety of unrelated molecular targets, the term "privileged substructure/moiety" is ascribed to this unique azole agent. It is believed that the interest in benzimidazole chemistry and as a scaffold/moiety in the discovery and development of drugs arose from the discovery of the rare and most prominent benzimidazole compound in nature, *N*-ribosyl-dimethylbenzimidazole (**2**) (**Figure 1**), which serves as an axial ligand for cobalt in vitamin B12 [7].

Although several benzimidazole derivatives have been approved for clinical use, including antiparasitic, antiulcer, antihypertensive, antihistaminic, and antiemetic drugs [1–4], only one anticancer drug, bendamustine (**3**) (**Figure 2**), has received FDA approval [8–10]. Two prominent benzimidazole agents, selumetinib (**4**) (**Figure 2**) [1, 6] and galeterone (**5**) (**Figure 2**) [11], that advanced to phase III clinical trials, but are yet to be approved as anticancer drugs, will also be discussed.

IntechOpen

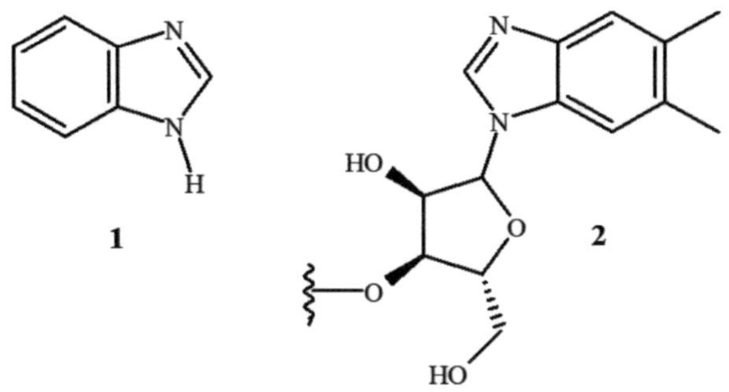

Figure 1.
Chemical structures of benzimidazole (1) and N*-ribosyl-dimethylbenzimidazole (2).*

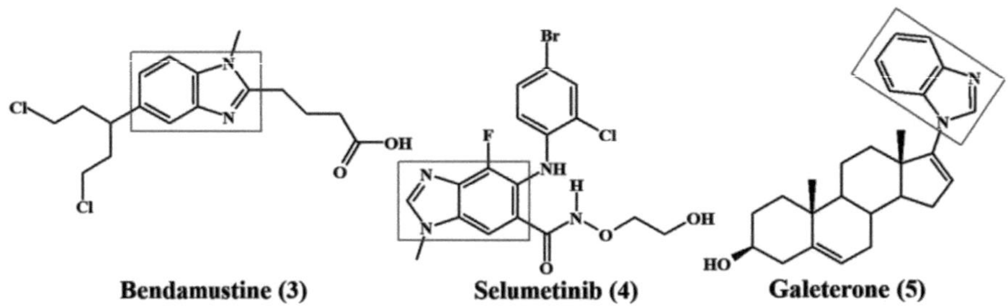

Figure 2.
Chemical structures of bendamustine (3), selumetinib (4), and galeterone (5).

2. Benzimidazole agents in the clinic and in clinical development

2.1 Bendamustine (3)

Bendamustine (**3**) (**Figure 2**) was discovered in a structure-activity relationship (SAR) campaign directed to obtain more effective and safer water-soluble analogs of chlorambucil (**6**) (**Figure 3**), a nitrogen mustard, which is used clinically against chronic lymphatic leukemia, lymphomas, and advanced ovarian and breast carcinomas [12]. The strategy was replacement of the benzene ring in compound **6** with purine-like *N*-methylbenzimidazole moiety in the hope of obtaining an anticancer agent with antimetabolite and DNA-alkylating activities. Although bendamustine was first synthesized in the early 1960s [13], it was approved under the trade name Treanda® by the US Food and Drug Administration (FDA) in 2008 for the treatment of chronic lymphocytic leukemia, multiple myeloma, and non-Hodgkin's lymphoma [10, 14–16].

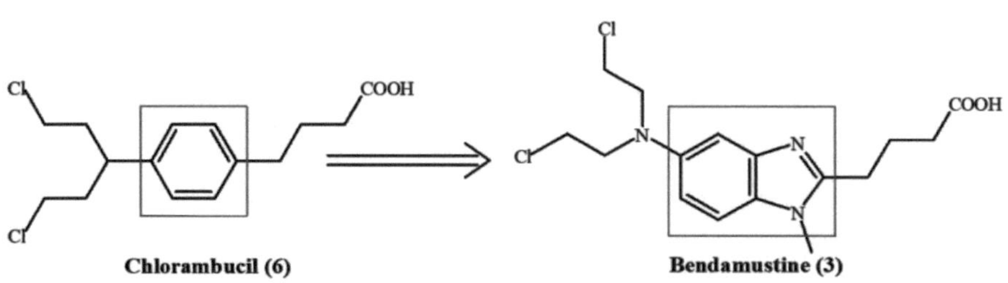

Figure 3.
Replacement of aromatic benzene ring of chlorambucil (6) to produce bendamustine (3).

2.1.1 Chemistry

Bendamustine (**3**) 4-{5-[bis(2-chloroethyl)amino]-1-methylbenzimidazol-2-yl} butanoic acid was first synthesized via eight synthetic steps with an overall yield of 12% [13, 17]. However, Chen and colleagues have developed a new, more efficient, and cost-effective route focused on the use of sustainable chemistry for the synthesis of bendamustine hydrochloride, with the overall yield improved from 12 to 45% as outlined in **Figure 4** [18]. This new synthesis is currently used for the commercial production of **3**.

2.1.2 Summary of bendamustine's preclinical and clinical pharmacology

Even though bendamustine is an alkylating agent, due to its ability to cause intra-strand and inter-strand cross-links between DNA bases, it has been reported that the DNA breaks induced by bendamustine are more extensive/durable than those induced by other alkylating agents, such as chlorambucil, cyclophosphamide, or carmustine [19–21]. In addition, the drug was shown to exhibit partial cross-resistance to other alkylating agents. These data suggested that bendamustine may possess additional mechanisms of action. Indeed, a comprehensive study by Leoni and colleagues clearly demonstrated that bendamustine exhibits a distinct pattern of activities unrelated to other alkylating drugs. Using a variety of lymphoid cancer cell lines, the study concluded that mechanisms of action include induction of mitotic catastrophe, inhibition of mitotic checkpoints, and activation of DNA-damage stress response and apoptosis. Compared to other alkylating agents, bendamustine was shown to activate the base excision DNA repair pathway rather than the alkyl transferase DNA repair mechanism [20].

Although bendamustine is approved for the treatment of a variety of lymphoid cancers, its activity has also been reported in several cancers, including cancers of small cell lung, breast, hepatic, bile duct, and head and neck. The studies by Chow and colleagues using leukemic cell lines in vitro or ex vivo cells from patients with

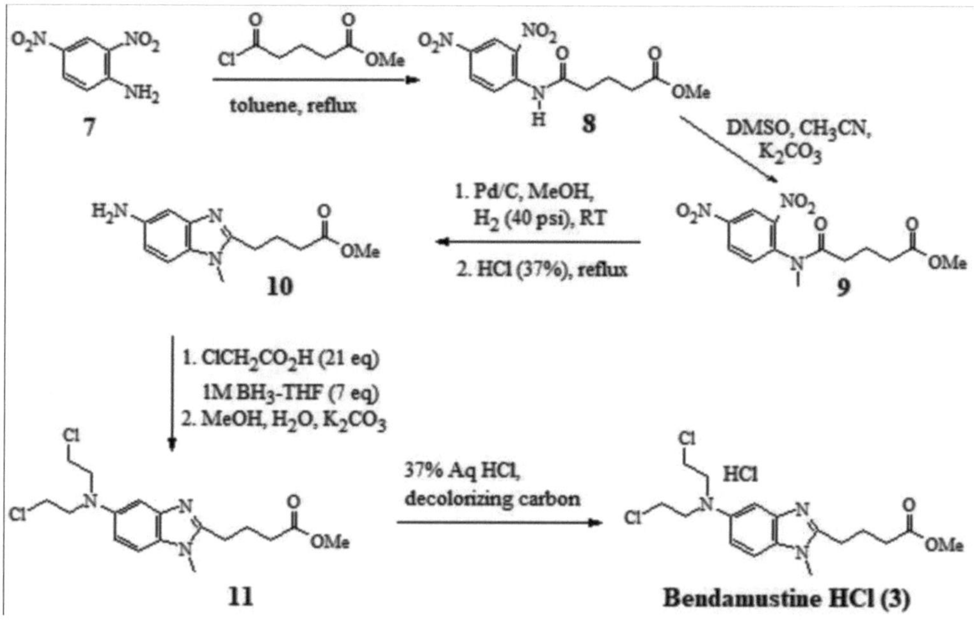

Figure 4.
New optimized synthesis of bendamustine hydrochloride (3).

leukemic progression to clarify interactions between bendamustine and other chemotherapeutic drugs unraveled synergy with cladribine, in contrast to observed antagonism with mitoxantrone or doxorubicin. The observation of synergism between bendamustine and rituximab (an anti-CD20 antibody) in in vitro CD20-positive DOHH-2 and WSU-NHL cell lines and ex vivo B-cell chronic lymphocytic leukemia (CLL) cells [22] and in mice with Daudi xenografts [23] provided the impetus for clinical trials combining these two drugs [24, 25].

Based on the discussion above, it is obvious that bendamustine is an "old drug rediscovered." For over 30 years, bendamustine was used in Eastern Germany as monotherapy for several cancers, including breast cancer, chronic lymphocytic leukemia (CLL), Hodgkin's lymphoma, non-Hodgkin's lymphoma (NHL), and multiple myeloma (MM) [26–37]. However, following the reunification of Germany, other countries initiated clinical trials of bendamustine as a single agent and in combination with other drugs. Bendamustine has achieved worldwide regulatory approval and is a standard-of-care drug for the treatment on many lymphoid malignancies. Several articles that provide comprehensive reviews of the discovery and development of this unique drug are available [8–10].

2.2 Selumetinib (4)

Selumetinib (**4**) (AZD6244: ARRY-142866) is an orally available, potent, selective inhibitor of mitogen-activated protein kinase (MAPK)/extracellular signal-related kinase (ERK) kinases 1 and 2 (MEK1 and MEK2) [6]. This agent has been extensively studied in many preclinical and clinical studies in several tumor types with mixed results. Here, we will summarize the chemistry, preclinical studies, and clinical studies.

2.2.1 Chemistry

Selumetinib (6-(4-bromo-2-chloroanilino)-7-fluoro-N-(2-hydroxyethoxy)-3-methyl benzimidazole-5-carboxamide) (**4**) is a diarylamine hydroxamide, containing mono-methylated benzimidazole subunit [38, 39]. It is a second-generation, orally active small molecule that acts as a selective and ATP-uncompetitive inhibitor of MEK1 and MEK2, binding to the allosteric binding site [38, 39]. The synthesis of selumetinib is yet to be reported in the literature.

2.2.2 Summary of selumetinib's preclinical and clinical pharmacology

Selumetinib inhibits the enzymatic activity of purified constitutively active MEK1 with a half maximal inhibitory concentration (IC_{50}) of 14 nM and was shown to be highly selective for inhibition of these targets compared to other related kinases [39]. Using several human cancer cell lines such as NSCLC, melanoma, and pancreatic and colorectal cell lines, it was shown that selumetinib was a potent antiproliferative agent. Analysis of the data revealed that cell lines with mutant BRAF and RAS were sensitive to selumetinib [40]. Selumetinib had little effect on the growth of Malme-3, the control cell line to the melanoma. Additional studies suggested that the growth inhibitory effects of selumetinib was not due to wide-ranging cytotoxicity [39], and it was also established that selumetinib effectively inhibits the phosphorylation of ERK 1 and ERK 2, which are substrates of MEK1 and MEK2 in the MAP kinase pathway. This mechanism of action was confirmed in tumor xenografts. Additionally, increased markers of apoptosis such as cleaved caspase 3 and decreased cell proliferation were seen in response to treatment with selumetinib in the xenograft models [39, 40].

The promising preclinical in vitro and in vivo data provided the rationale for multiple clinical trials in cancers with activated Raf-MEK-ERK signaling. In preparation for clinical evaluation of selumetinib, it was originally developed as a free base and administered as a liquid suspension, but subsequently a capsule formulation of the hydrogen sulfate salt was found to be more suitable for further development [38]. Several phase I and II clinical trials conducted against solid tumors to test the impact of selumetinib as a monotherapy were unsuccessful [41–44]. This led to the conduct of several clinical trials with selumetinib in combination with other cancer drugs. A notable trial was the randomized phase II study of selumetinib in combination with docetaxel, as a second-line treatment for patients with KRAS-mutant advanced NSCLC which showed very promising results [45]. The median progression-free survival was 5.3 months with selumetinib + docetaxel and 2.1 months with docetaxel alone. The objective response rate was 37% for selumetinib + docetaxel vs. 0% for docetaxel alone ($p < 0.001$), and the median overall survival was 9.4 months for selumetinib + docetaxel vs. 5.2 months for docetaxel alone (HR for death, 0.80 [80% CI, 0.56–1.14]; one-sided $p = 0.21$). Unfortunately, in a multinational 510 randomized patients with previously treated advanced KRAS-mutant NSCLC trial, the combination of selumetinib with docetaxel did not improve progression-free survival compared with docetaxel alone [46]. Clearly, addition clinical studies are required to realize the potential impact of selumetinib alone and in combination with other drugs for the treatment of a variety of cancers [47, 48].

2.3 Galeterone (5)

Galeterone (also called VN/124-1 or TOK-001) is an orally available anticancer agent. It was rationally designed as an inhibitor of androgen biosynthesis via inhibition of 17α-hydroxylase/17,20-lyase (CYP17), the key enzyme which catalyzes the biosynthesis of androgens from the progestins. Through extensive and rigorous preclinical studies, galeterone was shown to modulate two other targets in the androgen/androgen receptor (AR) signaling pathway [11] and shown to inhibit the eukaryotic initiation factor 4E (eIF4E) protein translational machinery [49]. Galeterone advanced successfully through phases I and II clinical trials in prostate cancer patients but was unsuccessful in the pivotal phase III clinical trial in men with castration-resistant prostate cancer (CRPC), harboring AR splice variants (e.g., AR-V7). We present a summary of the chemistry, preclinical studies, and clinical studies [50].

2.3.1 Chemistry

Galeterone, 3β-hydroxy-17-(1H-benzimidazole-1-yl)androsta-5,16-diene (**5**), is one of a series of novel Δ^{16}-17-azolyl steroid, which, unlike previously known 17-heteroaryl steroids, the azole moiety is attached to the steroid nucleus at C-17 via a nitrogen of the azole. The synthesis of galeterone from commercially available 3β-acetoxyandrosta-5-en-17-one (**12**) is presented in **Figure 5** [11, 51], and a facile and large-scale preparation (commercial process) of the compound has been developed but is yet to appear in the literature.

2.3.2 Summary of galeterone's preclinical and clinical pharmacology

Using intact CYP17 expressing *Escherichia coli*, galeterone was shown to be a potent inhibitor of the enzyme with an IC_{50} value of 300 nM and was shown to be more potent than abiraterone (IC_{50} value of 800 nM) [52]. Additional studies by

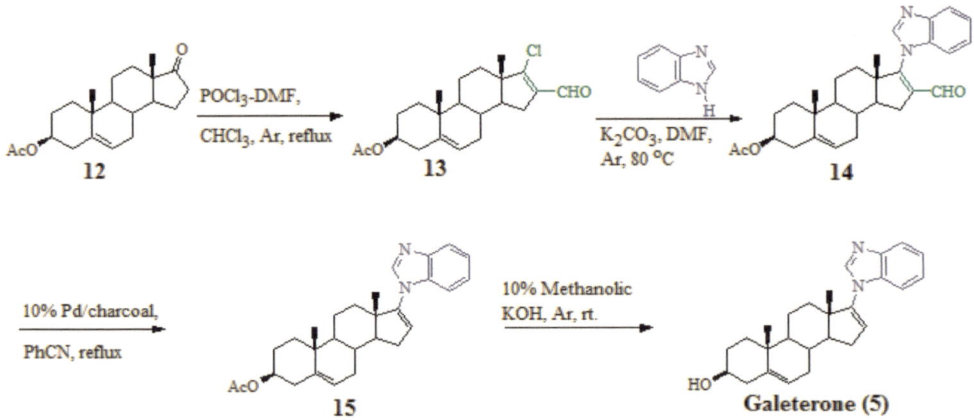

Figure 5.
Synthesis of galeterone.

our group revealed that galeterone could disrupt androgen signaling through multiple targets [51–54].

We strongly believe that the increased efficacy of galeterone in several prostate cancer models both in vitro and in vivo is due to its ability to downregulate the AR and block androgen binding to AR. Using well-established AR-competitive binding assays (against the synthetic androgen [^{3}H]R1881), galeterone was equipotent to Casodex in LNCaP cells but had a slightly higher affinity for the wild-type receptor in PC3-AR cells. In transcriptional activation assays (utilizing a luciferase reporter), galeterone was shown to be a pure AR antagonist of the wild-type AR and the T877A mutation found in LNCaP cells [53]. In prostate cancer cell lines, galeterone inhibited the growth of CRPCs, which had increased AR and were no longer sensitive to Casodex [53] and was also shown to inhibit the growth of AR-negative prostate cancer cells [54]. In addition, galeterone demonstrated superior synergy for growth inhibition in combination with everolimus or gefitinib compared with Casodex [55].

Recent in vitro studies have shown additional activities of galeterone, including proteasomal degradation of AR and its splice variants [56, 57] and inhibition of the eukaryotic initiation factor 4E (eIF4E) protein translational machinery via induction of proteasomal degradation of mitogen-activated protein kinase-interacting kinases 1 and 2 (Mnk1 and Mnk2) [58, 59].

Because of the short half-life ($t_{1/2}$ = ~40 min) in mice, galeterone was administered twice daily in our antitumor efficacy studies. Galeterone (0.13 mmol/kg twice daily) caused a 93.8% reduction (p = 0.00065) in the mean final LAPC-4 xenograft volume compared with controls, and this efficacy was significantly more effective than castration or our most potent CYP17 inhibitor, VN/85-1 [51]. In another antitumor efficacy study, treatment of galeterone (0.13 mmol twice daily) was very effective in preventing the formation of LAPC4 tumors (6.94 vs. 2410.28 mm^3 in the control group). Galeterone (0.13 mmol/kg twice daily) and VN/124-1 (0.13 mmol/kg twice daily) + castration induced *regression* of LAPC4 tumor xenografts by 26.55 and 60.67%, respectively [53]. Using castration-resistant prostate cancer (CRPC) HP-LNCaP tumor xenografts, we showed that galeterone + everolimus (m-TORC1 inhibitor) acted in concert to reduce tumor growth [60]. Utilizing the androgen-dependent LAPC-4 prostate cancer xenograft model, we have shown galeterone is more efficacious than the blockbuster prostate cancer drug abiraterone (Zytiga®) [61]. We also reported that galeterone potently inhibits the growth of CRPC CWR22Rv1 tumor xenografts [56].

Based on these impressive preclinical data, galeterone was licensed by the University of Maryland, Baltimore, to Tokai Pharmaceuticals, Inc., who initiated Androgen Receptor Modulation Optimized for Response 1 (ARMOR1) phase 1/ phase 2 trials in castrate-resistant prostate cancer patients on November 5, 2009 [11]. The ARMOR phase I and phase II studies conducted with galeterone demonstrated that galeterone is well tolerated with promising clinical activity in patients with CRCP [62, 63]. To determine whether galeterone has clinical activity in patients with C-terminal loss of the androgen receptor, circulating tumor cells were retrospectively tested for C-terminal loss. Of the seven patients identified, six had PSA50 responses. These promising phases I and II studies enabled the selection of galeterone 2500 mg/day dose for the pivotal phase III trial (ARMOR3-SV, NCT02438007). The retrospective data of patients with C-terminal loss of the androgen receptor supported the design of ARMOR3-SV pivotal trial in which patients with AR-V7 were randomized to receive either galeterone or enzalutamide. Regrettably, the trial was discontinued following review by the independent Data Monitoring Committee, though no safety concerns were cited regarding this recommendation [50]. Gratifyingly, Educational and Scientific LLC (ESL), Baltimore, announced (December 17, 2018) that the University of Maryland, Baltimore (UMB), has granted ESL an exclusive license for the development of galeterone for the treatment of patients with CRPC. We eagerly await the initiation of a new phase III clinical study of galeterone in men with prostate cancer.

3. Concluding remarks

Despite the enormous literature on the synthesis and preclinical evaluation of numerous benzimidazole-containing compounds, it is unclear why very few of this class of compounds have entered clinical trials for evaluation as potential anticancer drugs. Given the fact that many benzimidazole-containing drugs have achieved blockbuster status for other diseases, it may be reasonable to suggest that the researchers interested in the development of benzimidazole-contained anticancer drugs should carefully study the process that have resulted in successful non-cancer benzimidazole drugs. We hope that this review will stimulate research activities that would eventually produce new anticancer benzimidazole drugs.

Acknowledgements

This work was supported in part by a grant from the National Institutes of Health (NIH) and the National Cancer Institute (NCI) (R01CA224696) to VCON.

Conflict of interest

VCON is the lead inventor of galeterone; the patents and technologies thereof are owned by the University of Maryland, Baltimore. The other authors declare no potential conflict of interest.

Note added in proof

During the review of this manuscript, it was reported that an analog of Selumetinib called Binimetinib (Mektovi) in combination with a BRAF inhibitor

(Encorafenib, Braftovi) was approved by US Food and Drug Administration (FDA) for the treatment of unresectable or metastatic melanoma with BRAF mutations [64].

List of abbreviations

AME	apparent mineralocorticoid excess
AR	androgen receptor
ARMOR	androgen receptor modulation optimized for response
AR-V7	a type of androgen receptor splice variant
BRAF	a human gene that encodes a protein called B-Raf
CI	confidence interval
CLL	chronic lymphocytic leukemia
CRPC	castration-resistant prostate cancer
CWR22Rv1	a type of AR/AR-splice variants positive human prostate cancer cell line
CYP17	cytochrome P450 17α-hydroxylase/17,20-lyase
DNA	deoxyribonucleic acid
eIF4E	eukaryotic initiation factor 4E
ERK1 and ERK2	extracellular signal-regulated kinases 1 and 2
ESL	Educational and Scientific LLC
FDA	Food and Drug Administration
HIV	human immunodeficiency virus
HR	hazard ratio
IC_{50}	is the concentration of inhibitor required to inhibit the enzyme activity by 50%
KRAS	Kirsten retrovirus-associated DNA sequences
LAPC-4	a type of AR-positive human prostate cancer cell line
LNCaP	a type of AR-positive human prostate cancer cell line
MAPK	mitogen-activated protein kinase
MEK1 and MEK2	MAPK/ERK kinase
MM	multiple myeloma
Mnk1 and Mnk2	mitogen-activated protein kinase-interacting kinases 1 and 2
NHL	non-Hodgkin's lymphoma
NSCLC	non-small cell lung carcinoma
PC3-AR	a type of AR-negative prostate cancer cell line transfected with AR
PSA	prostate-specific antigen
RAF	rapidly accelerated fibrosarcoma
RAS	retrovirus-associated DNA sequences
SAR	structure-activity relationship
TB	tuberculosis
TOK-001	another code name of galeterone
UMB	University of Maryland, Baltimore
VN/85-1	code name for a CYP17 inhibitor/AR antagonist/AR degrader
VN/124-1	original code name of galeterone

Author details

Puranik Purushottamachar[1,2], Senthilmurugan Ramalingam[1,2]
and Vincent C.O. Njar[1,2,3*]

1 Department of Pharmacology, University of Maryland School of Medicine, Baltimore, MD, USA

2 Center for Biomolecular Therapeutics, University of Maryland School of Medicine, Baltimore, MD, USA

3 Marlene and Stewart Greenebaum Comprehensive Cancer Center, University of Maryland School of Medicine, Baltimore, MD, USA

*Address all correspondence to: vnjar@som.umaryland.edu

References

[1] Gaba M, Mohan C. Development of drugs based on imidazole and benzimidazole bioactive heterocycles: Recent advances and future directions. Medicinal Chemistry Research. 2016;**25**: 173-210

[2] Keri RS, Hiremathad A, Budagumpi S, Nagaraja BM. Comprehensive review in current developments of benzimidazole-based medicinal chemistry. Chemical Biology & Drug Design. 2015;**86**:19-65. DOI: 10.1111/cbdd.12462

[3] Shrivastava N, Naim MJ, Alam MJ, Nawaz F, Ahmed S, Alam O. Benzimidazole scaffold as anticancer agent: Synthetic approaches and structure-activity relationship. Archiv der Pharmazie. 2017;**350**:1-80. e1700040. DOI: 10.1002/ardp.201700040

[4] Yadav G, Ganguly S. Structure activity relationship (SAR) study of benzimidazole scaffold for different biological activities: A mini-review. European Journal of Medicinal Chemistry. 2015;**97**:419-443. DOI: 10.1016/j.ejmech.2014.11.053

[5] DeSimone RW, Currie KS, Mitchell SA, Darrow JW, Pippin DA. Privileged structures: Applications in drug discovery. Combinatorial Chemistry & High Throughput Screening. 2004;**7**: 473-494

[6] Gaba M, Singh S, Mohan C. Benzimidazole: An emerging scaffold for analgesic and anti-inflammatory agents. European Journal of Medicinal Chemistry. 2014;**76**: 494-505. DOI: 10.1016/j.ejmech.2014.01.030

[7] Barker HA, Smyth RD, Weissbach H, Toohey JI, Ladd JN, Volcani BE. Isolation and properties of crystalline cobamide coenzymes containing benzimidazole or 5,6-dimethylbenzimidazole. The Journal of Biological Chemistry. 1960;**235**:480-488

[8] Cheson BD, Brugger W, Damaj G, Dreyling M, Kahl B, Kimby E, et al. Optimal use of bendamustine in hematologic disorders: Treatment recommendations from an international consensus panel—An update. Leukemia & Lymphoma. 2016;**57**:766-782. DOI: 10.3109/10428194.2015.1099647

[9] Cheson BD, Rummel MJ. Bendamustine: Rebirth of an old drug. Journal of Clinical Oncology. 2009;**27**: 1492-1501. DOI: 10.1200/JCO.2008.18.7252

[10] Tageja N, Nagi J. Bendamustine: Something old, something new. Cancer Chemotherapy and Pharmacology. 2010;**66**:413-423. DOI: 10.1007/s00280-010-1317-x

[11] Njar VC, Brodie AM. Discovery and development of Galeterone (TOK-001 or VN/124-1) for the treatment of all stages of prostate cancer. Journal of Medicinal Chemistry. 2015;**58**: 2077-2087. DOI: 10.1021/jm501239f

[12] Nicolle A, Proctor SJ, Summerfield GP. High dose chlorambucil in the treatment of lymphoid malignancies. Leukemia & Lymphoma. 2004;**45**: 271-275

[13] Ozegowski W, Krebs D. IMET 3393, (−[1-methyl-5-bis-(−chloroethyl)-amino-benzimidazolyl-(2)]-butyric) acid hydrochloride, a new cytostatic agent from among the series of benzimidazole mustard compounds. Zentralblatt für die Pharmazie. 1971; **110**:1013-1019

[14] Knauf WU, Lissichkov T, Aldaoud A, Liberati A, Loscertales J, Herbrecht R, et al. Phase III randomized study of bendamustine compared with

chlorambucil in previously untreated patients with chronic lymphocytic leukemia. Journal of Clinical Oncology. 2009;**27**:4378-4384. DOI: 10.1200/JCO.2008.20.8389

[15] Rasschaert M, Schrijvers D, Van den Brande J, Dyck J, Bosmans J, Merkle K, et al. A phase I study of bendamustine hydrochloride administered once every 3 weeks in patients with solid tumors. Anti-Cancer Drugs. 2007;**18**:587-595. DOI: 10.1097/CAD.0b013e3280149eb1

[16] Rasschaert M, Schrijvers D, Van den Brande J, Dyck J, Bosmans J, Merkle K, et al. A phase I study of bendamustine hydrochloride administered day 1+2 every 3 weeks in patients with solid tumours. British Journal of Cancer. 2007;**96**:1692-1698. DOI: 10.1038/sj.bjc.6603776

[17] Werner W, Letsch G, Ihn W, Sohr R, Preiss R. Synthesis of a potential metabolite of the carcinostatic bendamustin (Cytostasen). Pharmazie. 1991;**46**:113-114

[18] Chen J, Przyuski K, Roemmele R, Bakale RP. Discovery of a novel, efficient, and scalable route to bendamustine hydrochloride: The API in Treanda. Organic Process Research and Development. 2011;**15**:1063-1072

[19] Hartmann M, Zimmer C. Investigation of cross-link formation in DNA by the alkylating cytostatic IMET 3106, 3393 and 3943. Biochimica et Biophysica Acta. 1972;**287**:386-389

[20] Leoni LM, Bailey B, Reifert J, Bendall HH, Zeller RW, Corbeil J, et al. Bendamustine (Treanda) displays a distinct pattern of cytotoxicity and unique mechanistic features compared with other alkylating agents. Clinical Cancer Research. 2008;**14**:309-317. DOI: 10.1158/1078-0432.CCR-07-1061

[21] Strumberg D, Harstrick A, Doll K, Hoffmann B, Seeber S. Bendamustine hydrochloride activity against doxorubicin-resistant human breast carcinoma cell lines. Anti-Cancer Drugs. 1996;**7**:415-421

[22] Chow KU, Sommerlad WD, Boehrer S, Schneider B, Seipelt G, Rummel MJ, et al. Anti-CD20 antibody (IDEC-C2B8, rituximab) enhances efficacy of cytotoxic drugs on neoplastic lymphocytes *in vitro*: Role of cytokines, complement, and caspases. Haematologica. 2002;**87**:33-43

[23] Kanekal S, Crain B., Elliott G. SDX-105 (Trenada) enhances the tumor inhibitory effect of rituximab in Daudi lymphoma xenografts. Blood 2004, 104: 229b, (abstr 4580)

[24] Robinson KS, Williams ME, van der Jagt RH, Cohen P, Herst JA, Tulpule A, et al. Phase II multicenter study of bendamustine plus rituximab in patients with relapsed indolent B-cell and mantle cell non-Hodgkin's lymphoma. Journal of Clinical Oncology. 2008;**26**: 4473-4479. DOI: 10.1200/JCO.2008.17.0001

[25] Rummel MJ, Al-Batran SE, Kim SZ, Welslau M, Hecker R, Kofahl-Krause D, et al. Bendamustine plus rituximab is effective and has a favorable toxicity profile in the treatment of mantle cell and low-grade non-Hodgkin's lymphoma. Journal of Clinical Oncology. 2005;**23**:3383-3389. DOI: 10.1200/JCO.2005.08.100

[26] Bottke D, Bathe K, Wiegel T, Hinkelbein W. Phase I trial of radiochemotherapy with bendamustine in patients with recurrent squamous cell carcinoma of the head and neck. Strahlentherapie und Onkologie. 2007;**183**:128-132. DOI: 10.1007/s00066-007-1597-1

[27] Eichbaum MH, Schuetz F, Khbeis T, Lauschner I, Foerster F, Sohn C, et al. Weekly administration of bendamustine as salvage therapy in metastatic breast

cancer: Final results of a phase II study. Anti-Cancer Drugs. 2007;**18**:963-968. DOI: 10.1097/CAD.0b013e328165d11a

[28] Hartmann JT, Mayer F, Schleicher J, Horger M, Huober J, Meisinger I, et al. Bendamustine hydrochloride in patients with refractory soft tissue sarcoma: A noncomparative multicenter phase II study of the German sarcoma group (AIO-001). Cancer. 2007;**110**:861-866. DOI: 10.1002/cncr.22846

[29] Kollmannsberger C, Gerl A, Schleucher N, Beyer J, Kuczyk M, Rick O, et al. Phase II study of bendamustine in patients with relapsed or cisplatin-refractory germ cell cancer. Anti-Cancer Drugs. 2000;**11**:535-539

[30] Koster W, Heider A, Niederle N, Wilke H, Stamatis G, Fischer JR, et al. Phase II trial with carboplatin and bendamustine in patients with extensive stage small-cell lung cancer. Journal of Thoracic Oncology. 2007;**2**:312-316. DOI: 10.1097/01.JTO.0000263714.46449.4c

[31] Koster W, Stamatis G, Heider A, Avramidis K, Wilke H, Koch JA, et al. Carboplatin in combination with bendamustine in previously untreated patients with extensive-stage small cell lung cancer (SCLC). Clinical Drug Investigation. 2004;**24**:611-618. DOI: 10.2165/00044011-200424100-00007

[32] Reichmann U, Bokemeyer C, Wallwiener D, Bamberg M, Huober J. Salvage chemotherapy for metastatic breast cancer: Results of a phase II study with bendamustine. Annals of Oncology. 2007;**18**:1981-1984. DOI: 10.1093/annonc/mdm378

[33] Schmidt-Hieber M, Schmittel A, Thiel E, Keilholz U. A phase II study of bendamustine chemotherapy as second-line treatment in metastatic uveal melanoma. Melanoma Research. 2004; **14**:439-442

[34] Schmittel A, Knodler M, Hortig P, Schulze K, Thiel E, Keilholz U. Phase II trial of second-line bendamustine chemotherapy in relapsed small cell lung cancer patients. Lung Cancer. 2007;**55**:109-113. DOI: 10.1016/j.lungcan.2006.09.029

[35] Schoppmeyer K, Kreth F, Wiedmann M, Mossner J, Preiss R, Caca K. A pilot study of bendamustine in advanced bile duct cancer. Anti-Cancer Drugs. 2007;**18**:697-702. DOI: 10.1097/CAD.0b013e32803d36e6

[36] von Minckwitz G, Chernozemsky I, Sirakova L, Chilingirov P, Souchon R, Marschner N, et al. Bendamustine prolongs progression-free survival in metastatic breast cancer (MBC): A phase III prospective, randomized, multicenter trial of bendamustine hydrochloride, methotrexate and 5-fluorouracil (BMF) versus cyclophosphamide, methotrexate and 5-fluorouracil (CMF) as first-line treatment of MBC. Anti-Cancer Drugs. 2005;**16**:871-877

[37] Zulkowski K, Kath R, Semrau R, Merkle K, Hoffken K. Regression of brain metastases from breast carcinoma after chemotherapy with bendamustine. Journal of Cancer Research and Clinical Oncology. 2002;**128**:111-113. DOI: 10.1007/s00432-001-0303-4

[38] Uehling DE, Harris PA. Recent progress on MAP kinase pathway inhibitors. Bioorganic & Medicinal Chemistry Letters. 2015;**25**:4047-4056. DOI: 10.1016/j.bmcl.2015.07.093

[39] Yeh TC, Marsh V, Bernat BA, Ballard J, Colwell H, Evans RJ, et al. Biological characterization of ARRY-142886 (AZD6244), a potent, highly selective mitogen-activated protein kinase kinase 1/2 inhibitor. Clinical Cancer Research. 2007;**13**:1576-1583. DOI: 10.1158/1078-0432.CCR-06-1150

[40] Davies BR, Logie A, McKay JS, Martin P, Steele S, Jenkins R, et al. AZD6244 (ARRY-142886), a potent inhibitor of mitogen-activated protein

kinase/extracellular signal-regulated kinase 1/2 kinases: Mechanism of action *in vivo*, pharmacokinetic/pharmacodynamic relationship, and potential for combination in preclinical models. Molecular Cancer Therapeutics. 2007;**6**:2209-2219. DOI: 10.1158/1535-7163.MCT-07-0231

[41] Beloueche-Babari M, Jamin Y, Arunan V, Walker-Samuel S, Revill M, Smith PD, et al. Acute tumour response to the MEK1/2 inhibitor selumetinib (AZD6244, ARRY-142886) evaluated by non-invasive diffusion-weighted MRI. British Journal of Cancer. 2013; **109**:1562-1569. DOI: 10.1038/bjc.2013.456

[42] El-Hoss J, Kolind M, Jackson MT, Deo N, Mikulec K, McDonald MM, et al. Modulation of endochondral ossification by MEK inhibitors PD0325901 and AZD6244 (selumetinib). Bone. 2014; **59**:151-161. DOI: 10.1016/j.bone.2013.11.013

[43] Jain N, Curran E, Iyengar NM, Diaz-Flores E, Kunnavakkam R, Popplewell L, et al. Phase II study of the oral MEK inhibitor selumetinib in advanced acute myelogenous leukemia: A University of Chicago phase II consortium trial. Clinical Cancer Research. 2014;**20**:490-498. DOI: 10.1158/1078-0432.CCR-13-1311

[44] Ma BB, Lui VW, Cheung CS, Lau CP, Ho K, Hui EP, et al. Activity of the MEK inhibitor selumetinib (AZD6244; ARRY-142886) in nasopharyngeal cancer cell lines. Investigational New Drugs. 2013;**31**:30-38. DOI: 10.1007/s10637-012-9828-4

[45] Janne PA, Shaw AT, Pereira JR, Jeannin G, Vansteenkiste J, Barrios C, et al. Selumetinib plus docetaxel for KRAS-mutant advanced non-small-cell lung cancer: A randomised, multicentre, placebo-controlled, phase II study. The Lancet Oncology. 2013;**14**: 38-47. DOI: 10.1016/S1470-2045(12)70489-8

[46] Janne PA, van den Heuvel MM, Barlesi F, Cobo M, Mazieres J, Crino L, et al. Selumetinib plus docetaxel compared with docetaxel alone and progression-free survival in patients with KRAS-mutant advanced non-small cell lung cancer: The SELECT-1 randomized clinical trial. JAMA. 2017; **317**:1844-1853. DOI: 10.1001/jama.2017.3438

[47] Cheng Y, Tian H. Current development status of MEK inhibitors. Molecules. 2017;**22**:1-20. DOI: 10.3390/molecules22101551

[48] Ciombor KK, Bekaii-Saab T. Selumetinib for the treatment of cancer. Expert Opinion on Investigational Drugs. 2015;**24**:111-123. DOI: 10.1517/13543784.2015.982275

[49] Ramamurthy VP, Ramalingam S, Kwegyir-Afful AK, Hussain A, Njar VC. Targeting of protein translation as a new treatment paradigm for prostate cancer. Current Opinion in Oncology. 2017;**29**: 210-220. DOI: 10.1097/CCO.0000000000000367

[50] McKay RR, Werner L, Fiorillo M, Roberts J, Heath EI, Bubley GJ, et al. Efficacy of therapies after galeterone in patients with castration-resistant prostate cancer. Clinical Genitourinary Cancer. 2017;**15**:463-471. DOI: 10.1016/j.clgc.2016.10.006

[51] Handratta VD, Vasaitis TS, Njar VC, Gediya LK, Kataria R, Chopra P, et al. Novel C-17-heteroaryl steroidal CYP17 inhibitors/antiandrogens: Synthesis, *in vitro* biological activity, pharmacokinetics, and antitumor activity in the LAPC4 human prostate cancer xenograft model. Journal of Medicinal Chemistry. 2005;**48**: 2972-2984. DOI: 10.1021/jm040202w

[52] Clement OO, Freeman CM, Hartmann RW, Handratta VD, Vasaitis TS, Brodie AM, et al. Three dimensional pharmacophore modeling of human CYP17 inhibitors. Potential agents for

prostate cancer therapy. Journal of Medicinal Chemistry. 2003;**46**: 2345-2351. DOI: 10.1021/jm020576u

[53] Vasaitis T, Belosay A, Schayowitz A, Khandelwal A, Chopra P, Gediya LK, et al. Androgen receptor inactivation contributes to antitumor efficacy of 17 {alpha}-hydroxylase/17,20-lyase inhibitor 3beta-hydroxy-17-(1H-benzimidazole-1-yl)androsta-5,16-diene in prostate cancer. Molecular Cancer Therapeutics. 2008;**7**:2348-2357. DOI: 10.1158/1535-7163.MCT-08-0230

[54] Bruno RD, Gover TD, Burger AM, Brodie AM, Njar VC. 17alpha-hydroxylase/17,20 lyase inhibitor VN/124-1 inhibits growth of androgen-independent prostate cancer cells via induction of the endoplasmic reticulum stress response. Molecular Cancer Therapeutics. 2008;**7**:2828-2836. DOI: 10.1158/1535-7163.MCT-08-0336.

[55] Schayowitz A, Sabnis G, Njar VC, Brodie AM. Synergistic effect of a novel antiandrogen, VN/124-1, and signal transduction inhibitors in prostate cancer progression to hormone independence *in vitro*. Molecular Cancer Therapeutics. 2008;**7**:121-132. DOI: 10.1158/1535-7163.MCT-07-0581

[56] Kwegyir-Afful AK, Ramalingam S, Purushottamachar P, Ramamurthy VP, Njar VC. Galeterone and VNPT55 induce proteasomal degradation of AR/AR-V7, induce significant apoptosis via cytochrome c release and suppress growth of castration resistant prostate cancer xenografts *in vivo*. Oncotarget. 2015;**6**:27440-27460. DOI: 10.18632/oncotarget.4578.

[57] Purushottamachar P, Godbole AM, Gediya LK, Martin MS, Vasaitis TS, Kwegyir-Afful AK, et al. Systematic structure modifications of multitarget prostate cancer drug candidate galeterone to produce novel androgen receptor down-regulating agents as an approach to treatment of advanced prostate cancer. Journal of Medicinal Chemistry. 2013;**56**:4880-4898. DOI: 10.1021/jm400048v

[58] Kwegyir-Afful AK, Bruno RD, Purushottamachar P, Murigi FN, Njar VC. Galeterone and VNPT55 disrupt Mnk-eIF4E to inhibit prostate cancer cell migration and invasion. The FEBS Journal. 2016;**283**:3898-3918. DOI: 10.1111/febs.13895

[59] Kwegyir-Afful AK, Murigi FN, Purushottamachar P, Ramamurthy VP, Martin MS, Njar VCO. Galeterone and its analogs inhibit Mnk-eIF4E axis, synergize with gemcitabine, impede pancreatic cancer cell migration, invasion and proliferation and inhibit tumor growth in mice. Oncotarget. 2017;**8**:52381-52402. DOI: 10.18632/oncotarget.14154

[60] Schayowitz A, Sabnis G, Goloubeva O, Njar VC, Brodie AM. Prolonging hormone sensitivity in prostate cancer xenografts through dual inhibition of AR and mTOR. British Journal of Cancer. 2010;**103**:1001-1007. DOI: 10.1038/sj.bjc.6605882

[61] Bruno RD, Vasaitis TS, Gediya LK, Purushottamachar P, Godbole AM, Ates-Alagoz Z, et al. Synthesis and biological evaluations of putative metabolically stable analogs of VN/124-1 (TOK-001): Head to head anti-tumor efficacy evaluation of VN/124-1 (TOK-001) and abiraterone in LAPC-4 human prostate cancer xenograft model. Steroids. 2011;**76**: 1268-1279. DOI: 10.1016/j.steroids.2011.06.002

[62] McKay RR, Mamlouk K, Montgomery B, Taplin ME. Treatment with galeterone in an elderly man with castration-resistant prostate cancer: A case report. Clinical Genitourinary Cancer. 2015;**13**:e325-e328. DOI: 10.1016/j.clgc.2014.12.015

[63] Montgomery B, Eisenberger MA, Rettig MB, Chu F, Pili R, Stephenson JJ, et al. Androgen receptor modulation optimized for response (ARMOR) phase I and II studies: Galeterone for the treatment of castration-resistant prostate cancer. Clinical Cancer Research. 2016;**22**:1356-1363. DOI: 10.1158/1078-0432.CCR-15-1432

[64] Rose AAN. Drugs Today (Barc). 2019;**55**(4):247-264

Chapter 7

Benzimidazoles: From Antiproliferative to Multitargeted Anticancer Agents

Yousef Najajreh

Abstract

Benzimidazole derivatives are known to act against a range of biological targets and thus gained clinical applications in a wide spectrum of diseases. Few examples of multitargeted benzimidazole derivatives that were reported during the last decade will be described in this chapter. Multitargeting agents for serving the polypharmacology approach to combat shortcomings of the main one-drug-one target main dogma will be briefly explored. In that context, the multitargeting benzimidazole derivatives gain a special attention. This includes discovery (hit-to-lead), structure-activity relationship (SAR), and binding mode of at least one lead (or hit) in each group. Special attention will be given to two structures dovitinib and AT9283 that are reported to exhibit potent in vitro and in vivo activities against a group of kinases and non-kinase target (as shown recently for dovitinib).

Keywords: benzimidazole, selective, cytotoxic, inhibitor, multitargeting, multikinase, polypharmacology, antiproliferative, quinolinone, carbamate, quinolinone, pyrazole, urea, aniline, anilinobenzimidazolylpyrimidine, chloroacetamide, amidine, binding, mode

1. Introduction

1.1 Antiproliferative action of benzimidazoles

Benzimidazole, a heterocyclic moiety comprising six-membered benzene ring fused with five-membered imidazole ring, containing molecules, was known for its ability to induce antiproliferative effects (named as antineoplastic, anticancer, or antitumor agents). Numerous structures were reported as effective inhibitors of cell growth and division, thus acting as antiviral, antibacterial, antifungal, anthelmintic (or antihelminthics), and anticancer agents. Over the years, several published scripts have reviewed the synthetic approaches, medicinal chemistry, SAR, bioactivities, and preclinical and clinical studies of such "gifted" fragment [1–8].

1.2 Benzimidazoles act on numerous biological targets

A wide range of activities and medical situations benzimidazole containing compounds have been used for. That includes antihypertensive [9–12], anti-inflammatory [13–15], antibacterial [16–18], antiviral [19–21], antifungal [22–24], anti-helmintic [25–28], anticancer [29–32], antiulcer [33–35], antioxidant [36–38], and psychoactive drugs [39]. And proton pump inhibitors [8, 33], anticoagulants [40, 41], immunomodulators [42], hormone modulators [43, 44], antidepressants [45], lipid level modulators [46–49], and antidiabetics [50–52] are partial list of therapeutic effects of benzimidazole containing comprising compounds. Benzimidazole derivatives exert their actions by interacting with vital biological targets including β-tubulin [52–55], DNA minor groove [56–58], serotonin receptors (5-hydroxytryptamine receptors; 5-HT) [59–62], histamine receptors 4 (H4H) [63], dopamine receptor 2 (D2R) [64], chemokine receptor (CXCR3) [65], interleukin 2-inducible T-cell kinase (ITK) [66], lymphocyte tyrosine kinase (Lck) [67], phosphatidylinositol 3-kinase (PI3K) [68], activated protein kinase (MEK1) [69, 70], anaplastic lymphoma kinase (ALK) [71], polo-like kinase 1 (PLK1) [72, 73], breakpoint cluster region-Abelson kinase (BCR-Abl) [74], casein kinase 2 (CK2) [75], telangiectasia and Rad3-related protein kinase (ATR) [76], tyrosine kinase receptors [fibroblast growth factor receptors (FGFR-1/FGFR-2/FGFR-3)], vascular endothelial growth factor receptor (VEGFR-1/VEGFR-2/VEGFR-3), platelet-derived growth factor receptor (PDGFR-α/PDGFR-β), stem cell factor receptor (c-KIT), FMS-like tyrosine kinase 3 (FLT3) [77], poly(ADP-ribose)polymerase-1 (PARP-1) [78–82], dihydroorotate dehydrogenase (DHODH) [83], topoisomerase 1 (TOPO1) [84], DNA and RNA polymerases [85–89], histone deacetylase 2 (HDAC2) and sirtuin [3, 90], antagonism of angiotensin 1 [2], neuropeptide Y binding [91], inhibition of proton pumps [8], DNA intercalating agents [92], inhibition of cyclin-dependent kinases (CDK) activity [93–96], activation of the p53 protein [97], etc. to mention part of the asserted cellular targets.

1.3 Scope: benzimidazoles as emerging multitargeting agents

The profound success in bringing into clinical application several kinase inhibitors as anticancer drugs made "kinase targeting" a central branch of targetable biomolecules during the past two decades. Nevertheless, the emerging of resistant tumors kinase-directed therapeutics and adverse side effects turned such promising "targeted therapeutics" into challenging field. In addition, it was noticed that lack of response to kinase inhibitors is accompanied by changes in signaling network composition through adaptive kinome reprogramming. Such reprogramming is believed to allow the tumor to escape effects of the drug and manifest resistance. In contrast to the "one-drug-one-target" approach, the "bitopic, that is, two drugs acting on one target" or the "dual, that is, one drug acting on two targets," "polypharmacology" which refers to a novel paradigm that purposes at "simultaneous modulation of more than two biological targets by a single drug" has been emerging as strategy to improve the efficacy and durability of clinical responses to therapies. In cancer treatment, polypharmacology is a result of the ability of "one drug" to simultaneously inhibit multiple cancer-driving targets. However, discovering inhibitors with an appropriate multitarget profile is a challenging task that necessitates a systemic deeper investigation accompanied by major clinical developments. Therefore, a strategy is required to identify single polypharmacological agents with the ability to target multiple cancer-promoting or -sustaining pathways that does not necessarily rely on inhibiting multiple kinases [98]. As a matter of fact, high ratio (~30%) of the FDA-approved kinome-targeting drugs were reported be multitargeted ones [99]. Actually, the first kinase inhibitor imatinib was approved as multitarget agent in a later stage (in addition to its primary target BCR-Abl, it inhibits stem cell factor receptor (c-KIT) and platelet-derived growth factor receptors A and B (PDGFRα and PDGFRβ) tyrosine kinases and human quinone reductase 2

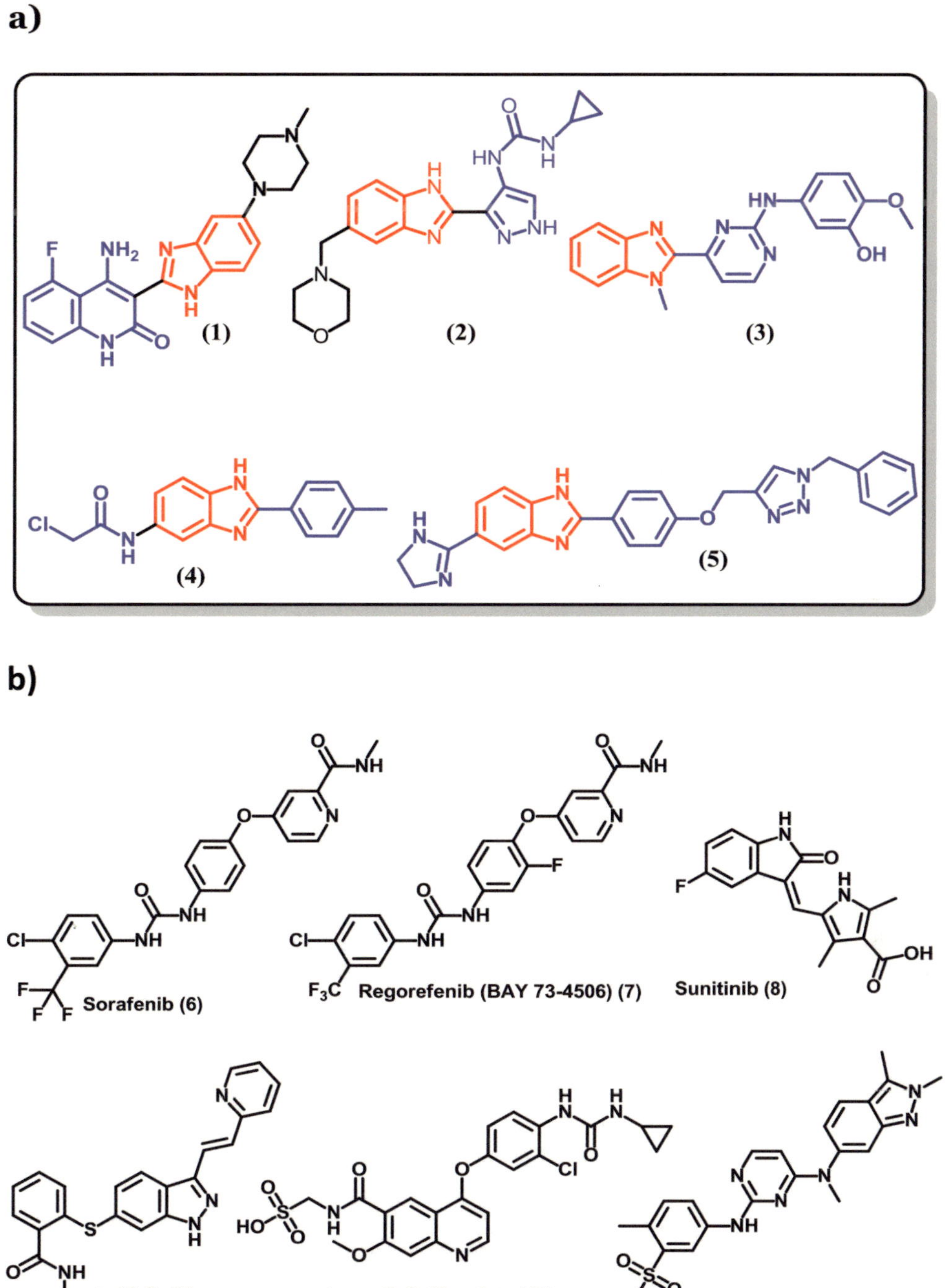

Figure 1.
Multitargeting anticancer agents. (a) Multitargeting cytotoxic benzimidazole-based structures. 3-Benzimidazol-2-ylhydroquinolin-2-one based dovitinib [TKI258, CHIR258; (1)], N-cyclopropyl-N'-[3-[5-(4-morpholinylmethyl)-1H-benzimidazol-2-yl]-1H-pyrazol-4-yl]Urea [AT9283 (2)], 2-anilino-4-(benzimidazol-2-yl)pyrimidine based [2-anilino-4-(benzimidazol-2-yl)-pyrimidine 2-methoxy-5-{[4-(1-methyl-1H-benzimidazol-2-yl)pyrimidin-2-yl]amino}phenol (3)], α-haloacetamidebenzimidazole-based [2-chloro-N-(2-(p-tolyl)-1H-benzo[d]imidazol-5-yl)acetamide (4)], and amidine-benzimidazole based [2-(4-((1-benzyl-1H-1,2,3-triazol-4-yl)methoxy)phenyl)-5-(4,5-dihydro-1H-imidazol-2-yl)-1H-benzo[d]imidazole (5)]; (b) FDA-approved multitargeting anticancer agents [sorafenib (6), regorafenib (7) sunitinib (8), axitinib (9), and lenvatinib (10) and pazopanib (11)].

(NQO2)). Thus the question of how efficacious are selective and specific one-drug-one-target-approved agents in treating advanced and metastatic cancer is still under evaluation [100–102].

This chapter will concisely provide a deeper insight into the benzimidazole-containing structures that exhibit action on multiple cellular targets. Special focus will be drawn to the identification and discovery, the structural activity relationship, proposed binding and interaction, and mechanism of action of each group of compounds. Detailed synthetic procedures and preclinical and clinical studies are out of scope of the current chapter. The focus of this chapter will be on groups of compounds that had been unveiled as concurrently antagonizing multiple targets. Instead, this chapter will focus on five groups of compounds reported to possess cytotoxic activities by acting on multiple (see **Figure 1a**) compounds (1, 2, 3, 4, and 5) holding the potential to be administered as "polytherapies."

2. Benzimidazole scaffold for multitargeting of cancer

Multitargeting agent is defined as "a single chemical entity exerting action as a result of direct interactions on multiple biomolecular targets" [103]. Such agents can be beneficial in overcoming single (or dual)-targeting limitations including compromised effectiveness, severe side effects, emergence of resistant target mutants, and target non-related mutations. In addition, the efficacy of single-molecular-targeted FDA-approved agents in treating brutal and mortal cancers (breast, colorectal lung, pancreatic, and prostate) is limited. Most tumors escape from the inhibition of any single chemotherapeutic agent, and thus one possible therapeutic strategy could be in (1) administering cocktails of highly selective inhibitors (combinational therapy) or (2) development of multitarget inhibitors that act on inhibiting concurrently multiple validated target in cancer cell initiating a concerted molecular response, leading to cell death. Multitargeting chemotherapeutics hold the potential of exhibiting synergistic or at least additive effects when compared to single-targeted ones. It is believed that advances in signaling cascades, networks and crosstalk, chemo- and bioinformatics, detailed three-dimensional structural information of target proteins, computational chemistry tools, proteomics, etc. will allow for designing novel multitarget inhibitors.

It has been realized that molecular targeted therapeutics are facing acquired resistance. Multitargeting approach is gaining increased attention especially when combating resistant cancer cells. Accumulated evidence showed that drug treatment aggravates "selective pressure" of evolutionary force exerted on tumor cells that leads to resistance.

Benzimidazole fragment is reported to be an integral part of multitargeted inhibitors. Such inhibitors challenge the dominant paradigm in drug discovery which deemed to design and develop bioactive agent with maximum selectivity and specificity to individual drug target. Such compounds hold the hope for a new avenue of combating disease cases that could not be cured with one inhibitor acting on single target such as cancer [104, 105].

2.1 Benzimidazolylquinolinone: a scaffold for targeting multiple biomolecules

2.1.1 Discovery of dovitinib (TKI258, CHIR258)

Dovitinib [(TKI258, CHIR258); 4-amino-5-fluoro-3-(5-(4-methylpiperazin-1-yl)-1*H*-benzo[d]imidazol-2-yl)quinolin-2(1*H*)-one (1)] was first designed and synthesized as vascular endothelial growth factor receptor (FEGFR) inhibitor in the context of developing targeted antiangiogenic treatments [106]. The compound was later reported as a multitargeted kinase inhibitor (by [107]) following the realization that its potent inhibitory effects on cancer cells are associated with action on other multiple key players in oncogenesis, development, and proliferation of cancer [107].

The commercially available 3-benzimidazol-2-ylhydroquinolin-2-one scaffold [benzimidazolylquinolinone for short from now on, **Figure 2** (13)] was identified using high-throughput screening (HTS) method and reported by Renhowe et al. (Novartis) as a potent (IC_{50} values < 0.1 μM) reversible ATP competitive inhibitor of VEGFR-2, FGFR-1, and PDGFRβ [106]. Due to desirable properties as low-molecular-weight compound exhibiting submicromolar activity, (12) was considered a good hit to start with. To overcome the undesirable physicochemical properties of (13) (low aqueous solubility), further optimization was needed that ended up with a drug-like compound (1). Determining the key structural features required for potent kinase inhibition, molecular modeling was employed. The assumption was that in quinolinone portion, both NH at position 1 and the carbonyl group, together with benzimidazole NH, form a donor-acceptor-donor motif that would most probably bind to the hinge region of the RTKs and should be preserved.

To test this hypothesis, a systematic study was conducted through which hydrogen bond donors were masked by methyl group (CH_3-) as shown in **Figure 3** (13a–c) and 14a and 14b. These changes led to significant loss in the potency against all three receptor tyrosine kinases (VEGFR-2, FGFR-1, and PDGFRβ RTKs). The dimethylated analogue (**Figure 3**, 14b) showed no kinase activity at a concentration as high as 25 μM. Interestingly, it was noticed that monomethylation seemed to affect the kinase selectivity profile as well. Introduction of a methyl on the benzimidazole NH (13b) had a more dramatic effect on VEGFR-2 affinity than the methylation at NH in position 1 of the hydroquinolin-2-one (13a). This underlies the importance

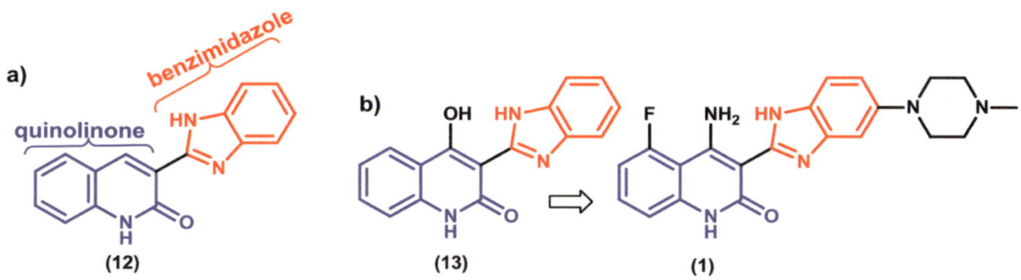

Figure 2.
Benzimidazolquinolinone-based multitargeting scaffold. (a) The basic skeleton of dovitinib (TKI258, CHIR258), 3-(1H-benzimidazol-2-yl)quinolin-2(1H)-one (12) with the two fragments quinolinone (blue) and benzimidazole (red) is indicated, (b) structures of commercially available starting "hit" (13) identified using HTS, and the "lead" approved as a multitargeting drug dovitinib (1).

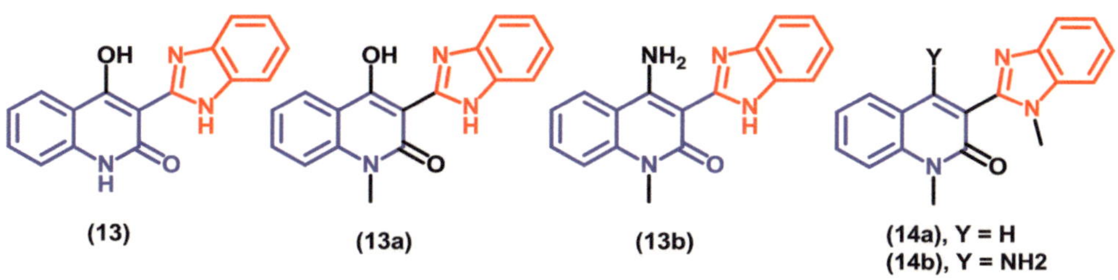

Figure 3.
Assessing the effect of the HBD and HBA on the activity of derivative of 3-benzimidazol-2-ylhydroquinolin-2-one. Methylated analogues of (13 and 14). The monomethylation caused a significant drop in the potency toward RTKs, while dimethylation aborted the RTKs' activity [106].

of preventing the hydrogen bond donor (HBD). An opposite effect was noticed for FGFR-1, which indicates that despite the high homology of the two ATP-binding sites in the tow targets, selectivity opportunities still exist that are likely due to small changes in the shape of binding site. Such change in the shape can influence the accessibility of alternate binding poses of the monomethylated ligands (13a–13b and 14a in **Figure 1**) [106].

2.1.1.1 Structure–activity relationship (SAR)

The scaffold (13) was annotated by four rings (A–D). Modifications were introduced in a systemic manner. Once the basic structural components needed for affinity to targets of interest were understood, a study of the structure–activity relationship around the periphery of central 3-benzimidazol-2-ylhydroquinolin-2-one (13) scaffold was undertaken. Besides electrophilicity, nucleophilicity, bulkiness, steric hindrance, HBD versus HBA, and basicity, C4 of ring A was used for incorporation of moieties that might impart favorable physicochemical properties.

SAR of ring B (C4): While removal of the hydroxyl group reduced the activity, its replacement with amine improved significantly affinity to RTK and also cell potency [EC_{50} of 0.078 μM (NH_2 > OH > H)], suggesting an importance of the HBD at C4 of the hydroquinolin-2-one fragment. Thus, incorporation of larger substituents on the C4-NH of the hydroquinolin-2-one was explored and found to be tolerated (see compounds 15b and 15c, **Figure 4**). Not only substantial improvement in the solubility was attained when the substituents carried an additional basic nitrogen were introduced to this position, it was noticed that this position modulates the selectivity profile of this class of compounds. It was reported that both derivatives (12a) and (12b) exhibited enhanced potency against PDGFR than VEGFR-1 (3000-fold) and FGFR (>1500-fold). Large basic amines like aminoquinuclidine potentiate the derivative (15d) against CHK-1 and GSK-3.

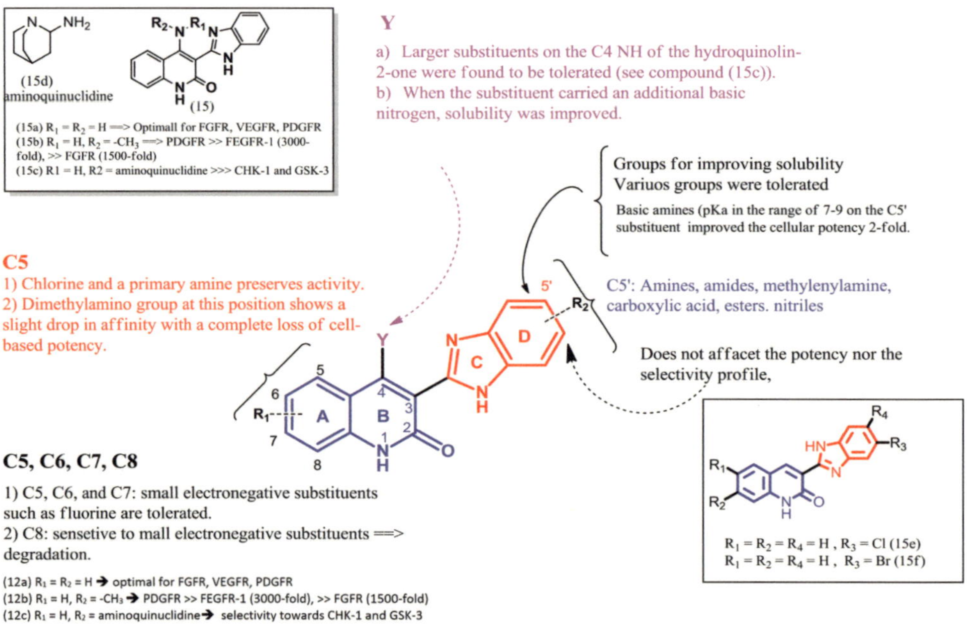

Figure 4.
Summary of structure–activity relationship (SAR) of 3-benzimidazol-2-ylhydroquinolin-2-one (1) [106].

In conclusion, substitution at C4 position was revealed as critical to the activity of the benzimidazolylhydroquinolinone scaffold; however for RTK inhibitor program, the NH_2 group was the optimal substituent at C4 as it avoided inhibition of these additional serine threonine kinases, which could complicate the pharmacological application of these agents.

SAR of ring D: The overall structure–activity relationship (SAR) is summarized in **Figure 4**. Medicinal chemistry efforts were concluded in the selection of compound (1) as a candidate for further development. The compound (1) displayed exceedingly potent inhibitory effect when assessed against receptor protein kinases VEGFR-2, FGFR-1 and FGFR-3, PDGFRβ, VEGFR-1, VEGFR-2 and VEGFR-3, c-KIT, CSF-1R, and FLT-3 with IC_{50} values between 3 and 27 nM. Such activity is translated into antiproliferative action on cells that are VEGF-, FGF-, SCF-, CSF-, or PDGF-driven. Mechanistically, it was also indicated that VEGF-mediated ERK phosphorylation was dipped in endothelial cells treated with (1).

In summary, dovitinib (1), an antineoplastic benzimidazolylquinolinone derivative, inhibits multiple growth factor receptor tyrosine kinases important for tumor angiogenesis and tumor growth. Dovitinib is well established as type III–V receptor tyrosine kinase (RTK) inhibitor. Though it potently inhibits fibroblast growth factor receptors (FGFR-1/FGFR-2/FGFR-3), the compound also inhibits vascular endothelial growth factor receptor (VEGFR-1/VEGFR-2/VEGFR-3), platelet-derived growth factor receptor (PDGFR-α/β), stem cell factor receptor (c-KIT), FMS-like tyrosine kinase 3 (FLT3), and colony-stimulating factor receptor 1 (CSFR-1) emphasizing the nonspecific action of the drug [108]. The orally bioavailable lactate salt of (1) strongly binds to fibroblast growth factor receptor 3 (FGFR3) and inhibits its phosphorylation, which may result in the inhibition of tumor cell proliferation and the induction of tumor cell death. The activation of the abovementioned RTK in singularity or together is associated with cell proliferation and survival in all cancer cell types.

Dovitinib (TKI258, 1) is a highly potent, novel multitargeting receptor tyrosine kinase inhibitor with IC_{50} of 1, 2, 10, 8, 27, and 36 nM for FLT3, c-KIT, VEGFR-1/VEGFR-2/VEGFR-3, PDGFRß, and CSFR –1, respectively. Due to its inhibitory effect of VEGFR1/VEGFR2, the compound displayed both antitumor and antiangiogenic activities in vivo.

Trudel and colleagues reported that in addition to inhibiting the abovementioned TRKs (types II, IV, V), (1) potently inhibits wild-type (WT) FGFR3, F384 L-FGFR3 (IC_{50} = 25 nM), and FGFR3 mutants (IC_{50} = 70–90 nM for the various mutations) driven by B9 cells [107]. Additionally, same group reported that (1) inhibited the proliferation of multiple myeloma (MM) cells. When assessing its antiproliferative effect against U266 and 8226 cells, (1) displayed a potent inhibitory effect (IC_{50} ~ 90 nM) against KMS11 (FGFR3-Y373C), OPM2 (FGFR3-K650E) cells and IC_{50} ~ 550 nM KMS18 (FGFR3-G384D) [109]. (1) Exhibited exceedingly potent antiproliferative effect against acute myelogenous leukemia (AML) cells MV4;11 (mutant FLT3-ITD) compared to AML RS4;11(FLT3 WT) cells [EC_{50} = 13 nmol/L and EC_{50} = 315 nmol/L for MV4;11 and RS4;11, respectively, i.e., (~24-fold decrease in potency for FLT3 WT cells)]. Such results indicated that (1) exhibited far more potent activity against cells that are dependent on constitutively active FLT3-ITD. A similar conclusion was affirmed by Heise et al. by the notion that (CHIR258, 1) inhibited the proliferation of MOLM13 and MOLM14 that are FLT3-ITD mutant cells with EC_{50} ~ 6 nmol/L similar to the ones with MV4;11 [109].

Besides the potent action of (1) against a wide range of RTK, its inhibitors' effect ion fibroblast growth factor receptors in a variety of tumor xenograft models in athymic mice, including acute myeloid leukemia, multiple myeloma, and colon- and prostate-derived models was promising.

Recent studies reported the comparative activities of dovitinib against 16 colorectal cancer (CRC) cell lines (among them, 10 were KRAS or BRAF mutants). Results showed the affectivity of the drug in inhibiting the proliferation of majority of the cell lines excluding the ones harboring KRAS or BRAF mutants. However, when assessing the efficacy of the drug in vivo, it reduced the tumor growth in vivo regardless of the KRAS and BRAF mutation status. The drug exerted significant reduction of the xenograft size of both resistant cell lines (KRAS mutant LoVo cells but not in BRAF mutant HT-29) but without a detectable effect in the resistant mutant cell BRAF mutant HT-29 in vitro on s. Such results were explained by the multitarget action of the drug in which by acting on FGFR and FGFR together with VEGFR has been able to interfere with resistance mechanisms emerging from the synergistic interaction between the various signaling cascades in promoting neovascularization that is believed to be one resistance factor in renal cell carcinoma or pancreatic cancer [110, 111].

Dovitinib was selected to proceed ahead for preclinical and clinical trials. Several clinical trials have been conducted, and others are also underway with the drug and alone or in combination with several chemotherapeutic agents [112–118].

2.1.2 Binding mode of dovitinib (CHIR258, 1) to FGFR-1

Based on FGFR-1 crystal structure (PDB 2FGI) in conjunction with the information received from the X-ray structure of (1) with CHK1, a homology model for (1) complexed with VEGRF2 was constructed [106]. The model was helpful in guiding for the important interactions of (1) with active site. It was concluded that (1) participated in three hydrogen bonds to the hinge domain (Glu917 and Cys919). In addition A-ring makes a VDW interaction with the hydrophobic gatekeeper Val916 and was engaged in an S-H/π interaction with Cys1045. Leu840, Val848 (both in the P-loop and ceiling of the purine pocket), Ala866 (ceiling of the purine pocket), Val 899 (floor of the purine pocket), Phe918 (part of the hinge), Lys920, Gly922 (both in the lower hinge region), and Leu1035 (floor of the purine pocket) took part in hydrophobic interaction with (1). In the following studies, the X-ray structures of (1) complexed with native and with mutant FGFR1 and with FGFR4 were reported [119–122].

2.1.2.1 Going beyond kinases

Although dovitinib binds to several kinases at nanomolar concentrations, recent studies reported its inhibitory effect against cancer-related targets including topoisomerase I and II (Topo I and II) [123] and human recombinant bone morphogenetic protein (BMP)-2, indicating that the cell growth inhibitory activity and the anticancer activity of dovitinib may result, in part, from its ability to target Topo I and II in addition to the ability to inhibit multiple kinases [124]. A study disclosed dovitinib inhibition of BMP-2 enhanced alkaline phosphatase (ALP) induction, which is a representative marker of osteoblast differentiation. Dovitinib also stimulated the translocation of phosphorylated Smad1/Smad5/Smad8 into the nucleus and phosphorylation of mitogen-activated protein kinases, including extracellular signal-regulated protein kinases 1 and 2 (ERK1/2) and p38 **Figure 5**. An increase in the expression of mRNA of BMP-4, BMP-7, ALP, and osteocalcin (OCN) was noticed following treatment with (1). It was also noted that the potent stimulating

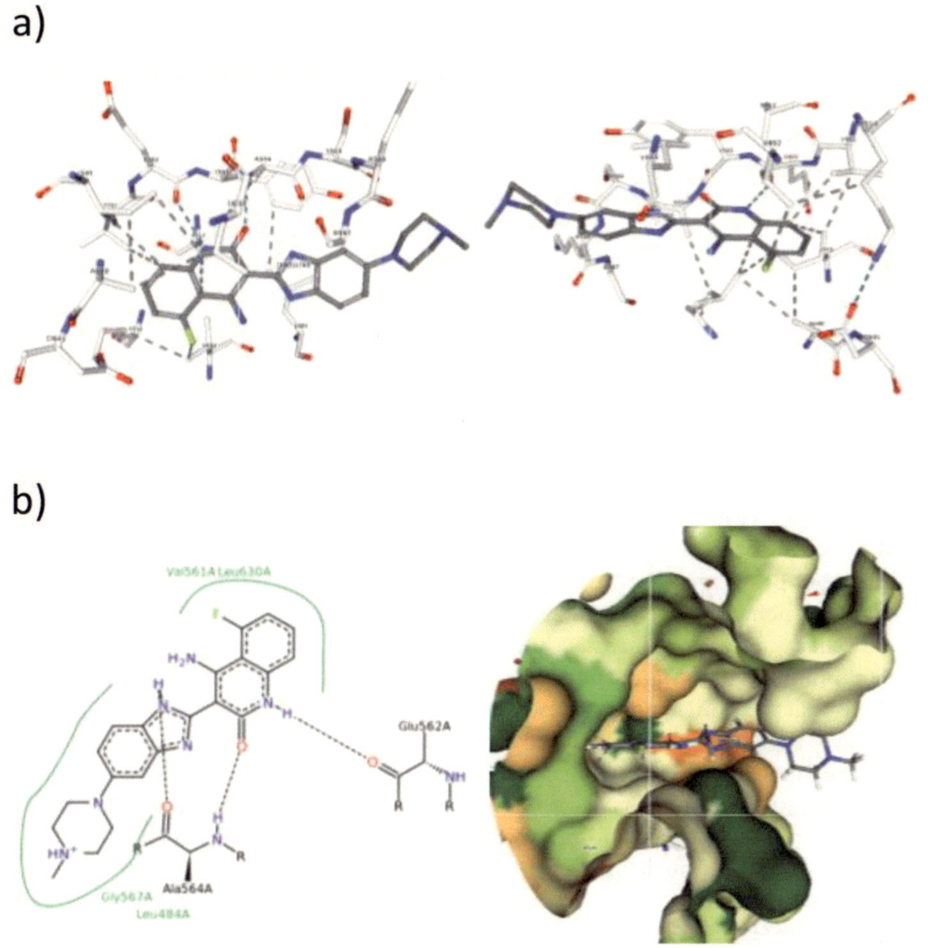

Figure 5.
(a) Cartoon representation of the crystallographic structure of complex of (1) to FGFR-1 (PDB 5 AM6); the binding site is depicted showing the kinase with residues interacting with the ligand in stick model, (b left) the main interactions between (1) and the kinase domain and (b right) the surface representation, with the surface colored by atom type (red, oxygen; blue, nitrogen; yellow, sulfur; gray, carbon/hydrogen). The donor-acceptor-donor motif is shown to interact with the hinge region while the 1-methylpiperazine substituent on C5′ points into solution [119].

effect of (1) on BMP-2-induced osteoblast differentiation suggests a potential repositioning for the use of (1) treatment of bone-related disorders [124]. In a recent study, Ye Zhang **et al**. initially used the central scaffold 3-(1*H*-benzimidazol-2-yl)quinolin-2(1*H*)-one (12) to explore that potential diversification of functional groups decorating (12). The compounds synthesized were assessed against HepG2 (human liver cancer cells), SK-OV-3 (human ovarian cancer cells), NCI-H460 (human large cell lung cancer cells), BEL-7404 (human liver cancer cells), and HL-7702 (human liver normal cells) cell lines. Initial studies showed that halogenated derivative [3-(6-chloro-1*H*-benzo[d]imidazol-2-yl)quinolin-2(1*H*)-one (15e) and 3-(6-bromo-1*H*-benzo[d]imidazol-2-yl)quinolin-2(1*H*)-one (15f) (see **Figure 4**)] exhibited better activity than 5-FU and cisplatin when assessed against HepG2, SKOV-3, NCI-H460, and BEL-7404 but not HL-7702. The authors postulated that (15e) and (15f) inhibit HepG2 proliferation by blocking the cells in G2/M stage through activation of p53 protein.

2.2 Pyrazolbenzimidazols as multikinase inhibitors

2.2.1 Discovery of AT9283a

In developing a selective potent aurora kinase inhibitor by employing fragment-based discovery method, the pyrazol-4-yl urea benzimidazole derivative (AT9283, 21) was identified as a multitargeting kinase inhibitor. The pyrazolebenzimidazole-based clinical candidate (21) was optimized by Steven Howard and his colleagues following efficient structure-guided fragment to hit IC_{50} as low as 3 nM activity as a dual potent inhibitor toward Aurora A/Aurora B [125]. AT9283 was identified starting from the pyrazole-benzimidazole fragment (16) that was previously identified during the endeavor of developing cyclin-dependent kinase (CDK) inhibitors. Subsequent structure-based approach using CDK2 crystallographic structure led to the identification of the benzamide analogue (18). Throughout the process of developing CDK2 inhibitors, pyrazole-benzimidazole derivative was identified to act with high potency toward Aurora A. Starting with fragment, (18) demonstrated superior ligand efficiency (LE = 0.59) for Aurora A compared to CDK1 and CDK2 and also sufficient potency to allow detection in a conventional enzyme bioassay [125].

Aiming at optimizing, the "hit" (18) on the way to end up with a lead SAR is performed on the benzamide analogues. The team was aided by polyploid phenotype in HCT116 cells, as a functional assay that differentiates for Aurora A and Aurora B inhibition, combined with potency when screening for further analogues. Guided by the hypothesis that introducing a basic motif into fragment (18) will improve the potency of the compound, modifications were introduced successfully to 5- or 6-position of the benzimidazole without causing any clashes with the protein. In a further step, the morpholinomethyl motif was functionalized at position 5. Details grasped from the X-ray crystal structure of (19) complexed with Aurora A revealed that the pyrazole-benzimidazole motif is positioned in an excellent complementarity with the narrow region of the ATP pocket. A result directed the steps to follow in the design of the optimized structure (**Figure 6**). While retaining the 5-morpholinomethyl on the pyrazole-benzimidazole motif, the benzamide portion was subjected to modifications. Keeping in mind the need to keep the molecular weight while introducing increased flexibility on the glycine region, the amide was converted to urea (20). This strategy was fruitful when comprehending that the urea analogue (20) exhibited reduced plasma protein binding while maintaining in vitro activity against Aurora kinases.

In the following step, the X-ray structure of (20) complexed with Aurora A was solved and iterated a similar binding mode to the hinge region. To resolve a twisted conformation of the phenyl plane in regard to pyrazole-benzimidazole portion of the molecule, a fully reduced cyclohexyl and difluorophenyl groups were also introduced (compound (20a) and (20b), respectively). Adsorption, disposition, metabolism, and excretion (ADME) considerations lead to proposing cyclopropyl derivative (21). As an alternative to introducing additional heterocyclic moiety, aiming at reducing the lipophilicity of (20a) for improving the ADME, the size of cyclohexyl ring was reduced to cyclopropyl analogue resulting in compound (21) that exhibits high enzyme and cellular potency still with reduced both the molecular weight (MW) and lipophilicity (log D7.4 = 2.1, MW = 381). Compound (21) demonstrated potent inhibition of HCT116 colony formation (IC_{50} = 12 nM), a clean CYP450 profile (IC_{50} > 10 μM for CYP3A4, 2D6, 1A2, 2C9, 2C19), acceptable mouse plasma protein binding (81.5%), and good thermodynamic solubility (2.0 mg/mL at pH = 7.0 and 13 mg/mL at pH = 5.5).

Later, AT9283 (21) was shown to bind and potently inhibit a number of kinases including the Aurora kinases A and B (serine–threonine kinases that are known to play essential roles in mitotic checkpoint control during mitosis at IC_{50} ~ 3 nM),

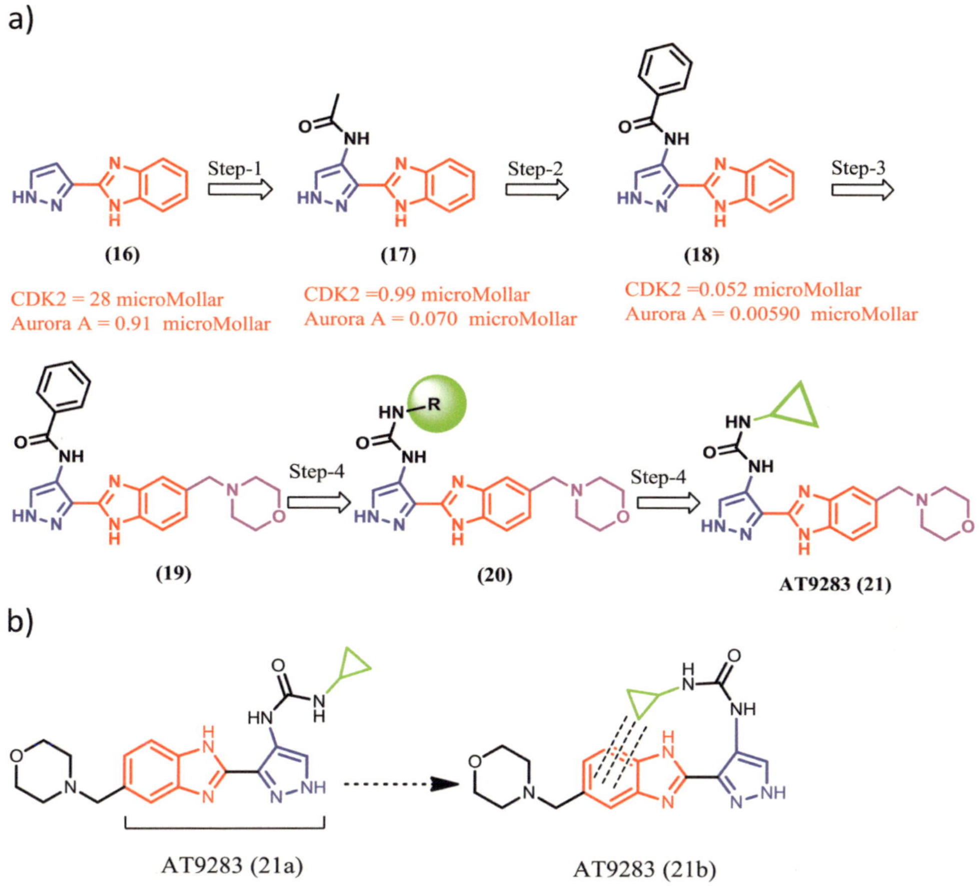

Figure 6.
(a) Main steps in the identification and discovery of pyrazolebenzimidazole-based multitargeting agent AT9283 (21) using fragment-based identification starting from fragment (16). (b) The structure N-cyclopropyl-N'-[3-[5-(4-morpholinylmethyl)-1H-benzimidazol-2-yl]-1H-pyrazol-4-yl] urea [AT9283 (21a)]. The "folded conformation" of (21b). Dotted lines to indicate the hydrophobic interaction between the cyclopropyl and benzimidazole motifs [125].

Janus kinase 2 (JAK2) and JAK3 (1.2 and 1.1 nM, respectively), breakpoint cluster region-Abelson (BCR-Abl) T315I (4 nM), and mitogen-activated protein kinase kinase kinase 2 (MEKK2) with IC_{50} values of lower nanomolar (4.7–18 nM). This set of known kinases is known to play key roles in mitotic progress in cell cycle, induction of proliferation, evasion of apoptosis and tumor growth and thus considered vital targets to chemotherapeutic agents (see **Table 1**). Therefore, AT9283 (21) is defined as multikinase (multitargeting) inhibitor [126].

AT-9283 inhibits effective proliferation of cancer cells both in vitro and in vivo with and its effect is enhanced by with other agents (see **Table 2**) [127]. Henceforth T9283 proceeded to clinical trials including in children with relapsed or refractory acute leukemia, imatinib-resistant BCR-Abl-positive leukemic cells, and patients with relapsed or refractory multiple myeloma. Accumulative results indicate a need for optimizing the pharmacological profile on the way to overcome faced challenges in clinical application of the compound [127, 128].

The activity in imatinib-resistant BCR-Abl chronic myelogeneous leukemia (CML) explained based on modeling which reiterated the assumption that AT-9283 is bound to the kinase domain in the "folded conformation" which allows the needed interactions with the hinge region without a clash between the cyclopropyl group and the isoleucine residue in the T315I mutant. The results obtained in

Enzyme	IC_{50} (nM)
Aurora A	52% I at 3.0 nM
Aurora B	58% I at 3.0 nM
JAK3	1.1
JAK2	1.2
Abl (T315I)	4
GSK3-β, FGFR2, VEGFR3 (Flt4), Mer, Ret, Rsk2, Rsk3, Tyk2, Yes	1–10
Abl(Q252H), DRAK1, FGFR1, FGFR1(V561 M), FGFR2(N549H), FGFR3, VEGFR1(Flt1), Flt-3, PDGFR-α(D842V), PDK1, PKCμ, Rsk4, SRC(T341 M), VEGFR2	10–30

Table 1.
The inhibitory concentration 50% (IC50 of the "lead" (21)) [126].

Inhibitory effect of AT9283 on tumor cell colony formation after 10–14 days treatment			
Origin	**Cell line**	**IC_{50} (nM)**	**p53 status**
Colon	HCT116	13	+
	HT-29	11	—
	SW620	14	—
Ovarian	A2780	7.7	+
Lung	A549	12	+
Breast	MCF7	20	+
Pancreatic	MIA-Pa-Ca-2	7.8	—
+ indicates expression of wild-type p53; – indicates no expression of p53 or that p53 is nonfunctional [126].			

Table 2.
IC_{50}s are the mean of two or more independent determinations.

refractory CML suggest that AT-9283 can be efficient in Ph + acute lymphoblastic leukemia (Ph + ALL). It is the distinct binding mode that allows AT-9283 in similar manner to MK-0457 and PHA-739358 to exhibit potent activity against imatinib-resistant T315I mutant [127, 129].

2.2.2 Binding mode of AT-9283 (21) to kinases

Currently, there exist 11 X-ray resolved crystallographic structures of AT-9283 complexed with target proteins that are documented at the Protein Data Bank. They include aurora A, aurora B, mutant of aurora B, JAK2, and protein kinase A mutants as surrogate model for Aurora B. A closer look clarifies that in a similar manner to the binding of dovitinib, the benzene portion in benzimidazole fragment is pointing in an orientation toward the solvents' exposed opening of the binding site. The pyrazole and urea fragments took part in multiple HBA and HBD interactions with the hinge region of the enzyme. The morpholine basic amine is oriented toward the solvent and enhanced significantly the solubility of the compound in physiological pH.

The crystal structure of compound (21) complexed with Aurora A is shown in **Figure 7** [130]. The molecule is positioned at the ATP-binding site of the kinase. It is revealed the urea linker adopts a *cis/trans* configuration that results in the molecule having a "folded conformation." This same conformation was also observed in the

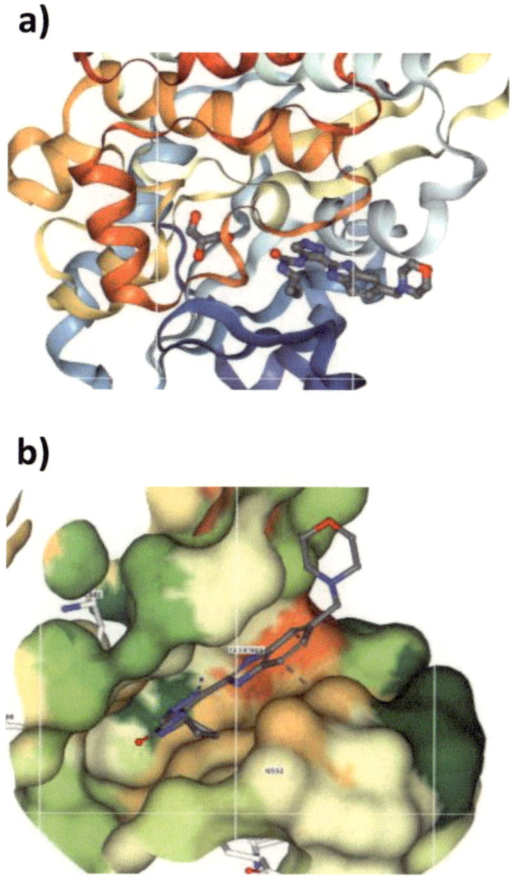

Figure 7.
(a) Cartoon representation of the crystallographic structure of complex pyrazol-4-yl urea AT9283 (21) complexed with JAK2 JH2 (PDB 5UT0); (b) the surface representation, with main interactions between (21) and the kinase domain, colored by atom type (red, oxygen; blue, nitrogen; yellow, sulfur; gray, carbon/hydrogen). The donor-acceptor-donor motif is shown to interact with the hinge region, while morpholine substituent on C5 points toward the solution [130].

crystal structure of (21B) alone (**Figure 8**) and in DMSO. Such "folded conformation" was confirmed by NMR measurement. An NOE was observed between H3b/H3b′ of the cyclopropyl ring and the H4 and H7 protons of the benzimidazole ring. This "folded conformation" was explained by the occurrence of additional stabilization due to a hydrophobic interaction between these two groups.

The crystallographic structures of complexes both dovitinib-FGFR-1 and AT-9283 –Aurora A, revealed that there is a co-planarity between the benzimidazole and the quinolin-2-one of dovitinib, and pyrazole motif in AT-9283. A tautomeric rearrangement of the double bond induces a restriction on the rotation around the connection between the two fragments in each case (see **Figure 8**). This indicate the favorite binding to the less rotatable conformer (21B).

Recently AT-9283 was phase I/phase II trial of AT9283, a selective inhibitor of Aurora kinase in children with relapsed or refractory acute leukemia: challenges to run early phase clinical trials for children with leukemia [131–137].

2.3 α-Haloacetamidebenzimidazole derivatives as multitargeting agents

Employing virtual screening methods of PubChem database as a first step, selected support vector machine (SVM) virtual hits were evaluated by Lipinski's

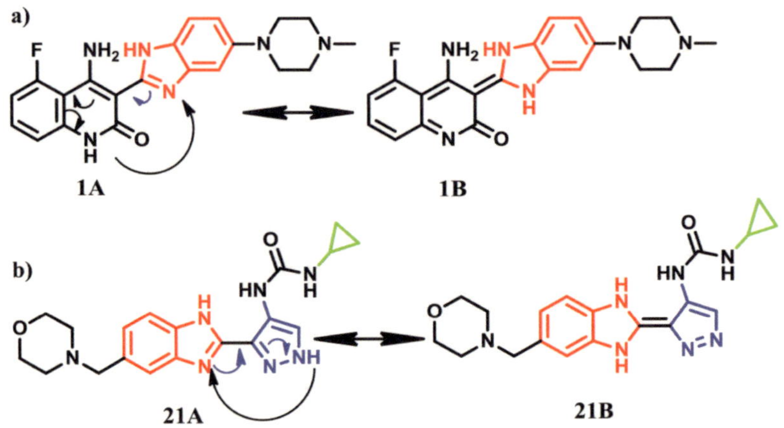

Figure 8.
Tautomeric rearrangement of multitarget inhibitors (1) and (21). (a) benzimidazole quinolin-2-one heterocyclic, and (b) benzimidazole pyrazole derivatives.

rule of five. The compounds which passed Lipinski's rule of five were subject to further and more refined screening by using molecular docking. This sequential refinement led to the identification of 2-aryl benzimidazole group of derivatives as multitarget "EGFR, VEGFR-2, and PDGFR" inhibitors [138]. A mechanistic study reported by Jiang and colleagues displayed that (22) exhibited low to moderate micromolar IC_{50} against nine established breast cancer cell lines that are known to have variable expressing EGFR and HER2 (MDA-MB-468, BT-549, MDA-MB-231, HCC1937, T-47D, BT-474, MDA-MB-453, ZR-75-1, MCF-7, and MCF-10 A). Using the 3-(4,5-dimethylthiazol-2-yl)-2,5-diphenyl tetrazolium bromide (MTT) assay, (24 and 25) exerts moderate inhibitory effect on growth of panel of breast cancer cell lines (IC_{50} values of 2–9 μM) and was reported to be more potent than lapatinib against MDA-MB-468, BT-549, MDA-MB-231, ZR-75-1, and MCF-7. A correlation was observed between the level of HER2 and EGFR amplification and expression and the sensitivity toward (22). IC_{50} = 3.58 μM against BT-474 (high expression of HER2), whereas against MDA-MB-453 (lower levels of HER2 expression) IC_{50} = 4.91 μM. The activity against lower EGFR and HER2 expressing cell lines (ZR-75-1 and MCF-7), IC_{50} = 1.81–2.99 μM was explained by the assumption that (22) is able to act via other targets of EGFR and HER2 [139].

Docking the compounds into kinase domains revealed that (22) occupies the ATP-binding site of EGFR (PDB: 2J6M). The compound was able to form a hydrogen bond with amino acid MET 793 (N–H···O:2.485 Å), claimed to be an important binding site of EGFR. The difference in the activity between the two compounds against VEGFR2 was explained by the difference in hydrogen bonding using docking into VEGFR-2 kinase. It was shown that (22) formed two hydrogen bonds with amino acids CYS917 (N–H···Cl:2.484 Å) and ASP1044 (N–H···O:2.429 Å), whereas compound (23) formed only one hydrogen bond with ASP1044 (N–H···O:2.419 Å) [140].

The authors concluded that electron-withdrawing substituent residing at 2-aryl ring together with shorter aliphatic chain contributed to the cytotoxic potency and to the induction of apoptosis by such group of compounds in HepG-2 cell lines. Though reported as multitargeting agent, the activity of 2-chloro-N-(2-*p*-tolyl-1*H*-benzo[d]imidazol-5-yl)acetamide (22) exhibiting most potency could not be explained explicitly by docking alone. (22) encompasses a reactive alphahaloacetamide (see **Figure 9**) that is vulnerable to nucleophilic substitution by biological

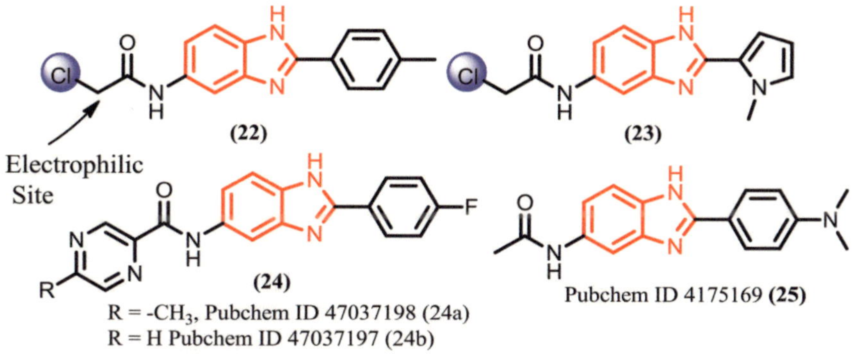

Figure 9.
α-Haloacetamidebenzimidazole derivatives as multitargeting agents. 2-chloro-N-(2-p-tolyl-1H-benzo[d]imidazol-5-yl)acetamide (21), a novel 2-arylbenzimidazole derivative exhibited remarkable activity toward breast cancer. In a study reported by Jiang et al. (22 and 23) were virtually identified as multikinase EGFR and VEGFR inhibitor while (22) was identified as EGFR inhibitor and (23) as PDGFR inhibitor [140, 142].

nucleophiles like thiols (-SH). Thus, a study to explore the formation of irreversible adducts with cellular proteins like kinases is recommended and hoped to uncover the principal mechanism of its wide action.

2.4 2-Anilino-4-(benzimidazol-2-yl)pyrimidines: a multikinase inhibitor scaffold

Anilinopyrimidines (**Figure 10**) displays a wide range of bioactivities. Asymmetric 2-anilinopyrimidines bearing 3-aminopropamides exhibit activity against epidermal growth factor receptor EGFR) [141]. 2-anilinopyrimidine derivatives bearing 4-piperidino substituents exhibited improved and selective activity against triple-negative breast cancer cell line MDA-MB-468 believed to be due to EGFR inhibition. Decorating the pyrimidine nucleus with different substituents at position 4 endowed the final derivatives (4-substituted-2-anilinopyrimidine) with activity as well as selective toward corticotropin-releasing factor (CRF) antagonists [142]. Having the anilino fragment at 2- together with thiazolyl at 4- of the pyrimidine core was reported to exert antagonistic effect of cyclin-dependent kinase-2 (CKD2) [143], and improved inhibitory activity toward CDK9 and (CDK2) [143–145].

Bis-anilinopyrimidine was reported as potent and selective PAK1 inhibitor and as highly selective group I p21-activated kinase (PAK1) inhibitor [146]. Additionally, N-phenyl-N′-[4-(pyrimidin-4-ylamino)phenyl] urea derivatives (see (27) at **Figure 10**) exhibit selective inhibition to class III receptor tyrosine kinase subfamily [147]. Other symmetric 4,6-dianilinopyrimidines induce selective EGFR inhibitions [148].

Notably, introducing the benzimidazolyl moiety at position 4 of the 2-anilinopyrimidine core to produce 2-anilino-4-(benzimidazol-2-yl)-pyrimidines renders such group of compounds' activity against a wider range of kinases (see **Figure 10**).

Renate Determann et al. reported the synthesis and in vitro activity of a small library of compounds that are based on the 2-anilino-4-(benzimidazol-2-yl)-pyrimidine scaffold (**Figure 10**, (30)) [142]. The most potent derivative exhibited antiproliferative activity for several cancer cell lines of the NCI panel in submicromolar concentrations. SAR study was concluded in indicating a basic correlation with the anilinopyrimidine fragment and the substitution pattern at the aniline moiety. It is worth mentioning that 2-anilinopyrimidine fragment (**Figure 10**, (30)) is found in a range of kinase inhibitors.

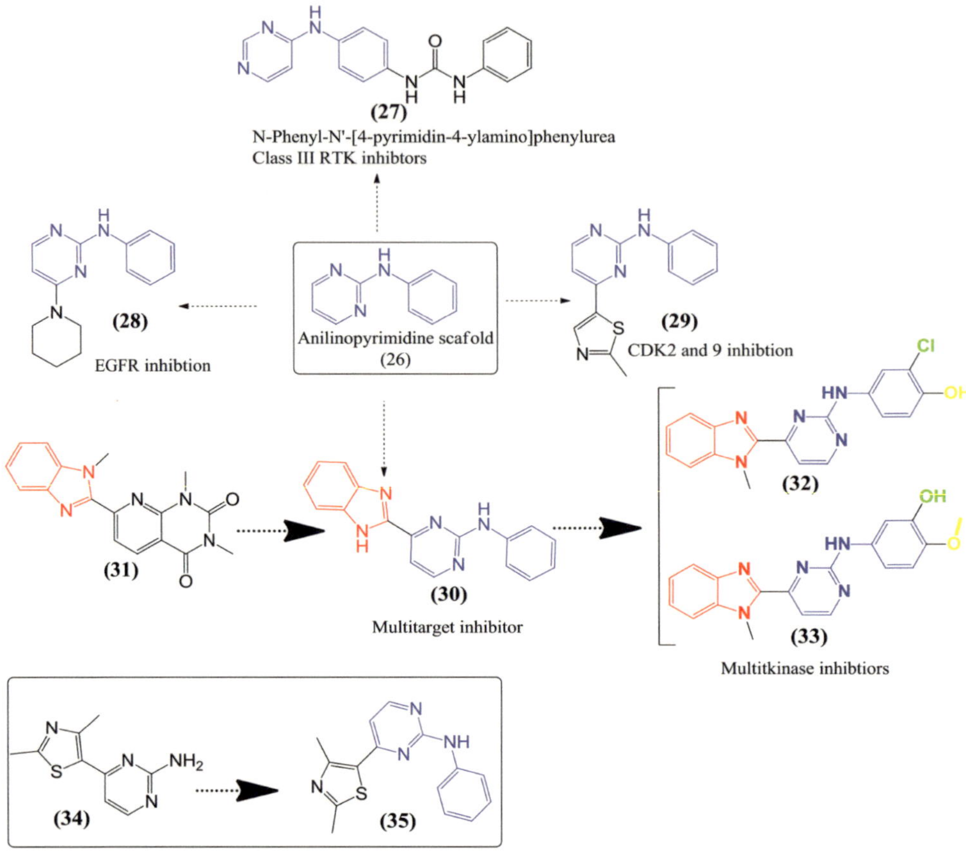

Figure 10.
Development of multitargeting 2-anilino-4-(benzimidazol-2-yl)-pyrimidine scaffold (30) starting from hinge binding compound 1,3-dimethyl-7-(1-methyl-1H-benzimidazol-2-yl)pyrido[2,3-d]pyrimidine-2,4(1H,3H)-dione (31). 2-anilino-4-(benzimidazol-2-yl)-pyrimidine derivatives 2-methoxy-5-{[4-(1-methyl-1H-benzimidazol-2-yl)pyrimidin-2-yl]amino}phenol (33) most potent compound and 2-anilino-4-(benzimidazol-2-yl)-pyrimidine derivatives 2-hydroxy-5-{[4-(1-methyl-1H-benzimidazol-2-yl) pyrimidin-2-yl]amino}phenol (32))), 4-(2,4-dimethyl-thiazol-5-yl)pyrimidin-2-ylamine (35), and 2-anilino-4-(thiazol-5-yl)pyrimidine (29).

Based on high-throughput screening method radiometric protein kinase assay (33PanQinase® Activity Assay) [149], 11 recombinant cancer-related protein kinases (AKT1, ARK5, Aurora B, AXL, FAK, IGF1-R, MET, PLK1, PRK1, SRC, VEGF-R2) were screened by a library of compounds. Interestingly, four kinases (Aurora B, FAK, PLK1, and VEGF-R2) proved to be of particular sensitivity to the tested compounds (**Table 3**). This group of four kinases is involved in oncogenesis and maintenance of vital processes of cancer. Thus it is believed that their concerted inhibition could be useful in the treatment of various malignancies. It is worth noting the infectivity of most of tested compounds, including the active ones against AKT1 (shown in **Table 3**).

2.4.1 2-Anilino-4-(benzimidazol-2-yl)pyrimidine-target interactions

Though the authors did not report a prudent SAR, however, docking compound (33) to ATP-binding pocket of PLK1 (PDB 2OWB) helped rationalize the initial observations [142]. One main reflection highlighted the positioning of the anilinopyrimidine fragment in the hinge region, forming a pair of hydrogen bonds to Cys133. Methoxy (CH_3O-) group at the position 2 of the aniline moiety forms a

	AKT1	Aurora B	FAK	PLK1	VEGF-R2
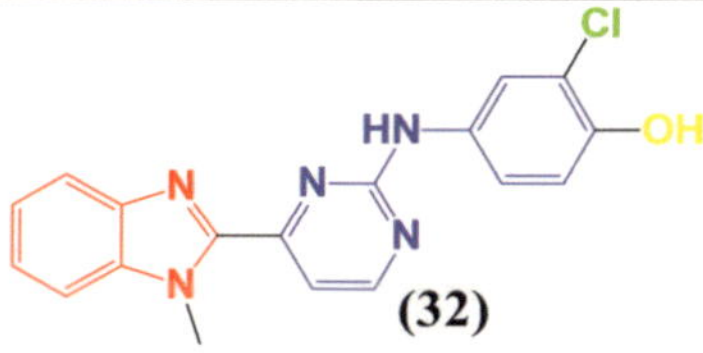 (32)	>100	7 ± 2.3	10.4 ± 2.7	6.0 ± 0.1	7.5 ± 2.0
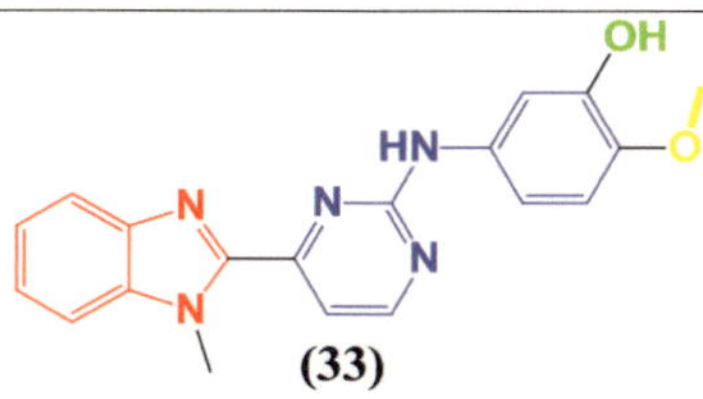 (33)	>100	6.0 ± 0.2	3.4 ± 0.8	1.2 ± 0.2	7.2 ± 0.3
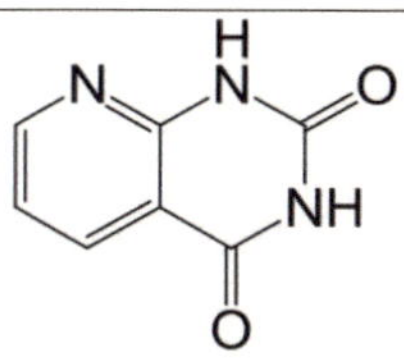 (39)	>100	>100	92	>100	85
Sorafenib	>10	1.8	>10	>10	0.022
Sunitinib	>10	1.5	1.6	>10	0.070

Compound (33) exhibited activities that range between IC_{50} = 1.2 and 7.2 μM [142].

Table 3.
Protein kinase inhibition by (32 and 33) compared pyrido[2,3-d]pyrimidine-2,4(1H,3H)-dione (39) to standard multitargeting FDA-approved agents sorafenib and sunitinib.

hydrogen bonding with the guanidine of Arg136 residing at the opening of the PLK1 ATP-binding pocket. The inactivity of derivatives with substituents bulkier than methoxy group (CH_3O-) was explained partially by the clash with Leu59 and Arg136 at the pocket entrance indicating limited tolerance to variation at this region.

2.5 Benzimidazolylamidines as multitargeting agents

Silvana Raić-Malića and colleagues reported the synthesis of a group of benzimidazole amidine derivatives [150]. Specifically, compound (**Figure 11**, (36)) abrogated the activity of several protein enzymes including tissue transglutaminase (TGM2) and kinases like CDK9, sphingosine kinase 1(SK1), and p38 mitogen-activated protein kinase (p38 MAPK), whereas compound (37) did not have profound effect on CDK9 and TGM2 but showed moderate downregulation of SK1 and significant reduction in p38 MAPK.

A small library comprising 27 compounds was screened for the potency. Two of them, *p*-chlorophenyl-substituted 1,2,3-triazolyl derivatives of amidine *N*-isopropyl amidine (36) and imidazoline amidine (37), exhibited remarkable antiproliferative activities with IC_{50} of 0.05 and 0.06 μM in non-small cell lung cancer cells A54 and was defined as multitarget inhibitors.

In their endeavor to look for potent inhibitors for treatment of non-small cell lung cancer, Silvana Raić-Malić and her team developed a group of benzimidazole amidine derivative that showed an inhibitory effect on several key players in cancer

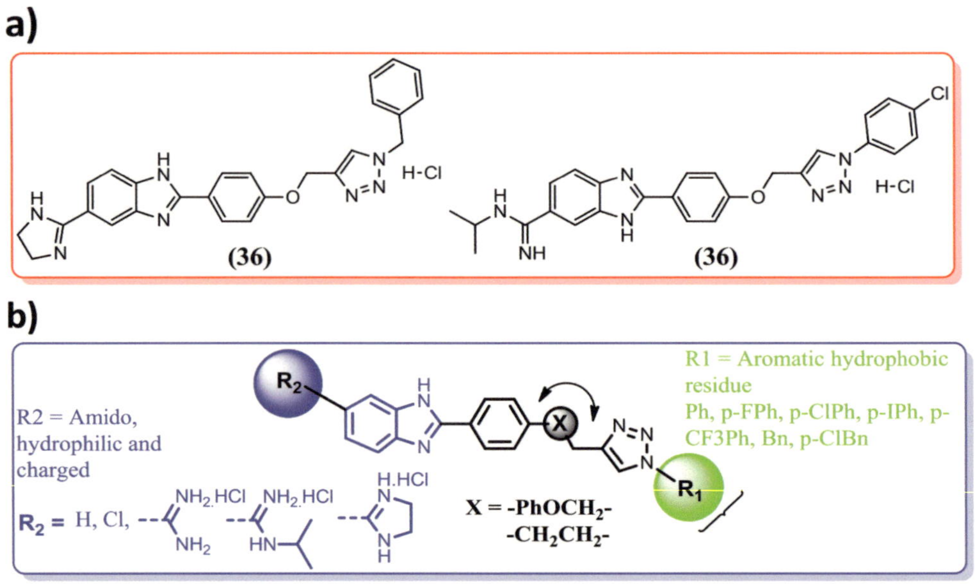

Figure 11.
(a) Hit compounds prepared and screened for multitarget action 2-(4-((1-Benzyl-1H-1,2,3-triazol-4-yl)methoxy)phenyl)-5-(4,5-dihydro-1H-imidazol-2-yl)-1H-benzo[d]imidazole hydrochloride (36),2-(4-((1-(4-chlorophenyl)-1H-1,2,3-triazol-4-yl)methoxy)phenyl)-N-isopropyl-1H-benzo[d]imidazole-6-carboximidamide (37); (b) summary of structure–activity relationship of benzimidazolylamidines [150].

proliferation [150]. A recent study reported that synthesis of amidino 2-substituted benzimidazoles linked to 1,4-disubstituted 1,2,3-triazoles by applying microwave and ultrasound irradiation in click reaction and subsequent condensation of thus obtained 4-(1,2,3-triazol-1-yl)benzaldehyde with *o*-phenylenediamines. The study concluded the improved cytotoxic effect (within the nanomolar range; IC_{50} of 50 and 60 nM) against hepatocellular carcinoma cells. A follow-up study affirms the conclusion that when benzimidazole is conjugated to 1,2,3-triazole moiety, the hybrid exerts potent and selective antiproliferative effect against a panel of cell lines [non-small cell lung cancer (A549), ductal pancreatic adenocarcinoma (CFPAC-1), cervical carcinoma (HeLa), and metastatic colorectal adenocarcinoma (SW620) as well as on normal human lung fibroblasts (WI38)] with 5-fluorouracil (5FU) as a positive control. Two hits (36) and (37) (**Figure 11a**) demonstrated a potent activity at nM range (IC_{50} of 50 and 60 nM) against non-small cell lung cancer (A549). Interestingly, benzyl-substituted 1,2,3-triazolyl analogue of imidazoline (36) exhibited a remarkable and selective activity (IC_{50} = 0.07 μM) on A549 cell line. A mechanistic study performed on A549 cell line using Western blotting reinforced the belief that nature of aromatic substituent of 1-(1,2,3-triazolyl) and amidino moiety at C-5 position of benzimidazole ring is critical to the cytostatic activity of this group of compounds. In silico analysis supported the conception that (36) is bound slightly better than (37) to ATP-binding site of p38 MAPK, which correlates with observed decrement in the expression level of phospho-p38 MAPK displayed by (36). The importance of triazole was referred to its ability to form one H-bond with Met109 in the hinge region. Aminobenzimidazole group forms a number of HB with polar amino acids Glu71, Hid148, and Asp168 in the linker region. Phenyl moieties found on the hybrid both are placed in the hydrophobic environment. The phenyl connected to the triazole is assumed to participate in a π-π stacking with Phe169 (see **Figure 11**). The study reported (36) as a multitarget inhibitor since it abrogated the activity of several protein kinases including TGM2, CDK9, SK1, and p38 MAPK.

3. Conclusion

3.1 Multitargeting and polypharmacology

According to the definition of Richard Morphy, the multitarget drugs are defined as "compounds that are designed to modulate multiple targets of relevance to a disease, with the overall goal of enhancing efficacy and/or improving safety" (Morphy, Rankovic, 2005) [151].

Modulating the function of numerous biological molecules is a well-established pharmacological approach in medicine practice. Paracetamol, a traditional therapeutic used worldwide, is believed to induce its effects via action on multiple targets. Several psychoactive, serotonergic, cholinergic, and adrenergic agonists or antagonists exercise their actions on a wider range of singular biomolecular target.

Apart from the alphachloroacetamidobenimidzoles (22), the groups of compounds reported so far in the literature as multitarget agents act in most cases on receptor tyrosine kinases (RTKs) as competitive ATP inhibitors. Those by virtue occupy the vicinity of ATP with the heteroaromatic system interactively buried in the purine portion pocket and interact with the hinge region of the kinase domain. The thiol (-SH)- π and the stacking π-π together with the hydrophobic interaction with the floor and the ceiling of the purine-binding regions are believed to do the required binding adjustment as kinase inhibitors. Crystallographic structure of dovitinib human FGFR1 revealed the occupancy of the purine-binding regions (part of the ATP-binding site) with the quinolinone-benzimidazole fragment, while the N-methylpiperazine attached to C5' at the phenyl part of the benzimidazole is pointing toward the opening and is exposed to the solution. Thus, it seems that benzimidazole portion is not interacting directly with the hinge region of the enzyme. Similar binding is noticed with AT9382. The pyrazolylbenzimidazole and the benzamide motif take part in HBD-HBA bridging with the hinge of the kinase domain.

In the case of 2-anilino-4-(benzimidazole-2-yl)pyrimidine, the benzimidazole portion looks immersed deep in the purine-binding regions of the ATP-binding site participating in direct interactions via hydrogen bonding and hydrophobic interactions, while the hydroxymethoxyaniline portion points towards the solvent exposed area.

3.2 Lessons learnt

3.2.1 Discovery methods

Despite the imbedded potential, the multitarget activity of the reported benzimidazole-based scaffolds was identified serendipitously. In other words, none of the benzimidazole anticancer multitargeting agents seem to be identified in unforeseen manner, and in many ways they emerge with no intention to be designed initially. While adhering to the development of selective and specific agents, results accumulated afterword revealed multitarget action. For example, 3-benzimidazol-2-ylhydroquinolin-2-one scaffold [benzimidazolylquinolinone (**Figure 4**, (12)] was identified using high-throughput screening (HTS). AT9283 (**Figure 6**, (21)) was identified following fragment-based structural approach with the initial aim to develop an Aurora selective inhibitor, and later it was reported to act as multitargeting agent.

It is hoped that the identification, discovery, and optimization of benzimidazole-based multitargeting anticancer agent will benefit from the "big data era" fueled by data available from public repositories.

3.2.2 Shift in the paradigm

Multitargeting can occur via three possible ways: acting on the same target, on different targets of the same pathway, or on different targets of different pathways. So far the benzimidazole derivatives that have been explored are reported to act as the third category "acting on different targets of different pathways." The focus has been so far on the kinome-relevant signaling key player with dovitinib widening the landscape to non-kinase targets. Broadening "multitargeting" concept to identify novel inhibitors with potency against key targets outside the human kinome necessitates treating complex diseases using "polypharmacology" gains special interest in resistant mutated spreadable cancers [151].

Despite the initial enthusiasm for the efficacy of molecular targeted therapeutics following the approval of imatinib, a small tyrosine kinase inhibitor targeting BCR-Abl, in chronic myeloid leukemia (CML) and trastuzumab, a monoclonal antibody against HER2, for treatment of metastatic breast cancer, scientists and clinicians were challenged by recurrent relapse due to cancer patients who developed drug resistance. In the case of RTKi, resistance can emerge as a result of selection for mutant sin in the target that renders the binding site inaccessible, reduced influx accompanied by enhance efflux, shift in metabolism and excretion of the drug, and the activation of alternative signaling pathways. Thus, the rationale for targeting drugs is shifting. In the last two decades, the main effort was aimed at developing highly specific inhibitors acting on single target. Now, there is a general agreement that molecules interfering simultaneously with multiple RTKs might be more effective than single-target agents. With the recent approval by the FDA of sorafenib, regorafenib, sunitinib, lenvatinib, and axitinib-targeting VEGFR, PDGFR, FLT-3, and c-KIT—more attention is drawn to broad-spectrum anticancer properties multikinase targeting drugs. Thus it is anticipated that more multitargeting agents will be getting into clinical trials and making their way to clinical application. It is hoped that identification, discovery, and optimization of benzimidazole-based multitargeting agents will benefit from the "big data era" fueled by the availability of big data, advances in technology, and artificial intelligence.

Acknowledgements

The author would like to acknowledge "Zamala" Program: University Fellowship Program in Palestinian Universities (http://www.taawon.org/ar/zamala) for the fund granted to make this research stay at London Regional Cancer Center, Western University Ontario, Canada, possible. Also, I would like to acknowledge my home institution Al-Quds University, Jerusalem-Palestine, for the support and encouragement. In addition, I wish to thank the editor and high value the editor's work, reading, and comments. The accomplishment of this chapter was made possible thanks to such support and help.

Conflict of interest

No conflict of interest exists.

Notes/thanks/other declarations

I would like to express my special thanks, gratitude and truthful appreciation to my dearest family: my wife Muna, my daughters Aseel and Layan and my sons

Mulham and Majd for all the love, compassion, and support I received from them all along.

This chapter is dedicated to the respectable memories of my mother Jaleelah and father Salem who died of old age and to the reminiscence of my dearest brother Mohammed who left this world due to leukemia. Peace Be Upon Them All.

Acronyms and abbreviations

MTT	3-(4,5-dimethylthiazol-2-yl)-2,5-diphenyl tetrazolium bromide
PRK1	actin-regulating kinase
MEK1	activated protein kinase
AML	acute myeloid leukemia
ADME	adsorption, disposition, metabolism, excretion
ARK5	AMPK-related kinase 5
ALK	anaplastic lymphoma kinase
ALP	alkaline phosphatase
AXL	"anexelekto" receptor tyrosine kinase
BCR-Abl	breakpoint cluster region-Abelson kinase
CK2	casein kinase 2
CXCR3	chemokine receptor
CML	chronic myelogeneous leukemia
DHODH	dihydroorotate dehydrogenase
D2R	dopamine receptor 2
FGFR-1/FGFR-2/FGFR-3	fibroblast growth factor receptors 1, 2, 3
FAK	focal adhesion kinase
IC_{50}	half maximal inhibitory concentration
HGF or MET	hepatocyte growth factor
H4H	histamine receptors 4
HDAC2	histone deacetylase 2
MDA-MB-468	human breast carcinoma cell lines
HER2	human epidermal growth factor receptor 2
HB	hydrogen bind
HBD	hydrogen bond donor
HBA	hydrogen bond acceptor
5-HTR	hydroxytryptamine receptors
CDK	inhibition of cyclin-dependent kinases
IGF-1R	insulin-like growth factor 1 receptor
ITK	interleukin 2-inducible T-cell kinase
JAK-1/JAK-2/JAK-3	Janus kinase-1/2/3
LE	ligand efficiency
Lck	lymphocyte tyrosine kinase
MEKK2	mitogen-activated protein kinase kinase kinase 2
nM	nanomolar
μM	micromolar
PAK1	p21-activated kinase
p38 MAPK	p38 mitogen-activated protein kinase
Ph + ALL	Ph + acute lymphoblastic leukemia
PI3K	phosphatidylinositol 3-kinase
PDGFR-α/β	platelet-derived growth factor receptor-α/β
PLK-1	polo-like kinase 1
PARP-1	poly(ADP-ribose)polymerase-1

ATP	adenosine triphosphate
PDB	Protein Data Bank
AKT1	RAC-alpha serine/threonine-protein kinase
PTKs	protein tyrosine kinase
PTP	protein tyrosine phosphatases
c-KIT	stem cell factor receptor
FLT3	FMS-like tyrosine kinase 3
FLT3-ITD	FMS-like tyrosine kinase 3 internal tandem duplication
SAR	structure–activity relationship
SVM	support vector machines
ATR	telangiectasia and Rad3-related protein kinase
TOPO1/TOPO2	topoisomerase 1/2
TKRs	tyrosine kinase receptors
VEGFR-1/VEGFR-2/VEGFR-3	vascular endothelial growth factor receptor-1, 2, 3
SRC	v-src sarcoma (Schmidt-Ruppin A-2) viral onco gene homolog (avian)

Author details

Yousef Najajreh
Anticancer Drugs Research Lab, Faculty of Pharmacy, Al-Quds University, Jerusalem, Palestine

*Address all correspondence to: y.s.najajreh@gmail.com

References

[1] Salahuddin, Shaharyar M, Mazumder A. Benzimidazoles: A biologically active compounds. Arabian Journal of Chemistry. 2017

[2] Bansal Y, Silakari O. The therapeutic journey of benzimidazoles: A review. Bioorganic & Medicinal Chemistry. 2012

[3] Yadav G, Ganguly S. Structure activity relationship (SAR) study of benzimidazole scaffold for different biological activities: A mini-review. European Journal of Medicinal Chemistry. 2015

[4] Walia R, Naaz SF, Iqbal K, Lamba HS. Benzimidazole derivatives—An overview. International Journal of Research in Pharmaceutical Chemistry. 2011

[5] Ahamad A, Pandurangan A, Rana K, Tiwari AK, Singh N, Anand P. Benzimidazole: A short review of their antimicrobial activities. International Current Pharmaceutical Journal. 2012

[6] Kathiravan MK, Salake AB, Chothe AS, Dudhe PB, Watode RP, Mukta MS, et al. The biology and chemistry of antifungal agents: A review. Bioorganic & Medicinal Chemistry. 2012

[7] Shah K, Chhabra S, Shrivastava SK, Mishra P. Benzimidazole: A promising pharmacophore. Medicinal Chemistry Research. 2013

[8] Sachs G, Shin JM, Howden CW. Review article: The clinical pharmacology of proton pump inhibitors. Alimentary Pharmacology and Therapeutics. 2006

[9] Sharma MC, Kohli DV, Sharmab S, Sharma AD. Synthesis and antihypertensive activity of some new benzimidazole derivatives of 4′-(6-methoxy-2-substituted-benzimidazole-1-ylmethyl)-biphenyl-2-carboxylic acid in the presences of BF3·OEt2. Der Pharmacia Sinica. 2010

[10] Shah DI, Sharma M, Bansal Y, Bansal G, Singh M. Angiotensin II-AT1 receptor antagonists: Design, synthesis and evaluation of substituted carboxamido benzimidazole derivatives. European Journal of Medicinal Chemistry. 2008

[11] Sakamoto H, Ojima M, Kubo K, Fuse H, Tanaka M, Kohara Y, et al. In vitro antagonistic properties of a new angiotensin type 1 receptor blocker, Azilsartan, in receptor binding and function studies. Journal of Pharmacology and Experimental Therapeutics. 2010

[12] Zhang J, Liu X, Wang SQ, Liu GY, Xu WR, Cheng XC, et al. Identification of dual ligands targeting angiotensin II type 1 receptor and peroxisome proliferator-activated receptor-γ by core hopping of telmisartan. Journal of Biomolecular Structure and Dynamics. 2017

[13] Achar KCS, Hosamani KM, Seetharamareddy HR. In-vivo analgesic and anti-inflammatory activities of newly synthesized benzimidazole derivatives. European Journal of Medicinal Chemistry. 2010

[14] Rathore A, Sudhakar R, Ahsan MJ, Ali A, Subbarao N, Jadav SS, et al. In vivo anti-inflammatory activity and docking study of newly synthesized benzimidazole derivatives bearing oxadiazole and morpholine rings. Bioorganic Chemistry. 2017

[15] Bukhari SNA, Lauro G, Jantan I, Chee CF, Amjad MW, Bifulco G, et al. Anti-inflammatory trends of new benzimidazole derivatives. Future Medicinal Chemistry. 2016

[16] Padalkar VS, Borse BN, Gupta VD, Phatangare KR, Patil VS, Umape PG, et al. Synthesis and antimicrobial activity of novel 2-substituted benzimidazole, benzoxazole and

benzothiazole derivatives. Arabian Journal of Chemistry. 2016

[17] Zhang HZ, Damu GLV, Cai GX, Zhou CH. Design, synthesis and antimicrobial evaluation of novel benzimidazole type of fluconazole analogues and their synergistic effects with chloromycin, norfloxacin and fluconazole. European Journal of Medicinal Chemistry. 2013

[18] Özkay Y, Tunali Y, Karaca H, Işikdağ I. Antimicrobial activity and a SAR study of some novel benzimidazole derivatives bearing hydrazone moiety. European Journal of Medicinal Chemistry. 2010

[19] Tonelli M, Simone M, Tasso B, Novelli F, Boido V, Sparatore F, et al. Antiviral activity of benzimidazole derivatives. II. Antiviral activity of 2-phenylbenzimidazole derivatives. Bioorganic & Medicinal Chemistry. 2010

[20] Budow S, Kozlowska M, Gorska A, Kazimierczuk Z, Eickmeier H, La Colla P, et al. Substituted benzimidazoles: Antiviral activity and synthesis of nucleosides. ARKIVOC. 2009

[21] Li YF, Wang GF, He PL, Huang WG, Zhu FH, Gao HY, et al. Synthesis and anti-hepatitis B virus activity of novel benzimidazole derivatives. Journal of Medicinal Chemistry. 2006

[22] Elnima EI, Zubair MU, Al-Badr AA. Antibacterial and antifungal activities of benzimidazole and benzoxazole derivatives. Antimicrobial Agents and Chemotherapy. 1981

[23] Bai YB, Zhang AL, Tang JJ, Gao JM. Synthesis and antifungal activity of 2-chloromethyl-1 H -benzimidazole derivatives against phytopathogenic fungi in vitro. Journal of Agricultural and Food Chemistry. 2013

[24] Singla P, Luxami V, Paul K. Benzimidazole-biologically attractive scaffold for protein kinase inhibitors. RSC Advances. 2014

[25] Prichard R. Anthelmintic resistance. Veterinary Parasitology. 1994

[26] Martin RJ. Modes of action of anthelmintic drugs. Veterinary Journal. 1997

[27] Gurvinder S, Maninderjit K, Mohan C. Benzimidazole: The latest information of biological activities. International Research Journal of Pharmacy. 2013

[28] Brown HD, Matzuk AR, Ilves IR, Peterson LH, Harris SA, Sarett LH, et al. Antiparasitic drugs. IV. 2-(4′-thiazolyl)-benzimidazole, a new anthelmintic. Journal of the American Chemical Society. 1961

[29] Nofal ZM, Soliman EA, Abd El-Karim SS, El-Zahar MI, Srour AM, Sethumadhavan S, et al. Novel benzimidazole derivatives as expected anticancer agents. Acta Poloniae Pharmaceutica. Drug Research. 2011

[30] Shaharyar M, Abdullah MM, Bakht MA, Majeed J. Pyrazoline bearing benzimidazoles: Search for anticancer agent. European Journal of Medicinal Chemistry. 2010

[31] Refaat HM. Synthesis and anticancer activity of some novel 2-substituted benzimidazole derivatives. European Journal of Medicinal Chemistry. 2010

[32] Paul K, Sharma A, Luxami V. Synthesis and in vitro antitumor evaluation of primary amine substituted quinazoline linked benzimidazole. Bioorganic & Medicinal Chemistry Letters. 2014

[33] Patil A, Ganguly S, Surana S. A systematic review of benzimidazole derivatives as an antiulcer agent. Rasayan Journal of Chemistry. 2008

[34] Sivakumar R, Pradeepchandran R, Jayaveera KN, Kumarnallasivan P, Vijaianand PR, Venkatnarayanan R. Benzimidazole: An attractive pharmacophore in medicinal chemistry. International Journal of Pharmaceutical Research. 2011

[35] Ganie AM, Dar AM, Khan FA, Dar BA. Benzimidazole derivatives as potential antimicrobial and antiulcer agents: A mini review. Mini Reviews in Medicinal Chemistry. 2018

[36] Gurer-Orhan H, Orhan H, Suzen S, Püsküllü MO, Buyukbingol E. Synthesis and evaluation of in vitro antioxidant capacities of some benzimidazole derivatives. Journal of Enzyme Inhibition and Medicinal Chemistry. 2006

[37] Mavrova AT, Yancheva D, Anastassova N, Anichina K, Zvezdanovic J, Djordjevic A, et al. Synthesis, electronic properties, antioxidant and antibacterial activity of some new benzimidazoles. Bioorganic & Medicinal Chemistry. 2015

[38] Rajasekaran S, Rao G, Chatterjee A. Synthesis, anti-inflammatory and anti-oxidant activity of some substituted benzimidazole derivatives. International Journal of Drug Development and Research. 2012

[39] Ueno H, Katoh S, Yokota K, Hoshi JI, Hayashi M, Uchida I, et al. Structure-activity relationships of potent and selective factor Xa inhibitors: Benzimidazole derivatives with the side chain oriented to the prime site of factor Xa. Bioorganic & Medicinal Chemistry Letters. 2004

[40] Nar H, Bauer M, Schmid A, Stassen JM, Wienen W, Priepke HWM, et al. Structural basis for inhibition promiscuity of dual specific thrombin and factor Xa blood coagulation inhibitors. Structure. 2001

[41] Janssen FW, Young EM, Kirkman SK, Sharma RN, Ruelius HW. Biotransformation of the immunomodulator, 3-(p-chlorophenyl)-2,3-dihydro-3-hydroxythiazolo[3,2a] benzimidazole-2-acetic acid, and its relationship to tyroid toxicity. Toxicology and Applied Pharmacology. 1981

[42] Oe T, Aramori I, Hosogai N, Mutoh S, Konishi S, Fujimura T, et al. FK614, a novel peroxisome proliferator-activated receptor γ modulator, induces differential transactivation through a unique ligand-specific interaction with transcriptional coactivators. Journal of Pharmacological Sciences. 2005

[43] Fujimura T, Sakuma H, Konishi S, Oe T, Hosogai N, Kimura C, et al. FK614, a novel peroxisome proliferator-activated receptor gamma modulator, induces differential transactivation through a unique ligand-specific interaction with transcriptional coactivators. Journal of Pharmacological Sciences. 2005

[44] Oh S, Ha H-J, Chi D, Lee H. Serotonin Receptor and Transporter Ligands - Current Status. Curr Med Chem. 2012

[45] Ramot Y, Mastrofrancesco A, Camera E, Desreumaux P, Paus R, Picardo M. The role of PPARγ-mediated signalling in skin biology and pathology: New targets and opportunities for clinical dermatology. Experimental Dermatology. 2015

[46] Huang THW, Teoh AW, Lin BL, DSH L, Roufogalis B. The role of herbal PPAR modulators in the treatment of cardiometabolic syndrome. Pharmacological Research. 2009

[47] Trémollières F, Lopes P. Specific estrogen receptor modulators (SERMS). Les Modul spécifiques des récepteurs œstrogéniques. 2002

[48] Gür ZT, Çalışkan B, Banoglu E. Drug discovery approaches targeting 5-lipoxygenase-activating protein (FLAP) for inhibition of cellular

leukotriene biosynthesis. European Journal of Medicinal Chemistry. 2018

[49] Aboul-Enein HY. Benzimidazole derivatives as antidiabetic agents. Med Chem (Los Angeles). 2015

[50] Bathini P, Kameshwari L, Vijaya N. Antidiabetic effect of 2 nitro benzimidazole in alloxan induced diabetic rats. International Journal of Basic & Clinical Pharmacology. 2013

[51] Spasov AA, Vassiliev PM, Lenskaya KV, Anisimova VA, Kuzmenko TA, Morkovnik AS, et al. Hypoglycemic potential of cyclic guanidine derivatives. Directed search, pharmacology, clinics. Pure and Applied Chemistry. 2017

[52] Robinson MW, McFerran N, Trudgett A, Hoey L, Fairweather I. A possible model of benzimidazole binding to β-tubulin disclosed by invoking an inter-domain movement. Journal of Molecular Graphics & Modelling. 2004

[53] Aguayo-Ortiz R, Méndez-Lucio O, Romo-Mancillas A, Castillo R, Yépez-Mulia L, Medina-Franco JL, et al. Molecular basis for benzimidazole resistance from a novel β-tubulin binding site model. Journal of Molecular Graphics & Modelling. 2013

[54] Silvestre A, Humbert JF. Diversity of benzimidazole-resistance alleles in populations of small ruminant parasites. International Journal for Parasitology. 2002

[55] Kwa MSG, Veenstra JG, Roos MH. Molecular characterisation of β-tubulin genes present in benzimidazole-resistant populations of *Haemonchus contortus*. Molecular and Biochemical Parasitology. 1993

[56] Pjura PE, Grzeskowiak K, Dickerson RE. Binding of Hoechst 33258 to the minor groove of B-DNA. Journal of Molecular Biology. 1987

[57] Stella S, Cascio D, Johnson RC. The shape of the DNA minor groove directs binding by the DNA-bending protein Fis. Genes & Development. 2010

[58] Nelson SM, Ferguson LR, Denny WA. Non-covalent ligand/DNA interactions: Minor groove binding agents. Mutation Research, Fundamental and Molecular Mechanisms of Mutagenesis. 2007

[59] Andrić D, Roglić G, Šukalović V, Šoškić V, Kostić-Rajačić S. Synthesis, binding properties and receptor docking of 4-halo-6-[2-(4-arylpiperazin-1-yl) ethyl]-1H-benzimidazoles, mixed ligands of D2 and 5-HT1A receptors. European Journal of Medicinal Chemistry. 2008

[60] Siracusa MA, Salerno L, Modica MN, Pittalà V, Romeo G, Amato ME, et al. Synthesis of new arylpiperazinylalkylthiobenzimidazole, benzothiazole, or benzoxazole derivatives as potent and selective 5-HT1A serotonin receptor ligands. Journal of Medicinal Chemistry. 2008

[61] Avila D, Frechilla D, Río JD, López-Rodríguez ML, Morcillo MJ, Schiapparelli L, et al. Design and synthesis of new benzimidazole-arylpiperazine derivatives acting as mixed 5-HT1A/5-HT3 ligands. Bioorganic & Medicinal Chemistry Letters. 2003

[62] Andrić D, Tovilović G, Roglić G, Šoškić V, Tomić M, Kostić-Rajačić S. 6-[2-(4-Arylpiperazin-1-yl)ethyl]-4-halo-1,3-dihydro-2H-benzimidazole-2-thiones: Synthesis and pharmacological evaluation. Journal of the Serbian Chemical Society. 2007

[63] Fernandes JPS, Pasqualoto KFM, Ferreira EI, Brandt CA. Molecular modeling and QSAR studies of a set of indole and benzimidazole derivatives as H4 receptor antagonists. Journal of Molecular Modeling. 2011

[64] Šukalović V, Andrić D, Roglić G, Kostić-Rajačić S, Schrattenholz A, Šoškić V. Synthesis, dopamine D2receptor binding studies and docking analysis of 5-[3-(4-arylpiperazin-1-yl)propyl]-1H-benzimidazole, 5-[2-(4-arylpiperazin-1-yl)ethoxy]-1H-benzimidazole and their analogs. European Journal of Medicinal Chemistry. 2005

[65] Hayes ME, Wallace GA, Grongsaard P, Bischoff A, George DM, Miao W, et al. Discovery of small molecule benzimidazole antagonists of the chemokine receptor CXCR3. Bioorganic & Medicinal Chemistry Letters. 2008

[66] Moriarty KJ, Takahashi H, Pullen SS, Khine HH, Sallati RH, Raymond EL, et al. Discovery, SAR and X-ray structure of 1H-benzimidazole-5-carboxylic acid cyclohexyl-methyl-amides as inhibitors of inducible T-cell kinase (Itk). Bioorganic & Medicinal Chemistry Letters. 2008

[67] Zhang G, Ren P, Gray NS, Sim T, Liu Y, Wang X, et al. Discovery of pyrimidine benzimidazoles as Lck inhibitors: Part I. Bioorganic & Medicinal Chemistry Letters. 2008

[68] Yaguchi SI, Fukui Y, Koshimizu I, Yoshimi H, Matsuno T, Gouda H, et al. Antitumor activity of ZSTK474, a new phosphatidylinositol 3-kinase inhibitor. Journal of the National Cancer Institute. 2006

[69] Yeh TC, Marsh V, Bernat BA, Ballard J, Colwell H, Evans RJ, et al. Biological characterization of ARRY-142886 (AZD6244), a potent, highly selective mitogen-activated protein kinase kinase 1/2 inhibitor. Clinical Cancer Research. 2007

[70] Bueno OF. The MEK1-ERK1/2 signaling pathway promotes compensated cardiac hypertrophy in transgenic mice. The EMBO Journal. 2002

[71] Antony SA, Sam Daniel Prabu D, Ramalakshmi N, Lakshmanan S, Thirumurugan K, Govindaraj D. Synthesis, characterization of benzimidazole carboxamide derivatives as potent anaplastic lymphoma kinase inhibitor and antioxidant activity. Synthetic Communications. 2019

[72] Long T, Neitz RJ, Beasley R, Kalyanaraman C, Suzuki BM, Jacobson MP, et al. Structure-bioactivity relationship for benzimidazole thiophene inhibitors of polo-like kinase 1 (PLK1), a potential drug target in *Schistosoma mansoni*. PLoS Neglected Tropical Diseases. 2016

[73] Blake DG, Green SR, Flynn CJ, Glover DM, Rogers NL, Emery A, et al. Abstract 4435: Discovery, biological characterization and oral antitumor activity of polo-like kinase 1 (Plk1) selective small molecule inhibitors. Cancer Research. 2011

[74] Hong S, Kim J, Yun SM, Lee H, Park Y, Hong SS, et al. Discovery of new benzothiazole-based inhibitors of breakpoint cluster region-abelson kinase including the T315i mutant. Journal of Medicinal Chemistry. 2013

[75] Sarno S, Reddy H, Meggio F, Ruzzene M, Davies SP, Donella-Deana A, et al. Selectivity of 4,5,6,7-tetrabromobenzotriazole, an ATP site-directed inhibitor of protein kinase CK2 ('casein kinase-2'). FEBS Letters. 2001

[76] Charrier JD, Durrant SJ, Golec JMC, Kay DP, Knegtel RMA, MacCormick S, et al. Discovery of potent and selective inhibitors of ataxia telangiectasia mutated and Rad3 related (ATR) protein kinase as potential anticancer agents. Journal of Medicinal Chemistry. 2011

[77] Mathias TJ, Natarajan K, Shukla S, Doshi KA, Singh ZN, Ambudkar SV, et al. The FLT3 and PDGFR

inhibitor crenolanib is a substrate of the multidrug resistance protein ABCB1 but does not inhibit transport function at pharmacologically relevant concentrations. Investigational New Drugs. 2015

[78] Penning TD, Zhu GD, Gandhi VB, Gong J, Liu X, Shi Y, et al. Discovery of the Poly(ADP-ribose) polymerase (PARP) inhibitor 2-[(R)-2-methylpyrrolidin-2-yl]-1H-benzimidazole-4-carboxamide (ABT-888) for the treatment of cancer. Journal of Medicinal Chemistry. 2009

[79] Abdullah I, Chee CF, Lee YK, Thunuguntla SSR, Satish Reddy K, Nellore K, et al. Benzimidazole derivatives as potential dual inhibitors for PARP-1 and DHODH. Bioorganic & Medicinal Chemistry. 2015

[80] Penning TD, Zhu GD, Gandhi VB, Gong J, Thomas S, Lubisch W, et al. Discovery and SAR of 2-(1-propylpiperidin-4-yl)-1H-benzimidazole-4-carboxamide: A potent inhibitor of poly(ADP-ribose) polymerase (PARP) for the treatment of cancer. Bioorganic & Medicinal Chemistry. 2008

[81] Zhou D, Chu W, Xu J, Jones LA, Peng X, Li S, et al. Synthesis, [^{18}F] radiolabeling, and evaluation of poly (ADP-ribose) polymerase-1 (PARP-1) inhibitors for in vivo imaging of PARP-1 using positron emission tomography. Bioorganic & Medicinal Chemistry. 2014

[82] Canan S, Webber SE, Newell DR, Thomas HD, Skalitzky D, Batey MA, et al. Preclinical selection of a novel poly(ADP-ribose) polymerase inhibitor for clinical trial. Molecular Cancer Therapeutics. 2007

[83] Singh A, Maqbool M, Mobashir M, Hoda N. Dihydroorotate dehydrogenase: A drug target for the development of antimalarials. European Journal of Medicinal Chemistry. 2017

[84] Bansal S, Sur S, Tandon V. Benzimidazoles: Selective inhibitors of topoisomerase I with differential modes of action. Biochemistry. 2019

[85] Ishida T, Suzuki T, Hirashima S, Mizutani K, Yoshida A, Ando I, et al. Benzimidazole inhibitors of hepatitis C virus NS5B polymerase: Identification of 2-[(4-diarylmethoxy)phenyl]-benzimidazole. Bioorganic & Medicinal Chemistry Letters. 2006

[86] Ryu K, Kim ND, Choi SI, Han CK, Yoon JH, No KT, et al. Identification of novel inhibitors of HCV RNA-dependent RNA polymerase by pharmacophore-based virtual screening and in vitro evaluation. Bioorganic & Medicinal Chemistry. 2009

[87] Adegboye AA, Khan KM, Salar U, Aboaba SA, Kanwal, Chigurupati S, et al. 2-Aryl benzimidazoles: Synthesis, In vitro α-amylase inhibitory activity, and molecular docking study. European Journal of Medicinal Chemistry. 2018

[88] Patil VM, R GK, Chudayeu M, Prakash Gupta S, Samanta S, Masand N, et al. Synthesis, in vitro and in silico NS5B polymerase inhibitory activity of benzimidazole derivatives. Medicinal Chemistry (Los Angeles). 2012

[89] Romero-Castro A, León-Rivera I, Ávila-Rojas LC, Navarrete-Vázquez G, Nieto-Rodríguez A. Synthesis and preliminary evaluation of selected 2-aryl-5(6)-nitro-1H-benzimidazole derivatives as potential anticancer agents. Archives of Pharmacal Research. 2011

[90] Yoon YK, Choon TS. Structural modifications of benzimidazoles via multi-step synthesis and their impact on sirtuin-inhibitory activity. Archiv der Pharmazie (Weinheim). 2016

[91] Tamura Y, Hayashi K, Omori N, Nishiura Y, Watanabe K, Tanaka N, et al.

Identification of a novel benzimidazole derivative as a highly potent NPY Y5 receptor antagonist with an anti-obesity profile. Bioorganic & Medicinal Chemistry Letters. 2013

[92] Song WJ, Lin QY, Jiang WJ, Du FY, Qi QY, Wei Q. Synthesis, interaction with DNA and antiproliferative activities of two novel Cu(II) complexes with norcantharidin and benzimidazole derivatives. Spectrochimica Acta Part A: Molecular and Biomolecular Spectroscopy. 2015

[93] Schaer DA, Beckmann RP, Dempsey JA, Huber L, Forest A, Amaladas N, et al. The CDK4/6 inhibitor abemaciclib induces a T cell inflamed tumor microenvironment and enhances the efficacy of PD-L1 checkpoint blockade. Cell Reports. 2018

[94] Fujiwara Y, Tamura K, Kondo S, Tanabe Y, Iwasa S, Shimomura A, et al. Phase 1 study of abemaciclib, an inhibitor of CDK 4 and 6, as a single agent for Japanese patients with advanced cancer. Cancer Chemotherapy and Pharmacology. 2016

[95] Lim JSJ, Turner NC, Yap TA. CDK4/6 inhibitors: Promising opportunities beyond breast cancer. Cancer Discovery. 2016

[96] Xu H, Yu S, Liu Q, Yuan X, Mani S, Pestell RG, et al. Recent advances of highly selective CDK4/6 inhibitors in breast cancer. Journal of Hematology & Oncology. 2017

[97] Deng Z, Yu L, Cao W, Zheng W, Chen T. Rational design of ruthenium complexes containing 2,6-bis(benzimidazolyl)pyridine derivatives with radiosensitization activity by enhancing p53 activation. ChemMedChem. 2015

[98] Melisi D, Piro G, Tamburrino A, Carbone C, Tortora G. Rationale and clinical use of multitargeting anticancer agents. Current Opinion in Pharmacology. 2013

[99] Li YH, Wang PP, Li XX, Yu CY, Yang H, Zhou J, et al. The human kinome targeted by FDA approved multi-target drugs and combination products: A comparative study from the drug-target interaction network perspective. PLoS One. 2016

[100] Krause DS, Van Etten RA. Tyrosine kinases as targets for cancer therapy. The New England Journal of Medicine. 2005

[101] Iqbal N, Iqbal N. Imatinib: A breakthrough of targeted therapy in cancer. Chemotherapy Research and Practice. 2014

[102] Cristofanilli M, Morandi P, Krishnamurthy S, Reuben JM, Lee BN, Francis D, et al. Imatinib mesylate (Gleevec®) in advanced breast cancer-expressing C-Kit or PDGFR-β: Clinical activity and biological correlations. Annals of Oncology. 2008

[103] Proschak E, Stark H, Merk D. Polypharmacology by design: A medicinal chemist's perspective on multitargeting compounds. Journal of Medicinal Chemistry. 2019

[104] Narayanan D, Gani OABSM, Gruber FXE, Engh RA. Data driven polypharmacological drug design for lung cancer: Analyses for targeting ALK, MET, and EGFR. Journal of Cheminformatics. 2017

[105] Hong WK, Kim ES, Lee JJ, Wistuba I, Lippman S. The landscape of cancer prevention: Personalized approach in lung cancer. Cancer Research. 2011

[106] Renhowe PA, Pecchi S, Shafer CM, Machajewski TD, Jazan EM, Taylor C, et al. Design, structure-activity relationships and in vivo characterization of 4-Amino-3-benzimidazol-2-ylhydroquinolin-2-ones:

Novel class of receptor tyrosine kinase inhibitors. Journal of Medicinal Chemistry. 2009

[107] Trudel S, Li ZH, Wei E, Wiesmann M, Chang H, Chen C, et al. CHIR-258, a novel, multitargeted tyrosine kinase inhibitor for the potential treatment of t(4;14) multiple myeloma. Blood. 2005

[108] Pyo K-H, Cho BC, Kim H, Moon YW, Jang KW, Kang HN, et al. Antitumor activity and acquired resistance mechanism of dovitinib (TKI258) in RET-rearranged lung adenocarcinoma. Molecular Cancer Therapeutics. 2015

[109] Lopes De Menezes DE, Peng J, Garrett EN, Louie SG, Lee SH, Wiesmann M, et al. CHIR-258: A potent inhibitor of FLT3 kinase in experimental tumor xenograft models of human acute myelogenous leukemia. Clinical Cancer Research. 2005

[110] Lee CK, Lee ME, Lee WS, Kim JM, Park KH, Kim TS, et al. Dovitinib (TKI258), a multi-target angiokinase inhibitor, is effective regardless of KRAS or BRAF mutation status in colorectal cancer. American Journal of Cancer Research. 2015

[111] Valverde A, Gomez-Espana A, Hernandez V, Jimenez J, Lopez-Sanchez LM, Cano MT, et al. The multi-targeted kinase inhibitor aee788 exerts anti-proliferative effects in braf mutated colorectal cancer cells. Annals of Oncology. 2010

[112] André F, Bachelot T, Campone M, Dalenc F, Perez-Garcia JM, Hurvitz SA, et al. Targeting FGFR with dovitinib (TKI258): Preclinical and clinical data in breast cancer. Clinical Cancer Research. 2013

[113] Angevin E, Lopez-Martin JA, Lin CC, Gschwend JE, Harzstark A, Castellano D, et al. Phase I study of dovitinib (TKI258), an oral FGFR, VEGFR, and PDGFR inhibitor, in advanced or metastatic renal cell carcinoma. Clinical Cancer Research. 2013

[114] Hahn NM, Bivalacqua TJ, Ross AE, Netto GJ, Baras A, Park JC, et al. A phase II trial of dovitinib in BCG-unresponsive urothelial carcinoma with FGFR3 mutations or overexpression: Hoosier Cancer Research Network trial HCRN 12-157. Clinical Cancer Research. 2017

[115] Musolino A, Campone M, Neven P, Denduluri N, Barrios CH, Cortes J, et al. Phase II, randomized, placebo-controlled study of dovitinib in combination with fulvestrant in postmenopausal patients with HR+, HER2- breast cancer that had progressed during or after prior endocrine therapy. Breast Cancer Research. 2017

[116] Schäfer N, Gielen GH, Kebir S, Wieland A, Till A, Mack F, et al. Phase I trial of dovitinib (TKI258) in recurrent glioblastoma. Journal of Cancer Research and Clinical Oncology. 2016

[117] Laurie SA, Hao D, Leighl NB, Goffin J, Khomani A, Gupta A, et al. A phase II trial of dovitinib in previously-treated advanced pleural mesothelioma: The Ontario Clinical Oncology Group. Lung Cancer. 2017

[118] Konecny GE, Finkler N, Garcia AA, Lorusso D, Lee PS, Rocconi RP, et al. Second-line dovitinib (TKI258) in patients with FGFR2-mutated or FGFR2-non-mutated advanced or metastatic endometrial cancer: A non-randomised, open-label, two-group, two-stage, phase 2 study. The Lancet Oncology. 2015

[119] Bunney TD, Wan S, Thiyagarajan N, Sutto L, Williams SV, Ashford P, et al. The effect of mutations on drug sensitivity and kinase activity of fibroblast growth factor receptors: A combined experimental and theoretical study. eBioMedicine. 2015

[120] Klein T, Vajpai N, Phillips JJ, Davies G, Holdgate GA, Phillips C, et al. Structural and dynamic insights into the energetics of activation loop rearrangement in FGFR1 kinase. Nature Communications. 2015

[121] Tucker JA, Klein T, Breed J, Breeze AL, Overman R, Phillips C, et al. Structural insights into FGFR kinase isoform selectivity: Diverse binding modes of AZD4547 and ponatinib in complex with FGFR1 and FGFR4. Structure. 2014

[122] Lesca E, Lammens A, Huber R, Augustin M. Structural analysis of the human fibroblast growth factor receptor 4 kinase. Journal of Molecular Biology. 2014

[123] Hasinoff BB, Wu X, Nitiss JL, Kanagasabai R, Yalowich JC. The anticancer multi-kinase inhibitor dovitinib also targets topoisomerase I and topoisomerase II. Biochemical Pharmacology. 2012

[124] A receptor tyrosine kinase inhibitor, dovitinib (TKI-258), enhances BMP-2-induced osteoblast differentiation in vitro. Molecules and Cells. 2016

[125] Howard S, Berdini V, Boulstridge JA, Carr MG, Cross DM, Curry J, et al. Fragment-based discovery of the pyrazol-4-yl urea (AT9283), a multitargeted kinase inhibitor with potent aurora kinase activity. Journal of Medicinal Chemistry. 2009

[126] Curry J, Angove H, Fazal L, Lyons J, Reule M, Thompson N, et al. Aurora B kinase inhibition in mitosis: Strategies for optimising the use of aurora kinase inhibitors such as AT9283. Cell Cycle. 2009

[127] Tanaka R, Squires MS, Kimura S, Yokota A, Nagao R, Yamauchi T, et al. Activity of the multitargeted kinase inhibitor, AT9283, in imatinib-resistant BCR-ABL-positive leukemic cells. Blood. 2010

[128] Santo L, Hideshima T, Cirstea D, Bandi M, Nelson EA, Gorgun G, et al. Antimyeloma activity of a multitargeted kinase inhibitor, AT9283, via potent Aurora kinase and STAT3 inhibition either alone or in combination with lenalidomide. Clinical Cancer Research. 2011

[129] Smyth T, Reule M, Yokota A, Ottmann OG, Nagao R, Tanaka R, et al. Activity of the multitargeted kinase inhibitor, AT9283, in imatinib-resistant BCR-ABL-positive leukemic cells. Blood. 2010

[130] Puleo DE, Kucera K, Hammarén HM, Ungureanu D, Newton AS, Silvennoinen O, et al. Identification and characterization of JAK2 pseudokinase domain small molecule binders. ACS Medicinal Chemistry Letters. 2017

[131] Moreno L, Marshall LV, Pearson ADJ, Morland B, Elliott M, Campbell-Hewson Q, et al. A phase I trial of AT9283 (a selective inhibitor of aurora kinases) in children and adolescents with solid tumors: A cancer research UK study. Clinical Cancer Research. 2015

[132] Qi W, Liu X, Cooke LS, Persky DO, Miller TP, Squires M, et al. AT9283, a novel aurora kinase inhibitor, suppresses tumor growth in aggressive B-cell lymphomas. International Journal of Cancer. 2012

[133] Vormoor B, Veal GJ, Griffin MJ, Boddy AV, Irving J, Minto L, et al. A phase I/II trial of AT9283, a selective inhibitor of aurora kinase in children with relapsed or refractory acute leukemia: Challenges to run early phase clinical trials for children with leukemia. Pediatric Blood & Cancer. 2017

[134] Foran JM, Ravandi F, O'Brien SM, Borthakur G, Rios M, Boone P, et al. Phase I and pharmacodynamic trial of AT9283, an aurora kinase inhibitor, in patients with refractory leukemia. Journal of Clinical Oncology. 2008

[135] Dent S, Chi K, Jonker D, Capier K, Simpson R, Chen E, et al. 512 NCIC CTG IND.181: Phase I study of AT9283 given as a weekly 24 hour infusion. European Journal of Cancer Supplements. 2010

[136] Jayanthan A, Cooper TM, Hoeksema KA, Lotfi S, Woldum E, Lewis VA, et al. Occurrence and modulation of therapeutic targets of Aurora kinase inhibition in pediatric acute leukemia cells. Leukemia & Lymphoma. 2013

[137] Duong JK, Griffin MJ, Hargrave D, Vormoor J, Edwards D, Boddy AV. A population pharmacokinetic model of AT9283 in adults and children to predict the maximum tolerated dose in children with leukaemia. British Journal of Clinical Pharmacology. 2017

[138] Li Y, Tan C, Gao C, Zhang C, Luan X, Chen X, et al. Discovery of benzimidazole derivatives as novel multi-target EGFR, VEGFR-2 and PDGFR kinase inhibitors. Bioorganic & Medicinal Chemistry. 2011

[139] Chu B, Liu F, Li L, Ding C, Chen K, Sun Q, et al. A benzimidazole derivative exhibiting antitumor activity blocks EGFR and HER2 activity and upregulates DR5 in breast cancer cells. Cell Death & Disease. 2015

[140] Yun CH, Boggon TJ, Li Y, Woo MS, Greulich H, Meyerson M, et al. Structures of lung cancer-derived EGFR mutants and inhibitor complexes: Mechanism of activation and insights into differential inhibitor sensitivity. Cancer Cell. 2007

[141] Han C, Wan L, Ji H, Ding K, Huang Z, Lai Y, et al. Synthesis and evaluation of 2-anilinopyrimidines bearing 3-aminopropamides as potential epidermal growth factor receptor inhibitors. European Journal of Medicinal Chemistry. 2014

[142] Determann R, Dreher J, Baumann K, Preu L, Jones PG, Totzke F, et al. 2-Anilino-4-(benzimidazol-2-yl) pyrimidines—A multikinase inhibitor scaffold with antiproliferative activity toward cancer cell lines. European Journal of Medicinal Chemistry. 2012

[143] Wang S, Meades C, Wood G, Osnowski A, Anderson S, Yuill R, et al. 2-Anilino-4-(thiazol-5-yl)pyrimidine CDK inhibitors: Synthesis, SAR analysis, X-ray crystallography, and biological activity. Journal of Medicinal Chemistry. 2004

[144] Shao H, Shi S, Huang S, Hole AJ, Abbas AY, Baumli S, et al. Substituted 4-(thiazol-5-yl)-2-(phenylamino) pyrimidines are highly active CDK9 inhibitors: Synthesis, X-ray crystal structures, structure-activity relationship, and anticancer activities. Journal of Medicinal Chemistry. 2013

[145] Wang S, Griffiths G, Midgley CA, Barnett AL, Cooper M, Grabarek J, et al. Discovery and characterization of 2-anilino-4-(thiazol-5-yl)pyrimidine transcriptional CDK inhibitors as anticancer agents. Chemistry & Biology. 2010

[146] McCoull W, Hennessy EJ, Blades K, Chuaqui C, Dowling JE, Ferguson AD, et al. Optimization of highly kinase selective bis-anilino pyrimidine PAK1 inhibitors. ACS Medicinal Chemistry Letters. 2016

[147] Gandin V, Ferrarese A, Dalla Via M, Marzano C, Chilin A, Marzaro G. Targeting kinases with anilinopyrimidines: Discovery of N-phenyl-N'-[4-(pyrimidin-4-ylamino) phenyl]urea derivatives as selective inhibitors of class III receptor tyrosine kinase subfamily. Scientific Reports. 2015

[148] Zhang Q, Liu Y, Gao F, Ding Q, Cho C, Hur W, et al. Discovery of EGFR selective 4,6-disubstituted pyrimidines

from a combinatorial kinase-directed heterocycle library. Journal of the American Chemical Society. 2006

[149] Von Ahsen O, Bömer U. High-throughput screening for kinase inhibitors. Chembiochem. 2005

[150] Bistrović A, Krstulović L, Harej A, Grbčić P, Sedić M, Koštrun S, et al. Design, synthesis and biological evaluation of novel benzimidazole amidines as potent multi-target inhibitors for the treatment of non-small cell lung cancer. European Journal of Medicinal Chemistry. 2018

[151] Morphy R, Rankovic Z. Designed multiple ligands. An emerging drug discovery paradigm. Journal of Medicinal Chemistry. 2005

[152] Ballas MS, Chachoua A. Rationale for targeting VEGF, FGF, and PDGF for the treatment of NSCLC. OncoTargets and Therapy. 2011

Chapter 8

Bisbenzimidazoles: Anticancer Vacuolar (H^+)-ATPase Inhibitors

Renukadevi Patil, Olivia Powrozek, Binod Kumar, William Seibel, Kenneth Beaman, Gulam Waris, Neelam Sharma-Walia and Shivaputra Patil

Abstract

Small molecule chemotherapeutic agents such as Imatinib, Gefitinib, and Erlotinib have played a significant role in the treatment of cancer. Although the unprecedented progress has been achieved in cancer treatment with these targeted agents, there is a strong demand for the development of selective and highly efficacious cancer drugs. V-ATPases are emerging as important target for the identification of novel therapeutic agents for cancer. Our screening and drug discovery processes have identified the bisbenzimidazole derivative (**RP-15**) as a potent anticancer V-ATPase inhibitor. In the present study, bisbenzimidazoles (compound-**25**, **RP-11** and **RP-15**) have been tested for proton-pump inhibition activity in human hepatoma cell line (Huh7.5). **RP-15** displayed comparable proton-pump inhibition activity to the standard Bafilomycin A1. We examined the antiproliferative activity of these analogs in two highly invasive and metastatic inflammatory breast cancer (IBC) cell lines (SUM 149PT and SUM190PT) along with Huh7.5. The compound-**25** (SUM190PT: IC_{50} = 0.43±0.11 μM) and its structural analog **RP-11** (SUM190PT: IC_{50} = 0.49±0.09 μM) have shown significant inhibition toward IBC cell lines. Additionally, **RP-11** and **RP-15** have demonstrated very good cytotoxicity toward the majority of cancer cell lines in the NCI 60 cell line panel.

Keywords: bisbenzimidazoles, anticancer, V-ATPase, proton-pump, inhibitors

1. Introduction

Since Paul Ehrlich's introduction of the concept of chemotherapy, development of chemotherapeutic agents for cancer over the past several decades has seen marvelous records of accomplishments [1, 2]. Cancer is one of the major health problems globally and is second leading cause of death in the USA [3, 4]. Cancer is a very complex disease and our understanding towards it has been advanced tremendously over the last six decades since the first human cancer cell line HeLa identified in 1952 [5]. Over the past few years, the search for new anticancer drugs has changed dramatically. Advances in the molecular nature of drug action, new technology and more recently market considerations have produced new approaches to cancer drug discovery [6]. Recent advances in molecular biology, high throughput screening

(HTS), computer-aided drug design (CADD), and combinatorial chemistry technologies have allowed a combination of both knowledge around the drug receptor and large library screening to be used for anticancer drug discovery today [7–10].

As the understanding of human biology and new technologies progressed, the discovery and development process moved from a random pattern to a more predictable one. The development of a molecularly targeted anticancer drug has gained importance in recent years [11]. One of the important small molecule targeted therapy, Imatinib (Gleevec®), a tyrosine kinase inhibitor, achieved incredible advancement in cancer treatment [12–14]. Imatinib's success stimulated the scientists to develop variety of targeted anticancer agents including Gefitinib (Iressa™) and Erlotinib (Terceva®) for the treatment of different types of cancer patients (**Figure 1**). Targeted agents represented significant developments in cancer treatment and have increased the life expectancy of patients [15–18]. Despite the unprecedented progress achieved, the anticancer drug discovery research remains highly challenging and there is strong demand for the development of highly efficacious and safe anticancer drugs which can overcome cancer metastasis, and drug resistance.

Recent studies suggest that an acidic microenvironment in the tumor is responsible for cancer development, progression, and metastasis. Novel drugs that specifically target the mechanism by which V-ATPase lowers the pH of the tumor microenvironment are essential for cancer chemotherapy. Among the key regulators of the tumor, acidic microenvironment V-ATPases plays an important role in the regulation of the pH gradient. V-ATPases play a vital role in the maintenance of the tumor acidic microenvironment and are overexpressed in many types of metastatic cancers including breast cancer. V-ATPases are functionally expressed in plasma membranes of tumor cells and they have specialized functions in metastasis [19]. Recent research has demonstrated that the preferential expression of V-ATPase at the cell surface is important for the acquisition of invasiveness and the metastasis of breast cancer cells [19]. Therefore, V-ATPase is a potential target to investigate for metastatic breast cancer therapy. Discovery and development of easily synthesized,

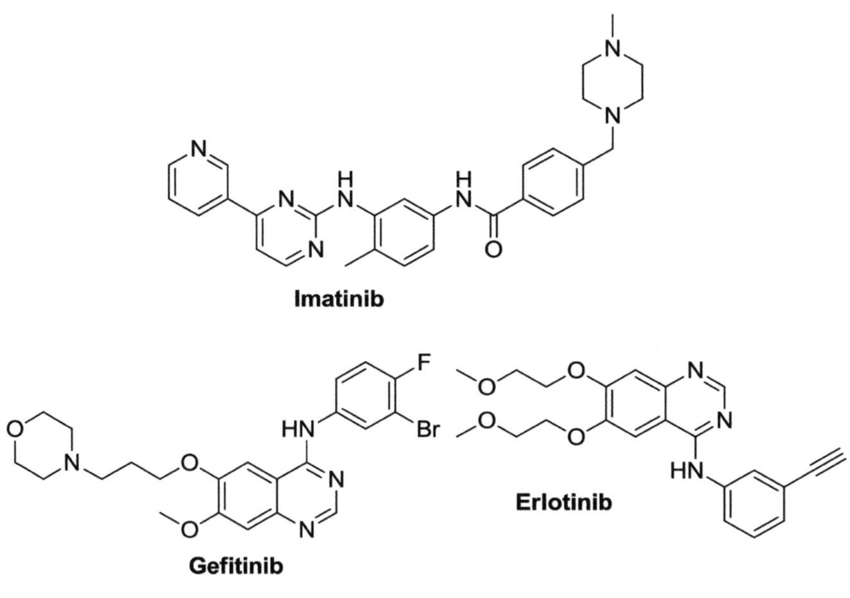

Figure 1.
Molecularly targeted clinically successful chemotherapeutic agents.

cost-effective, and potent small molecule drugs targeting V-ATPase are needed to evaluate the therapeutic potential of V-ATPase inhibitors in metastatic breast cancer.

The V-ATPases are a family of ATP-driven proton pumps that couple ATP hydrolysis with translocation of protons across membranes. The V-ATPase proton pump is a macromolecular complex composed of at least 14 subunits organized into two functional domains, V1 is responsible for ATP hydrolysis and V0 provides the transmembrane proton channel [20–23]. The V-ATPases have been associated with cancer invasion, metastasis and drug resistance [19, 24–27]. Several pre-clinical studies have reported the anticancer effects of V-ATPase inhibitors [28–32]. V-ATPase inhibitors will be beneficial for cancer patients given either in combination with cytotoxic agents or dual-acting (anticancer and V-ATPase inhibitor) agents. Thus, V-ATPases are emerging as an important target for the identification of potential novel chemotherapeutic agents. Despite the clear involvement of V-ATPases in cancer, to date, therapeutic use of V-ATPase targeting small molecules have not reached the clinic. Natural products macrolide antibiotics, such as bafilomycin and concanamycin, potently inhibit V-ATPases [33–37] (**Figure 2**), but their use is complicated by non-specific effects on other targets. Moreover, these molecules have been difficult to synthesize in large quantities. Despite huge efforts by both academic and pharmaceutical industry medicinal chemists, development of useful V-ATPase inhibitors has been limited because of the complicated chemical structures of existing natural inhibitors.

We have been actively involved in the design and development of novel small molecular agents for different types of cancers. Past few years, we have reported the chromene-, chromenopyridine-, and imidazoquinoline-based pharmacophores as initial lead anticancer drug candidates through screening and drug development process [38–40]. Notably, we have identified the highly potent microtubule targeting anticancer agent (**SP-6-27**) for ovarian cancer [41]. Since then our laboratory has been active in identifying anticancer agents with different mechanisms of action. In continuation of our drug discovery research, we recently initiated a collaborative effort on the V-ATPases as anti-cancer targets. Successful identification of new lead small molecule drugs for ovarian cancer by screening and drug development processes [41] inspired us to screen the library of compounds based on the literature of known V-ATPase inhibitors. We identified the bisbenzimidazole scaffold from screening process. Bisbenzimidazoles are nitrogen heterocycles with wide spectrum of biological activities. We

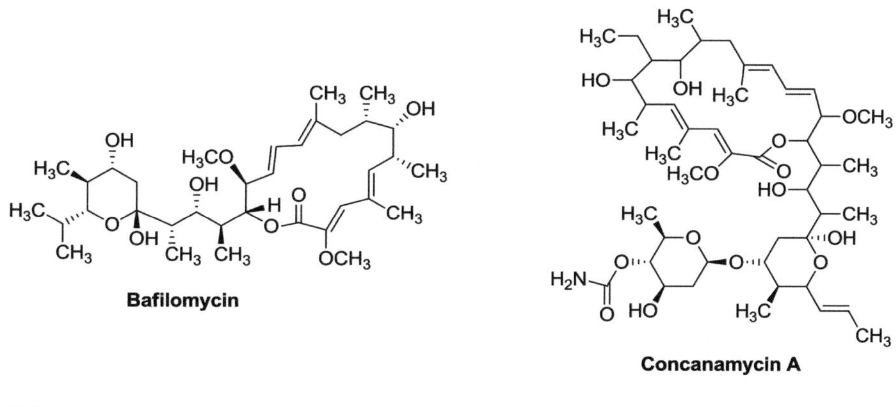

Figure 2.
Natural potent V-ATPase inhibitors.

reported the focused set of bisbenzimidazoles as anticancer V-ATPase agents (**Figure 3**) [42]. Bisbenzimidazole derivatives (**RP-3–RP-15**) have been screened in selected human breast cancer (MDA-MB-231, MDA-MB-468, MCF-7) and ovarian cancer (cisplatin-sensitive A2780, cisplatin-resistant Cis-A2780 and PA-1) cell lines. Among this small set of bisbenzimidazoles, **RP-15** demonstrated high potency towards the epidermal growth factor receptor (EGFR) over expressed triple negative breast cancer (TNBC) cell line, MDA-MB-468 (IC_{50} = 0.04 ± 0.02 μM). Very interestingly, **RP-15** is not toxic to normal breast epithelial cells. It is nearly 40 times less toxic in the normal breast epithelial cell line, MCF10A (IC_{50} = 1.62 ± 0.14 μM). Furthermore, the bisbenzimidazole derivatives (Compound-**25**, **RP-11** and **RP-15**) have demonstrated encouraging proton pump inhibition activity in MDA-MB-231. In particular our most efficacious anticancer analog **RP-15** has shown comparable proton pump inhibition activity to standard agent Bafilomycin A1.

In the present study, we selected and screened top two bisbenzimidazole derivatives (**RP-11** and **RP-15**) along with initial hit (compound **25**) for proton pump inhibition activity in human hepatoma cells, Huh7.5 using pH indicator Lysosensor Yellow/Blue DND-160. These compounds have also been screened for their antiproliferative activity using BrDU incorporation assay in selected inflammatory breast cancer (IBC) cell lines (SUM149PT and SUM190PT) along with Huh7.5 human hepatoma cancer cell line. Additionally, **RP-11** and

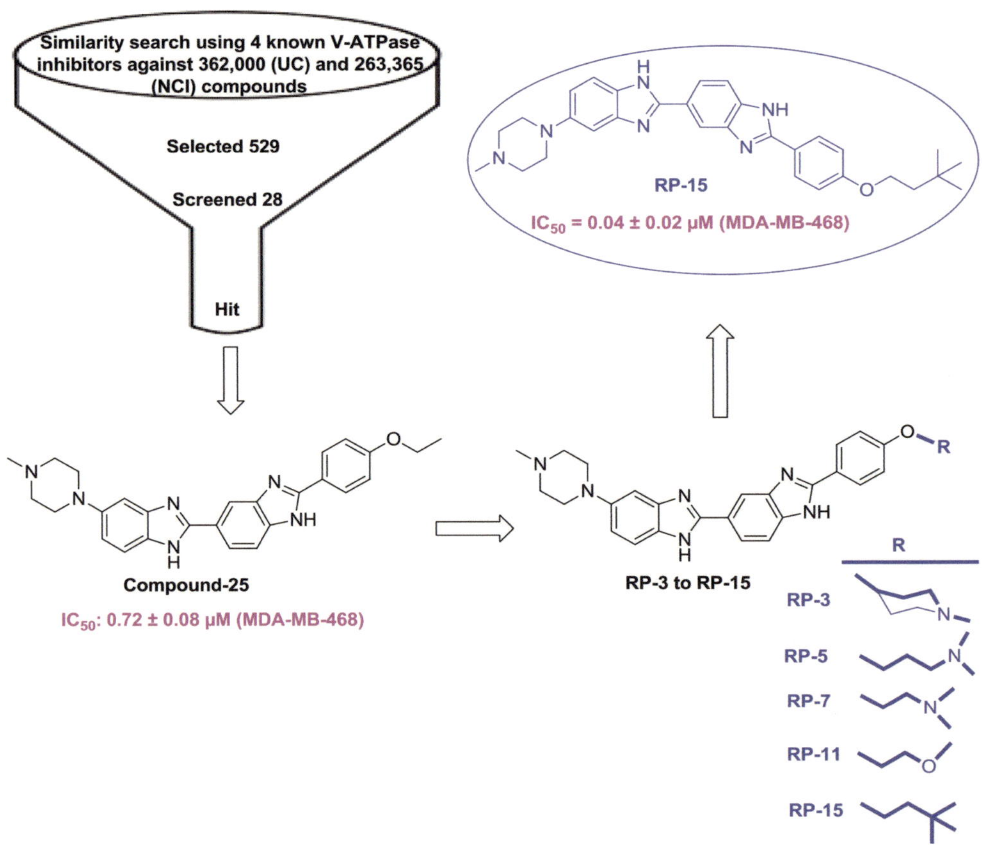

Figure 3.
Bisbenzimidazoles derivatives.

RP-15 have been tested in NCI Developmental Therapeutics Program (DTP) nine major (leukemia, non-small cell lung cancer, colon cancer, CNS cancer, melanoma, ovarian cancer, renal cancer, prostate cancer and breast cancer) 60 human cancer cell lines.

2. Methods

2.1 Chemical synthesis

We recently reported the synthesis and detailed characterization of all these new bisbenzimidazoles [42]. In brief, we developed a fast and efficient synthetic one pot procedure to prepare all these analogs (**RP-3**–**RP-15**). Condensation of 4-(6-(4-methylpiperazin-1-yl)-1H, 30H-[2, 50-bibenzo [d]imidazol]-20-yl) phenol with substituted alkyl halides in the presence of cesium carbonate in dimethyl formamide (DMF). For the more detailed synthesis and spectral and analytical characterization of all these compounds please see Ref. [42].

2.2 Proton pump inhibition activity in human hepatoma (Huh7.5) cell line

We used Huh7.5 cell line for proton pump activity. Briefly, the Huh7.5 cells were cultured in DMEM media supplemented with 10% serum to a confluency of 80%. The Huh7.5 cells were treated with the compounds (Compound-**25**, **RP-11** and **RP-15**) at a concentration of 12 μM for 20 minutes followed by incubation with Lysosensor Yellow/Blue DND-160 (10 μM) for 10 minutes at 37°C. The cells were visualized under the microscope.

2.3 Antiproliferative activity

Cell proliferation ELISA BrdU colorimetric (assay no. 11647229001; Roche, Basel, Switzerland) was used to quantify cell proliferation by the measurement of BrdU incorporated during DNA synthesis. Cells from a 90% confluent T-25 flask were seeded 100 μL/well of 96-well plates and incubated overnight. Dimethyl Sulfoxide (DMSO) stock solutions of the compounds (Compound-**25**, **RP-11** and **RP-15**) were diluted in pure F-12 media and exposed to different concentrations for 24 and 48 hours. Each concentration and controls were done in triplicates. The mean ± standard deviation (S.D.) was calculated and shown on the graph with untreated cells serving as a negative control, 20 minutes after adding the substrate, the absorbance was read at 370 nm. The compound concentration that inhibited cell growth by 50% of the untreated control (IC_{50}) was calculated from the dose response curves constructed by normalizing the data to percentages based of the negative control and a nonlinear regression analysis in GraphPad Prism Software 7 (GraphPad Software, San Diego, CA, USA). For the Huh7.5 cell line we used CellTiter-Glo Luminescent Cell Viability Assay kit (Promega, Madison, WI, USA).

2.4 The NCI 60 cell lines *in vitro* screening

The bisbenzimidazoles (**RP-11** and **RP-15**) have been tested for growth inhibition against 60 human cancer cell lines from the NCI's anticancer screening program. The NCI's screening procedure has been given in detail elsewhere [43–47] and presently DTP uses the sulforhodamine B (SRB) assay.

3. Results and discussion

Inhibition of V-ATPase has shown the link between cell biophysical properties and proliferative signaling selectively in malignant hepatocellular carcinoma (HCC) cells, which provides a new strategy to combat HCC [48]. HCC is the third most common cause of cancer-related deaths worldwide. HCC is accounting for almost 90% of primary malignant hepatic tumors in adults. In continuation of our work on V-ATPase inhibition, we used Huh7.5 cells for the proton pump inhibition activity. We have performed proton pump inhibitory activity of selected bisbenzimidazole derivatives (Compound-**25**, **RP-11** and **RP-15**) in Huh7.5 cells using Lysosensor Yellow/Blue DND-160 protocol [49]. The DND-160 is a pH indicator and cellular compartments with acidic pH elicit yellow fluorescence when stained, while the destabilized compartments with higher pH elicit blue fluorescence.

The compound **RP-15** displayed maximum inhibition of the proton-pump activity of V-ATPase followed by compound-**25** and **RP-11**. The untreated cells showed the strong intensity of yellow fluorescence (converted to pseudo-green in the **Figure 4A**) while the cells treated with bisbenzimidazoles (Compound-**25**, **RP-11** and **RP-15**) showed the strong intensity of blue fluorescence representing varying degree of destabilization of pH due to impaired vacuolar ATPase activity (**Figure 4A** and **B**). Additionally, these compounds have been tested for their cytotoxicity towards Huh7.5 cells using the CellTiter-Glo Luminescent Cell Viability Assay. The IC_{50} were calculated based on the results obtained for these compounds treated for 24 hours only for Huh7.5 cells compared to breast and ovarian cancer cell lines where we treated all test compounds for 48 hours. Bisbenzimidazoles, **RP-11** and **RP-15** have demonstrated very moderate antiproliferative activity towards Huh7.5 cells for 24 hours (**Table 1**).

High potency of bisbenzimidazole analog (**RP-15**) against the EGFR over expressed TNBC cell line (MDA-MB-468) inspired us to explore the selected bisbenzimidazoles in other breast cancer cell lines for anticancer activity. We selected two IBC cell lines (triple negative SUM149PT and Het2 positive SUM190PT) for the *in vitro* screening process [50]. Both SUM149 and SUM190 cell lines have been established from primary IBC tumors. IBC is one of the highly invasive, metastatic and lethal variant of human breast cancer. Development of therapeutic targets and agents for IBC is still in very early stage and it represents an opportunity for medicinal chemists to develop novel (pre) clinical drug candidates.

In vitro screening of the bisbenzimidazoles (Compound-**25**, **RP-11** and **RP-15**) towards these inflammatory cell lines has shown encouraging results (**Figure 5** and **Table 1**). Very interestingly our initial hit, compound-**25** (SUM149PT: IC_{50} = 0.80 ± 0.08 μM; SUM190PT: IC_{50} = 0.43 ± 0.11 μM) and its structural analog **RP-11** (SUM149PT: IC_{50} = 0.91 ± 0.15 μM; SUM190PT: IC_{50} = 0.49 ± 0.09 μM) have shown very good inhibition, whereas our TNBC lead **RP-15** (SUM149PT: IC_{50} = 1.77 ± 0.08 μM; SUM190PT: IC_{50} = 2.08 ± 0.56 μM) has demonstrated moderate inhibition towards these IBC cell lines. The high potency shown by compound-**25** and **RP-11** towards IBC has given us more insights to develop new anticancer agents for it and we plan to explore the structure-activity relationship (SAR) studies based on the bisbenzimidazole scaffold in very near future.

The Development Therapeutic Program (DTP) of the National Cancer Institute's 60 human tumor cell lines screen was developed as an *in vitro* drug discovery tool. We submitted both compounds (**RP-11** and **RP-15**) to the NCI Developmental Therapeutics Program (DTP) anticancer drug screen. Both of them have been first tested for three cell lines (MCF-7 breast cancer; NCI-H460 large-cell lung cancer; SF-268 glioma) to advance to the 60 cell line screen. This pre-screen process eliminates the inactive compounds but preserves active agents for 60 cell line screening.

DOI: http://dx.doi.org/10.5772/intechopen.85231

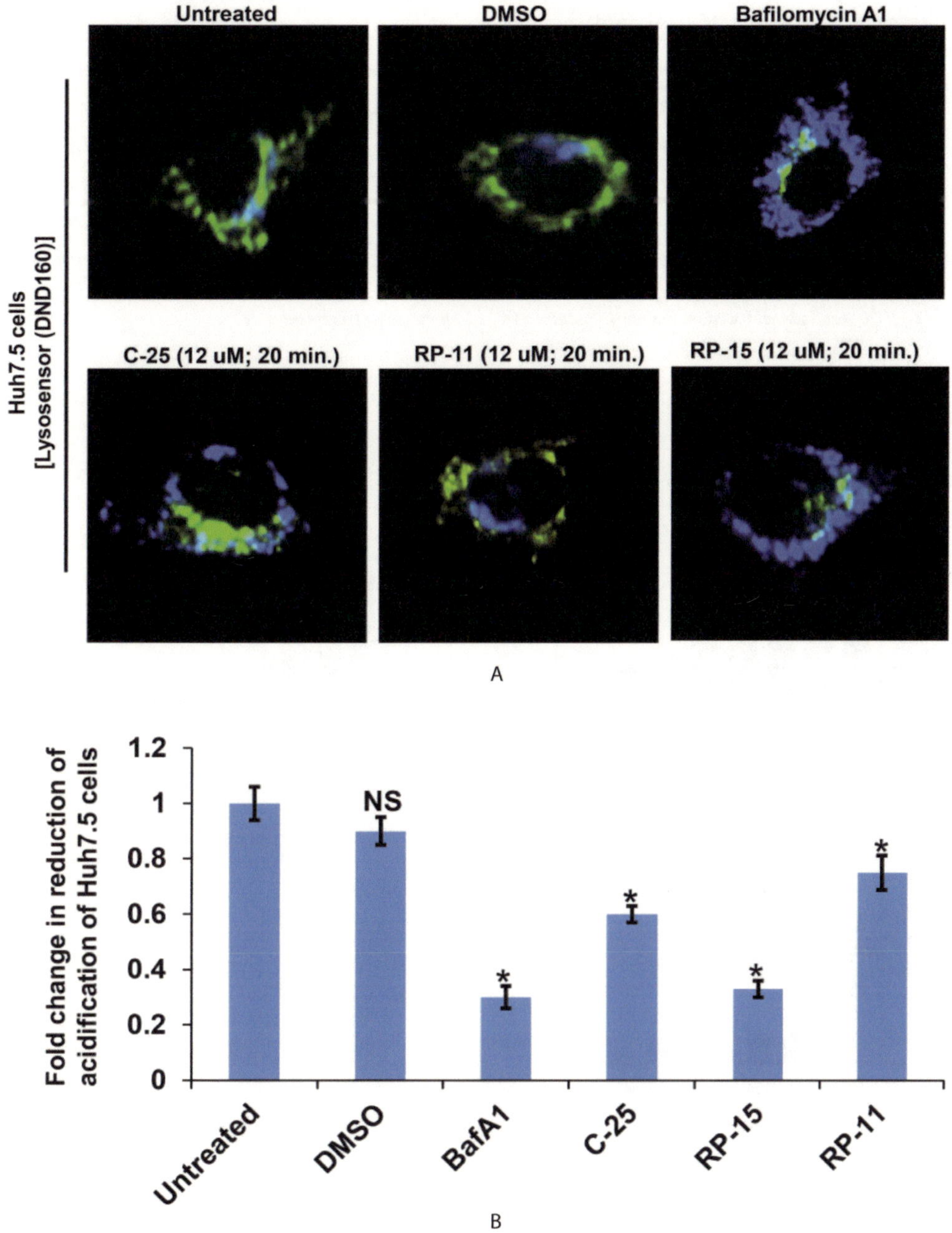

Figure 4.
*(A) Staining of acidic compartments: Yellow signal (converted to pseudo-green) represents acidic pH, while the blue color represents slightly acidic to neutral pH. Huh7.5 cells were treated with the compounds (Compound-**25**, **RP-11** and **RP-15**) at a concentration of 12 μM and standard Bafilomycin A1 at concentration of 2 μM for 20 minutes followed by incubation with Lysosensor Yellow/Blue DND-160 (10 μM) for 10 minutes at 37°C. The cells were visualized under the microscope. The DND-160 is a pH indicator and cellular compartments with acidic pH elicit yellow fluorescence when stained, while the destabilized compartments with higher pH elicit blue fluorescence. The expected yellow color showed yellowish green of the filters available in the microscope. (B) Fold change in overall acidification of Huh7.5 cells upon treatment with bisbenzimidazoles (Compound-**25**, **RP-11** and **RP-15**) along with positive control BafilomycinA1.*

Both compounds have been advanced to 60 cell lines representing nine major cancers (leukemia, non-small cell lung, central nervous system, colon, melanoma, ovarian, renal, prostate, and breast). Compounds have been tested over a broad range of concentrations against every cell line in the panel (five 10 fold dilutions starting

Compd.	IC_{50} ± SD (μM)					
	SUM149PT	SUM190PT	MDA-MB-468‡	MCF10A‡	Cis-A2780‡	Huh7.5†
C-25	0.80 ± 0.08	0.43 ± 0.11	0.72 ± 0.08	1.14 ± 0.13	3.95 ± 0.33	17.1 ± 0.85
RP-11	0.91 ± 0.15	0.49 ± 0.09	0.56 ± 0.05	1.55 ± 0.04	3.03 ± 0.18	17.0 ± 0.78
RP-15	1.77 ± 0.08	2.08 ± 0.56	0.04 ± 0.02	1.62 ± 0.14	1.34 ± 0.14	16.4 ± 0.65
Baf A1	ND	ND	ND	0.036 ± 0.04	0.008 ± 0.01	ND

ND: not determined.
‡Data from Ref. [42].
†The IC50 is calculated based on the results obtained from 24 hours drug treatment only.

Table 1.
Half maximal inhibitory concentration of novel bisbenzimidazole analogs in different cancer cell lines.

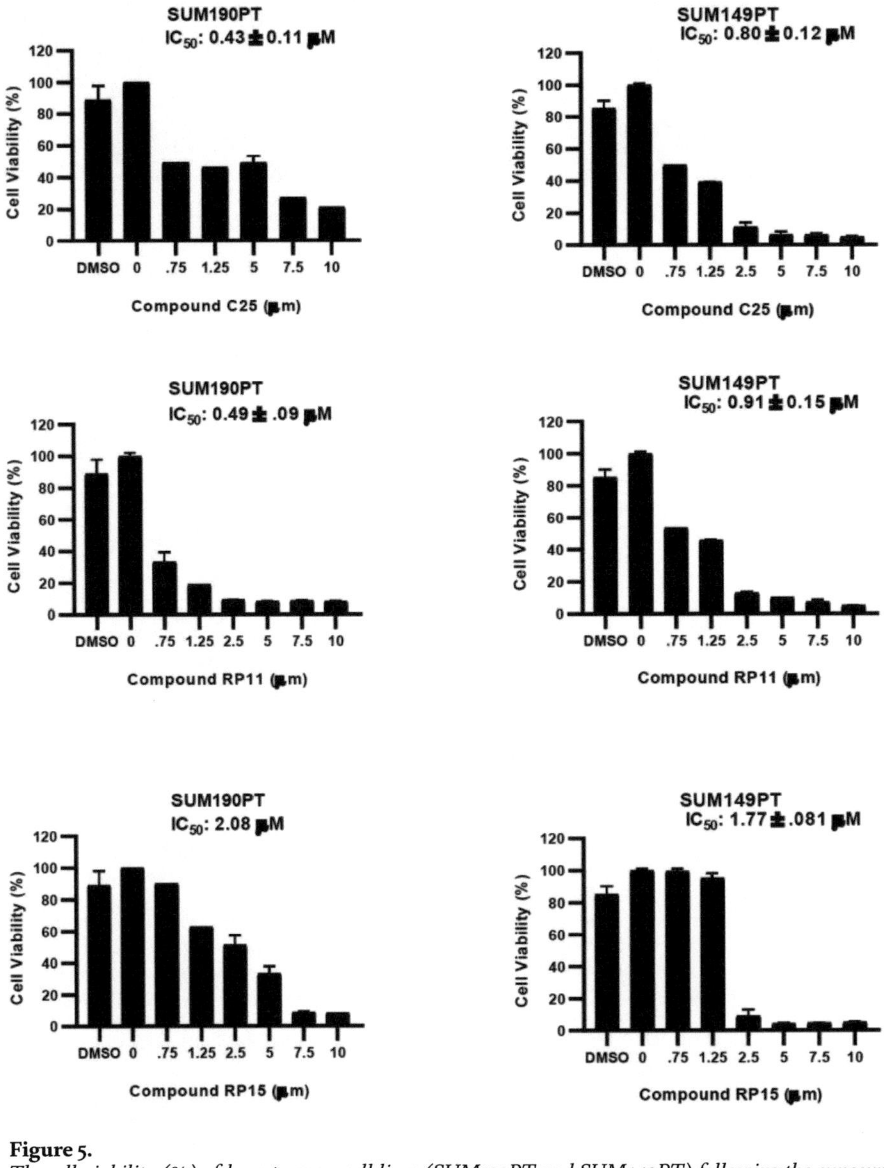

Figure 5.
*The cell viability (%) of breast cancer cell lines (SUM190PT and SUM149PT) following the exposure of various concentrations of bisbenzimidazoles (Compound-25, **RP-11** and **RP-15**) for 48 hours.*

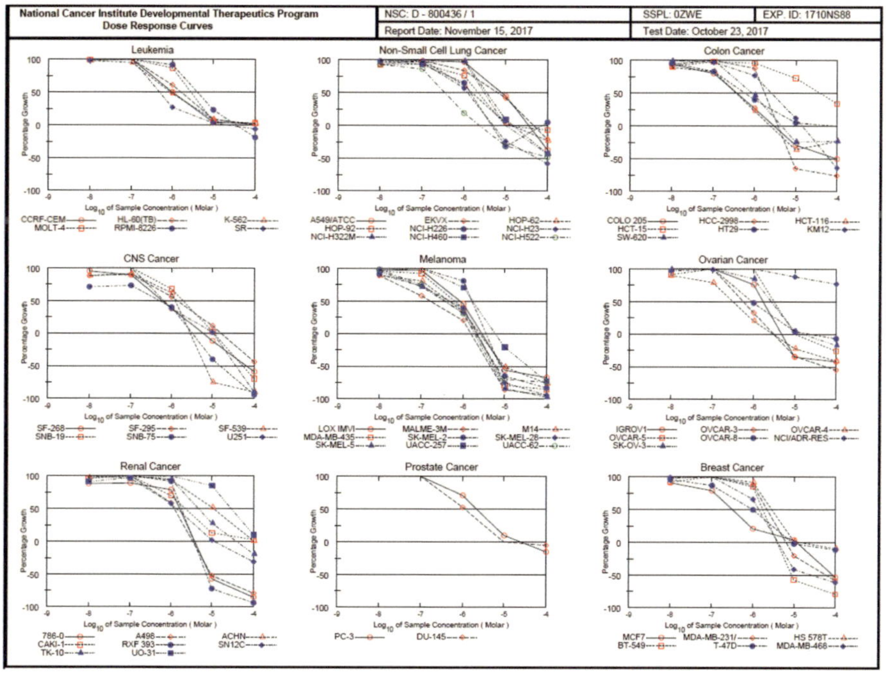

Figure 6.
Dose response curves derived from screening of compound **RP-11** *(NSC: D-800436) in 60 cell line screen using nine major human cancer cell lines (leukemia, non-small cell lung cancer, colon cancer, CNS cancer, melanoma, ovarian cancer, renal cancer, prostate cancer and breast cancer).*

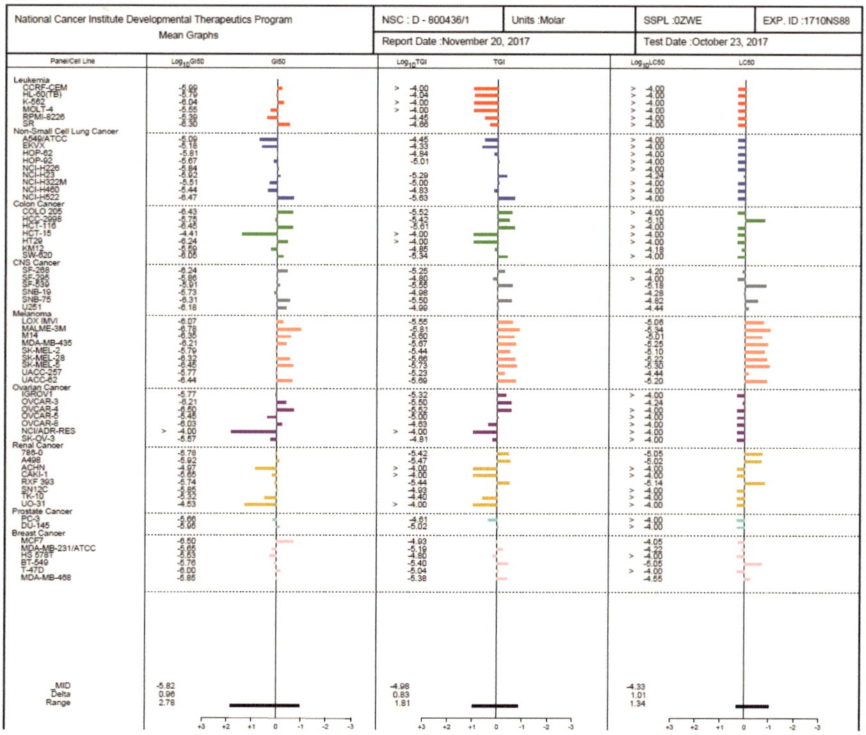

Figure 7.
The mean graph representation of antitumor effects of compound **RP-11** *(NSC: D-800436). The GI_{50} (50% of growth inhibition), TGI (total growth inhibition) and LC_{50} (50% of lethal concentration) mean graphs are derived from the dose response curves using* ***Figure 6*** *from the initial screening.*

from 10^{-4} M concentration). **Figures 6** and **8** describe the dose response curves for compounds **RP-11** (NSC: D-800436) and **RP-15** (NSC: D-800437) respectively. From these dose response curves three end points were calculated (GI_{50}: 50% of growth inhibition; TGI: total growth inhibition; LC_{50}: 50% of lethal concentration). **Figures** 7 and **9** demonstrate mean graph patterns for compound **RP-11** and **RP-15** respectively. Mean graphs are created for GI_{50}, TGI, and LC_{50} by plotting positive and negative values termed as deltas generated from dose response curves. More sensitive cell lines are displayed as bars that project to the right of the mean, whereas the less sensitive cell lines are displayed with bars projected to the left. The length of each bar is proportional to the relative sensitivity of the agent with the mean determination.

Both bisbenzimidazole analogs, **RP-11** and **RP-15** demonstrated very good cytotoxicity towards the majority of cancer cell lines in the 60 cell line panel. Compound **RP-11** displayed growth inhibition and total growth inhibition to low micromolar range and is moderate towards LC_{50} for MCF7 (GI_{50}: 0.32 μM, TGI: 11.8 μM and LC_{50}: 88.7 μM), MDA-MB-468 (GI_{50}: 1.42 μM, TGI: 4.16 μM and LC_{50}: 28.2 μM) and MDA-MB-231 (GI_{50}: 2.25 μM, TGI: 6.49 μM and LC_{50}: 60.4 μM). Interestingly, it showed low micromolar range effects against other cell lines such as SR (GI_{50}: 0.50 μM); NCI-H522 (GI_{50}: 0.34 μM); COLO 205 (GI_{50}: 0.37 μM); SF-268 (GI_{50}: 0.58 μM); OVCAR-3 (GI_{50}: 0.62 μM) and MDA-MB-435 (GI_{50}: 0.62 μM) (**Table 2**). Compound **RP-15** shows similar behavior as **RP-11**. Compound **RP-15** exhibited GI_{50}: 1.91 μM, TGI: 4.13 μM and LC_{50}: 8.91 μM for the MDA-MB-468 cell line, whereas a similar trend is observed for the MDA-MB-231 cell line (GI_{50}: 2.85 μM, TGI: 5.83 μM and LC_{50}: 21.3 μM). Similarly, low micromolar growth inhibition was observed for other cell lines such as MDA-MB-435 (GI_{50}: 1.97 μM), RXF

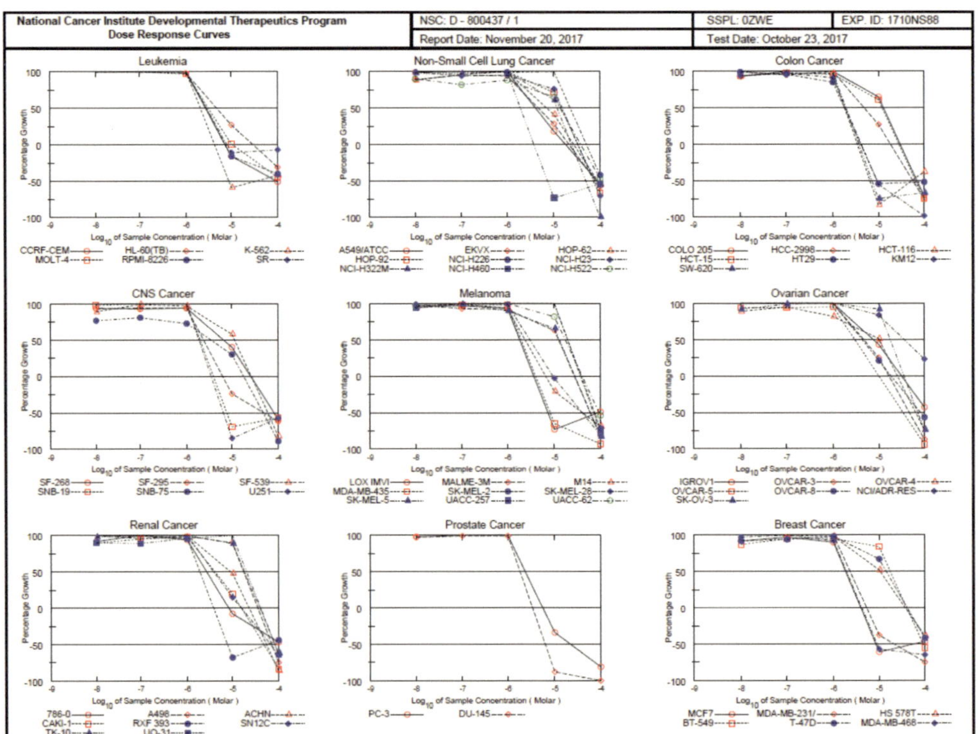

Figure 8.

*Dose response curves derived from screening of compound **RP-15** (NSC: D-800437) in 60 cell line screen using nine major human cancer cell lines (leukemia, non small cell lung cancer, colon cancer, CNS cancer, melanoma, ovarian cancer, renal cancer, prostate cancer and breast cancer).*

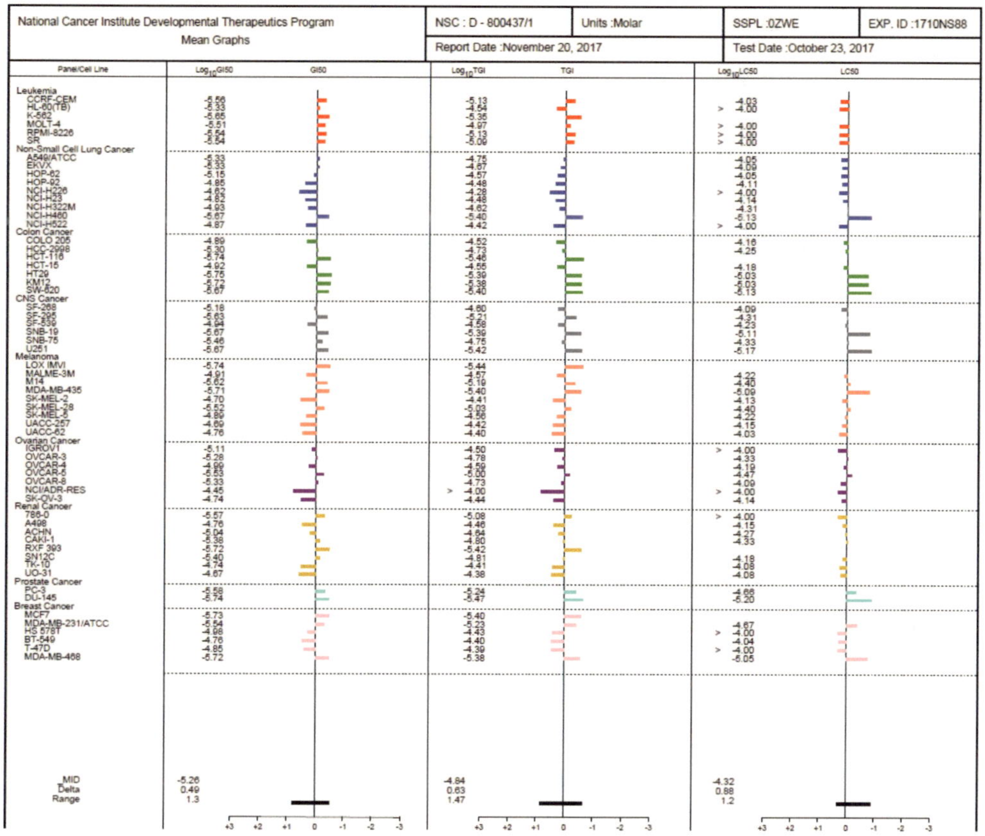

Figure 9.
*The mean graph representation of antitumor effects of compound RP-15 (NSC: D-800437). The GI_{50} (50% of growth inhibition), TGI (total growth inhibition) and LC_{50} (50% of lethal concentration) mean graphs are derived from the dose response curves using **Figure 8** from the initial screening.*

Cell line	RP-11 (μM)			RP-15 (μM)			Cell line	RP-11 (μM)			RP-15 (μM)		
	GI_{50}	TGI	LC_{50}	GI_{50}	TGI	LC_{50}		GI_{50}	TGI	LC_{50}	GI_{50}	TGI	LC_{50}
Leukemia							MDA-MB-435	0.62	2.14	5.66	1.97	3.99	8.10
CCRF-CEM	1.02	>100	>100	2.72	7.34	92.8	SK-MEL-2	1.64	3.59	7.89	19.7	3.5	75.0
HL-60(TB)	1.62	91.8	>100	4.71	29.1	100	SK-MEL-28	0.478	2.21	6.08	3.03	9.43	40.1
K-562	0.91	>100	>100	2.24	4.45	–	SK-MEL-5	0.351	1.88	5.06	12.7	27.6	59.8
MOLT-4	2.81	>100	>100	3.12	10.7	100	UACC-257	1.71	5.92	35.9	20.5	38.3	71.4
RPMI-8226	4.11	35.5	>100	2.86	7.44	100	UACC-62	0.36	2.02	6.37	17.3	40.1	93.1
SR	0.50	21.9	>100	2.90	8.05	100	Ovarian cancer						
Non-small cell Lung							IGROV1	1.71	4.81	>100	7.69	31.6	100
A549/ATCC	8.05	35.2	>100	4.69	17.7	88.6	OVCAR-3	0.62	3.13	57.0	5.22	16.8	46.7
EKVX	6.56	46.3	>100	4.67	21.4	81.4	OVCAR-4	0.313	2.99	>100	10.1	25.5	64.0
HOP-62	1.56	14.5	>100	7.10	26.8	89.0	OVCAR-5	3.59	9.96	>100	2.96	9.96	33.8
HOP-92	2.16	9.71	>100	14.3	33.2	77.0	OVCAR-8	0.931	23.4	>100	4.69	18.6	81.5
NCI-H226	1.44	–	>100	23.8	52.2	100	NCI/ADR-RES	>100	>100	>100	35.7	>100	>100
NCI-H23	1.21	5.09	57.2	15.1	33.0	71.9	SK-OV-3	2.70	15.6	>100	18.0	36.0	72.0

Cell line	RP-11 (μM)			RP-15 (μM)			Cell line	RP-11 (μM)			RP-15 (μM)		
	GI_{50}	TGI	LC_{50}	GI_{50}	TGI	LC_{50}		GI_{50}	TGI	LC_{50}	GI_{50}	TGI	LC_{50}
NCI-H322M	3.07	9.93	>100	11.6	23.8	48.8	Renal cancer						
NCI-H460	3.63	14.7	>100	2.12	3.98	7.44	786.0	1.64	3.81	8.86	2.70	8.33	>100
NCI-H522	0.34	2.36	>100	13.6	38.1	>100	A498	1.21	3.39	9.54	17.5	35.1	70.3
Colon cancer							ACHN	10.7	>100	>100	9.20	22.8	54.2
COLO 205	0.37	3.05	>100	12.9	30.0	69.8	CAKI-1	2.25	>100	>100	4.15	15.7	47.2
HCC-2998	1.78	3.78	8.03	5.06	18.5	56.7	RXF 393	1.80	3.63	7.32	1.91	3.84	–
HCT-116	0.36	2.46	>100	1.82	3.44	–	SN12C	1.42	11.8	>100	4.00	15.4	66.2
HCT-15	38.7	>100	>100	12.1	28.1	65.4	TK-10	4.74	40.2	>100	18.2	38.8	83.0
HT29	0.58	>100	>100	1.78	4.07	9.30	UO-31	29.4	>100	>100	21.3	41.8	82.3
KM12	2.58	14.3	65.4	1.91	4.20	9.26	Prostate cancer						
SW-620	0.90	10.2	36.5	2.12	3.96	7.37	PC-3	2.19	24.5	>100	2.63	5.81	21.8
CNS cancer							DU-145	1.13	9.59	>100	1.83	3.40	6.29
SF-268	0.58	5.66	62.9	6.58	25.4	81.0	Breast cancer						
SF-295	1.38	15.7	>100	2.34	6.22	48.4	MCF7	0.32	11.8	88.7	1.85	3.96	–
SF-539	1.22	2.83	6.59	11.5	26.1	59.5	MDA-MB-231/ATCC	2.25	6.49	60.4	2.85	5.83	21.3
SNB-19	1.87	10.5	52.9	2.15	4.09	7.79	HS578T	2.94	15.9	>100	10.5	37.1	>100
SNB-75	0.49	3.14	15.2	3.46	18.0	47.0	BT-549	1.76	3.96	8.91	17.5	40.1	91.7
U251	0.66	10.2	36.5	2.16	3.81	6.74	T-47D	1.00	9.09	>100	14.2	41.1	>100
Melanoma							MDA-MB-468	1.42	4.16	28.2	1.91	4.13	8.91
LOX IMVI	0.85	2.83	8.66	1.84	3.67	–							
MALME-3 M	0.16	1.55	4.61	12.2	27.2	60.4							
M14	0.45	2.50	9.80	2.39	6.53	40.0							

Table 2.
The NCI 60 cancer cell line screening results.

393 (GI_{50}: 1.91 μM), HT29 (GI_{50}: 1.78 μM), LOXIMVI (GI_{50}: 1.84 μM), DU-145 (GI_{50}: 1.83 μM) and KM12 (GI_{50}: 1.91 μM) (**Table 2**). Overall, the NCI 60 cell line results are encouraging for both new bisbenzimidazole derivatives.

4. Conclusions and future directions

In summary, our screening and drug discovery processes have identified the bisbenzimidazole (**RP-15**) as a potent anticancer V-ATPase inhibitor for TNBC and **RP-11** as initial lead for the IBC. The compound **RP-15** showed maximum inhibition of the proton-pump activity which is comparable to our standard agent Bafilomycin A1. The *in vitro* antiproliferative activity of these bisbenzimidazole analogs (Compound-**25**, **RP-11** and **RP-15**) towards IBC cell lines revealed that compound-**25** and its structural analog **RP-11** could be possibly considered for further exploration in other IBC cell lines. Bisbenzimidazoles **RP-11** (NSC: D-800436) and **RP-15** (NSC: D-800437) have demonstrated very good cytotoxicity towards the majority of cancer cell lines in the NCI 60 cell line panel. Overall, our research identified efficacious and selective anticancer V-ATPase inhibitors for TNBC and

IBC. We will continue to explore the SAR with this exciting pharmacophore to identify the highly selective and potent V-ATPase inhibitors which will ultimately lead to the generation of investigational new drug (IND) candidates for the clinical testing in TNBC and IBC patients.

Acknowledgements

The screening of **RP-11** and **RP-15** against 60 human cancer cell lines of NCI's development therapeutic program (DTP) is greatly acknowledged. Rosalind Franklin University of Medicine and Science University start-up grant to NSW. National Institute of Health grant (DK106244) to GW.

Conflict of interest

The authors declare no conflict of interest, financial or otherwise.

Author details

Renukadevi Patil[1], Olivia Powrozek[2], Binod Kumar[2], William Seibel[3], Kenneth Beaman[2], Gulam Waris[2], Neelam Sharma-Walia[2] and Shivaputra Patil[1*]

1 Pharmaceutical Sciences Department, College of Pharmacy, Rosalind Franklin University of Medicine and Science, North Chicago, IL, USA

2 Department of Microbiology and Immunology, Chicago Medical School, Rosalind Franklin University of Medicine and Science, North Chicago, IL, USA

3 Division of Oncology, Cincinnati Children's Hospital Medical Center, Cincinnati, OH, USA

*Address all correspondence to: shivaputra.patil@rosalindfranklin.edu

References

[1] Kaufmann SH. Paul Ehrlich: Founder of chemotherapy. Nature Reviews. Drug Discovery. 2008;**7**(5):373-373. DOI: 10.1038/nrd2582

[2] Xu J, Mao W. Overview of research and development for anticancer drugs. Journal of Cancer Therapy. 2016;**7**: 762-772. DOI: 10.4236/jct.2016.710077

[3] Siegel RL, Miller KD, Jemal A. Cancer statistics. A Cancer Journal for Clinicians. 2017;**67**:7-30. DOI: 10.3322/caac.21387

[4] Noone AM, Howlader N, Krapcho M, Miller D, Brest A, Yu M, et al., editors. SEER Cancer Statistics Review. Bethesda, MD: National Cancer Institute; 1975-2015. Available from: https://seer.cancer.gov/csr/1975_2015/

[5] Gey GO, Coffman WD, Kubicek MT. Tissue culture studies of the proliferative capacity of cervical carcinoma and normal epithelium. Cancer Research. 1952;**12**:264-265

[6] Prasad V, De Jesús K, Mailankody S. The high price of anticancer drugs: Origins, implications, barriers, solutions. Nature Reviews. Clinical Oncology. 2017;**14**(6):381-390. DOI: 10.1038/nrclinonc.2017.31

[7] Belfield GP, Delaney SJ. The impact of molecular biology on drug discovery. Biochemical Society Transactions. 2006;**34**(2):313-316. DOI: 10.1042/BST20060313

[8] Liu B, Li S, Hu J. Technological advances in high-throughput screening. American Journal of Pharmacogenomics. 2004;**4**(4):263-276. DOI: 10.2165/00129785-200404040-00006

[9] Ooms F. Molecular modeling and computer aided drug design. Examples of their applications in medicinal chemistry. Current Medicinal Chemistry. 2000;**7**:141-158. DOI: 10.2174/0929867003375317

[10] Hogan JC Jr. Combinatorial chemistry in drug discovery. Nature Biotechnology. 1997;**15**:328-330. DOI: 10.1038/nbt0497-328

[11] Aggarwal S. Targeted cancer therapies. Nature Reviews. Drug Discovery. 2010;**9**(6):427-428. DOI: 10.1038/nrd3186

[12] Baselga J. Targeting tyrosine kinases in cancer: The second wave. Science. 2006;**312**:1175-1178. DOI: 10.1126/science.1125951

[13] Iqbal N, Iqbal N. Imatinib: A breakthrough of targeted therapy in cancer. Chemotherapy Research and Practice. 2014;**2014**:357027. DOI: 10.1155/2014/357027

[14] Jones RL, Judson IR. The development and application of imatinib. Expert Opinion on Drug Safety. 2005;**4**(2):183-191. DOI: 10.1517/14740338.4.2.183

[15] Herbst RS, Fukuoka M, Baselga J. Gefitinib–A novel targeted approach to treating cancer. Nature Reviews. Cancer. 2004;**4**(12):956-965. DOI: 10.1038/nrc1506

[16] Sanford M, Scott LJ. Gefitinib: A review of its use in the treatment of locally advanced/metastatic non-small cell lung cancer. Drugs. 2009;**69**(16):2303-2328. DOI: 10.2165/10489100-000000000-00000

[17] Dowell J, Minna JD, Kirkpatrick P. Erlotinib hydrochloride. Nature Reviews. Drug Discovery. 2005;**4**(1): 13-14. DOI: 10.1038/nrd1612

[18] Blackhall FH, Rehman S, Thatcher N. Erlotinib in non-small cell lung

cancer: A review. Expert Opinion on Pharmacotherapy. 2005;**6**(6):995-1002. DOI: 10.1517/14656566.6.6.995

[19] Sennoune SR, Bakunts K, Martínez GM, Chua-Tuan JL, Kebir Y, Attaya MN, et al. Vacuolar H^+-ATPase in human breast cancer cells with distinct metastatic potential: Distribution and functional activity. American Journal of Physiology. Cell Physiology. 2004;**286**(6):C1443-C1452. DOI: 10.1152/ajpcell.00407.2003

[20] Nishi T, Forgac M. The vacuolar (H^+)-ATPases—nature's most versatile proton pumps. Nature Reviews. Molecular Cell Biology. 2002;**3**(2): 94-103. DOI: 10.1038/nrm729

[21] Finbow ME, Harrison MA. The vacuolar H^+-ATPase: A universal proton pump of eukaryotes. The Biochemical Journal. 1997;**324**(Pt 3):697-712. DOI: 10.1042/bj3240697

[22] Yokoyama K, Imamura H. Rotation, structure, and classification of prokaryotic V-ATPase. Journal of Bioenergetics and Biomembranes. 2005;**37**(6):405-410. DOI: 10.1007/s10863-005-9480-1

[23] Wang Y, Cipriano DJ, Forgac M. Arrangement of subunits in the proteolipid ring of the V-ATPase. The Journal of Biological Chemistry. 2007;**282**(47):34058-34065. DOI: 10.1074/jbc.M704331200

[24] Fais S, De Milito A, You H, Qin W. Targeting vacuolar H^+-ATPases as a new strategy against cancer. Cancer Research. 2007;**67**(22):10627-10630. DOI: 10.1158/0008-5472.CAN-07-1805

[25] Sennoune SR, Luo D, Martinez-Zaguilan R. Plasmalemmal vacuolar-type H^+-ATPase in cancer biology. Cell Biochemistry and Biophysics. 2004;**40**(2):185-206. DOI: 10.1385/CBB:40:2:185

[26] Capecci J, Forgac M. The function of vacuolar ATPase (V-ATPase) a subunit isoforms in invasiveness of MCF10a and MCF10CA1a human breast cancer cells. The Journal of Biological Chemistry. 2013;**288**(45):32731-32741. DOI: 10.1074/jbc.M113.503771

[27] Rofstad EK, Mathiesen B, Kindem K, Galappathi K. Acidic extracellular pH promotes experimental metastasis of human melanoma cells in athymic nude mice. Cancer Research. 2006;**66**(13):6699-6707. DOI: 10.1158/0008-5472.CAN-06-0983

[28] Ohta T, Arakawa H, Futagami F, Fushida S, Kitagawa H, Kayahara M, et al. Bafilomycin A1 induces apoptosis in the human pancreatic cancer cell line Capan-1. Journal of Pathology. 1998;**185**:324-330. DOI: 10.1002/(SICI)1096-9896(199807)185:3<324::AID-PATH72>3.0.CO;2-9

[29] Lee JC, Lee CH, Su CL, Huang CW, Liu HS, Lin CN, et al. Justicidin a decreases the level of cytosolic Ku70 leading to apoptosis in human colorectal cancer cells. Carcinogenesis. 2005;**26**:1716-1730. DOI: 10.1093/carcin/bgi133

[30] Schneider LS, von Schwarzenberg K, Lehr T, Ulrich M, Kubisch-Dohmen R, Liebl J, et al. Vacuolar-ATPase inhibition blocks iron metabolism to mediate therapeutic effects in breast cancer. Cancer Research. 2015;**75**: 2863-2874. DOI: 10.1158/0008-5472.CAN-14-2097

[31] Nakashima S, Hiraku Y, Tada-Oikawa S, Hishita T, Gabazza EC, Tamaki S, et al. Vacuolar H^+-ATPase inhibitor induces apoptosis via lysosomal dysfunction in the human gastric cancer cell line MKN-1. Journal of Biochemistry. 2003;**134**:359-364. DOI: 10.1093/jb/mvg153

[32] Spugnini EP, Citro G, Fais S. Proton pump inhibitors as anti

vacuolar-ATPases drugs: A novel anticancer strategy. Journal of Experimental & Clinical Cancer Research. 2010;**29**:44. DOI: 10.1186/1756-9966-29-44

[33] Bowman EJ, Graham LA, Stevens TH, Bowman BJ. The bafilomycin/concanamycin binding site in subunit c of the V-ATPases from Neurospora crassa and *Saccharomyces cerevisiae*. The Journal of Biological Chemistry. 2004;**279**(32):33131-33138. DOI: 10.1074/jbc.M404638200

[34] Scheidt KA, Bannister TD, Tasaka A, Wendt MD, Savall BM, Fegley GJ, et al. Total synthesis of (−)-bafilomycin A1. Journal of the American Chemical Society. 2002;**124**(24):6981-6990. DOI: 10.1021/ja017885e

[35] Lim JH, Park JW, Kim MS, Park SK, Johnson RS, Chun YS. Bafilomycin induces the p21-mediated growth inhibition of cancer cells under hypoxic conditions by expressing hypoxia-inducible factor-1alpha. Molecular Pharmacology. 2006;**70**(6):1856-1865. DOI: 10.1124/mol.106.028076

[36] Hayashi Y, Katayama K, Togawa T, Kimura T, Yamaguchi A. Effects of bafilomycin A1, a vacuolar type H^+ ATPase inhibitor, on the thermosensitivity of a human pancreatic cancer cell line. International Journal of Hyperthermia. 2006;**22**(4):275-285. DOI: 10.1080/02656730600708049

[37] Huss M, Ingenhorst G, König S, Gassel M, Dröse S, Zeeck A, et al. Concanamycin a, the specific inhibitor of V-ATPases, binds to the V(o) subunit c. The Journal of Biological Chemistry. 2002;**277**(43):40544-40548. DOI: 10.1074/jbc.M207345200

[38] Patil SA, Wang J, Li XS, Chen J, Jones TS, Hosni-Ahmed A, et al. New substituted 4H-chromenes as anticancer agents. Bioorganic & Medicinal Chemistry Letters. 2012;**22**(13):4458-4461. DOI: 10.1016/j.bmcl.2012.04.074

[39] Patil R, Ghosh A, Sun Cao P, Sommer RD, Grice KA, Waris G, et al. Novel 5-arylthio-5H-chromenopyridines as a new class of anti-fibrotic agents. Bioorganic & Medicinal Chemistry Letters. 2017;**27**(5):1129-1135. DOI: 10.1016/j.bmcl.2017.01.089

[40] Patil SA, Pfeffer SR, Seibel WL, Pfeffer LM, Miller DD. Identification of imidazoquinoline derivatives as potent antiglioma agents. Medicinal Chemistry. 2015;**11**(4):400-406. DOI: 10.2174/1573406410666140914162701

[41] Kulshrestha A, Katara GK, Ibrahim SA, Patil R, Patil SA, Beaman KD. Microtubule inhibitor, SP-6-27 inhibits angiogenesis and induces apoptosis in ovarian cancer cells. Oncotarget. 2017;**8**(40):67017-67028. DOI: 10.18632/oncotarget.17549

[42] Patil R, Kulshrestha A, Tikoo A, Fleetwood S, Katara G, Kolli B, et al. Identification of novel bisbenzimidazole derivatives as anticancer vacuolar (H^+)-ATPase inhibitors. Molecules. 2017;**22**(9):pii: E1559. DOI: 10.3390/molecules22091559

[43] Boyd MR, Paull KD. Some practical considerations and applications of the National Cancer Institute in vitro anticancer drug discovery screen. Drug Development Research. 1995;**34**:91-109. DOI: 10.1002/ddr.430340203

[44] Holbeck SL, Collins JM, Doroshow JH. Analysis of Food and Drug Administration-approved anticancer agents in the NCI60 panel of human tumor cell lines. Molecular Cancer Therapeutics. 2010;**9**(5):1451-1460. DOI: 10.1158/1535-7163.MCT-10-0106

[45] Covell DG, Huang R, Wallqvist A. Anticancer medicines in development: Assessment of bioactivity profiles within the National Cancer

Institute anticancer screening data. Molecular Cancer Therapeutics. 2007;**6**(8):2261-2270. DOI: 10.1158/1535-7163.MCT-06-0787

[46] Skehan P, Streng R, Scudiero D, Monks A, McMahon J, Vistica D, et al. New colorimetric cytotoxicity assay for anticancer-drug screening. The Journal of the National Cancer Institute. 1990;**82**:1107-1112. DOI: 10.1093/jnci/82.13.1107

[47] Monks A, Scudiero D, Skehan P, Shoemaker R, Paull K, Vistica D, et al. Feasibility of a high-flux anticancer drug screen using a diverse panel of cultured human tumor cell lines. Journal of the National Cancer Institute. 1991;**11**:757-766. DOI: 10.1093/jnci/83.11.757

[48] Bartel K, Winzi M, Ulrich M, Koeberle A, Menche D, Werz O, et al. V-ATPase inhibition increases cancer cell stiffness and blocks membrane related Ras signaling–a new option for HCC therapy. Oncotarget. 2017;**8**(6):9476-9487. DOI: 10.18632/oncotarget.14339

[49] Asleh R, Ward J, Levy NS, Safuri S, Aronson D, Levy AP. Haptoglobin genotype-dependent differences in macrophage lysosomal oxidative injury. The Journal of Biological Chemistry. 2014;**289**(23):16313-16325. DOI: 10.1074/jbc.M114.554212

[50] Forozan F, Veldman R, Ammerman CA, Parsa NZ, Kallioniemi A, Kallioniemi OP, et al. Molecular cytogenetic analysis of 11 new breast cancer cell lines. British Journal of Cancer. 1999 Dec;**81**(8):1328-1334. DOI: 10.1038/sj.bjc.6695007

Section 4

Benzimidazole Derivatives in Chemistry of Materials

Chapter 9

Optical Sensing (Nano)Materials Based on Benzimidazole Derivatives

Ema Horak, Robert Vianello and Ivana Murković Steinberg

Abstract

Benzimidazole derivatives are well-known biologically active substances, and therefore, they are mostly synthesised for therapeutic purposes. However, such heteroaromatic molecular systems own structure-related properties that enable a variety of applications, especially in optical science. Multifunctionality of the benzimidazole unit, such as electron accepting ability, π-bridging, chromogenic pH sensitivity/switching and metal-ion chelating properties, makes it an exceptional structural candidate for the design of optical chemical sensors and functional materials. Development of smart molecular sensors and novel (nano)materials is the emerging trend observed in materials and optical sensing science in general, in which the benzimidazole molecular systems strongly contribute and participate. In this chapter, we summarised recent advances in optical sensing (nano)materials that incorporate the benzimidazole structural moiety. Solid-state optical sensing systems, including self-assembled molecular materials based on benzimidazoles, are reviewed and discussed. In addition, immobilisation of benzimidazole derivatives onto or into various substrates and matrices, such as organic and inorganic polymers, bulk membranes and nanoparticles, utilising different chemical and physical methods, is presented and analysed.

Keywords: benzimidazole, functional materials, optical sensor, solid-state, absorbance, fluorescence, aggregation-induced emission

1. Introduction

Optical chemical sensors are widely applied in chemical science and technology, as well as in other disciplines such as biology, medicine and environmental science. They enable continuous monitoring of the target analytes and exhibit high sensitivity and fast response time. The biggest advantages of optical chemical sensors, in comparison to other sensing devices, are the economic production, ease of operation and the possibility of on-site application without reference devices, which are preferred in chemical and biological applications. Performance of every chemical sensor is primarily determined by the sensing chemistry that operates in the background, that is, the recognition unit—receptor. The receptor, the core of every optical chemical sensor, is the sensing molecule that selectively responses to the presence of the target analyte by changing the photophysical properties of the observed molecular system. Fluorescence techniques are most commonly applied for the generation and transfer of the analytical signal, while providing high sensitivity and selectivity. Therefore, fluorescent sensing molecules are the most promising candidates for the chemical sensing. Their design often starts from the heterocyclic molecular skeleton, due to its excellent spectral properties and the ability to detect diverse analytes. Heterocyclic chromophores and fluorophores are the most investigated classes of optical sensing molecules; hence, the interest for benzimidazole as a structural block of novel molecular systems is constantly increasing. Although benzimidazole derivatives are primarily known as biologically and therapeutically active substances [1], such heteroaromatic molecules have structure-related properties that enable a variety of applications in optoelectronics and non-linear optics (NLO) [2, 3], photovoltaics [4, 5], sensing [6] and bioimaging [7, 8]. Indeed, multifunctionality of the benzimidazole unit, such as electron accepting ability, π-bridging, chromogenic pH sensitivity/switching and metal-ion chelating properties, makes it an exceptional structural candidate for the design of optical chemical sensors [9, 10]. From the chemical point of view, the benzimidazole ring possesses a high degree of stability. Benzimidazole, for example, is not affected by concentrated sulphuric acid and is quite resistant to reduction. Oxidation cleaves its benzene ring, yet only under vigorous conditions. The two imidazole nitrogens are different from one another in their nature, which makes the properties of the ring system diverse in character. The hydrogen attached to the nitrogen can easily tautomerise to the other nitrogen atom. With the pK_a values 5.3 and 12.3, benzimidazoles are weakly basic, being somewhat less basic than the imidazoles and sufficiently acidic to make them usually more soluble in polar environments and less soluble in organic solvents. Benzimidazole, for example, is soluble in hot water but difficultly soluble in ether and insoluble in benzene, all of which can be modified upon the substitution. The acid/base properties of benzimidazoles are due to the stabilisation of the charged ion by the resonance effect.

However, development of optical chemical sensors is much more complicated than designing a sensing molecule (recognition unit), since the process combines molecular recognition, material science and device implementation. Employing the sensing chemistry in a form of optical sensing material is perhaps the key step towards the ultimate goal, since its implementation can directly result in a functional sensor. Although there is a large number of fluorescent indicators and sensing molecules presented in literature, many of them lose their selectivity upon the implementation in functional devices, which makes the design of optical sensing materials a very challenging task [11, 12].

Recently, developments in optical sensing molecular systems that incorporate benzimidazole structural unit are reviewed and discussed [13]. As can be deduced from a given review, molecular sensors based on benzimidazole derivatives are mainly applied in solution, while materials for optical sensing are still rare, yet very promising. Development of novel (nano)materials and especially 'smart' molecular sensors, some of which include nanotubes, nanowires and nanoparticles, is the emerging trend observed in materials and optical sensing science. Although the growth of scientific interest in benzimidazole-based materials is evident in the last decade, such systems are indeed untapped potential in the field of optical chemical sensing.

In this chapter, we summarised the recent advances in optical solid state sensing systems and (nano)materials that incorporate the benzimidazole structural moiety. Immobilisation of benzimidazole derivatives in bulk membranes, polymers, sol-gel materials, as well as self-assembled (nano)materials for optical sensing are reviewed and discussed. Representative examples have been selected and commented in next sections, based on the type of applied material.

2. Polymer-based sensing materials

Polymers are the most commonly used support for optical chemical sensors [14]. They can be utilised to immobilise the sensing component, but can also directly participate in the sensing mechanism. In general, most commonly used polymers for analyte sensing are cellulose derivatives and hydrogels because of their excellent mechanical properties, stability at broad temperature and pH ranges, as well as high permeability towards water, ions and undissolved gases. In addition, polymers like polyurethane and pHEMA are biocompatible, a fact that enables new possibilities of their application.

The simplest form of the polymeric sensing material is the polymer membrane in which the chemosensor molecule is physically entrapped [11, 12]. Such dye-impregnated polymers are widely used in sensing chemistry, due to economic and simple methods of preparation. The choice of polymer depends on its permeability towards a specific analyte, its stability, availability and potential for immobilisation. Still, the development of such membranes is a challenging task because the polymer microenvironment has a strong effect on spectral characteristics of the immobilised sensing molecule, its acid-base equilibrium, selectivity towards the analyte and fluorescence lifetime. Ion-selective optodes are well-known examples of such optical sensing systems, where the bulk membranes are mostly formed from plasticised PVC [15].

An alternative for dye-impregnated polymers are the polymers with covalently attached fluorescent molecules. Stability of covalently bonded systems provides many advantages and can significantly improve analytical performance of chemical sensor. Another class of materials for optical sensing is the luminescent polymers. Design and synthesis of novel conjugated or coordination polymers is a constantly growing area of research, due to the enormous potential for the application of these kinds of functional materials.

The polymer-based materials for optical sensing that incorporate benzimidazole unit are summarised in **Table 1**, while the representative examples have been selected and discussed in next sections.

2.1 Dye-impregnated polymers

Polymer-based sensing materials incorporating physically entrapped benzimidazole-based receptors are very often used when relying on the electrochemical detection [16, 17]. However, examples utilising optical sensing techniques are not that common. For example, novel fluorescent sensors are developed by the immobilisation of benzimidazole-based ionophores in plasticised PVC, resulting in ion-selective optode for mercury [18] and silver detection [19]. Presented ion-selective optodes are complex systems, where a number of parameters, including lipophilicity, polarity and microviscosity affect the heterogeneous ion-exchange equilibrium. The same sensing mechanism is presented for benzimidazole-based acrylonitrile derivatives [20] and Schiff bases [21], where novel colorimetric and fluorimetric sensing materials are applied for detecting the acidity changes. Moreover, immobilisation of this class of compounds into polymer matrices is demonstrated as a convenient way to overcome certain problems of organic fluorophores occurring in aqueous solution, such as hydrolysis of imino-bond or low quantum yields. For instance, a reversible spectroscopic response to pH is achieved because protonation of the immobilised benzimidazole Schiff bases occurs on the stable benzimidazole moiety (electron acceptor), while the imino bond of the Schiff base remains preserved [21]. At the same time, spectral properties of fluorescent sensing molecules are significantly altered due to the interactions between molecules in bulk, that is, in novel environment, where the molecular system becomes more rigid with partially disabled *cis-trans* isomerisation. Optical properties of developed materials

Material	Analyte	BI-based sensing molecule	Detection method	Limit of detection (mol L^{-1})	Ref.
PVC	Hg^{2+}	Crown-based ionophore	Fluorimetric	3.5×10^{-13}	Firooz et al. [18]
PVC	Ag^{+}	Crown-based ionophore	Fluorimetric	2.8×10^{-12}	Firooz et al. [19]
PVC	pH	Acrylonitrile derivative	Colorimetric	—	Horak et al. [20]
PVC	pH	Schiff bases	Fluorimetric	—	Horak et al. [21]
Photocrosslinked membrane	Hg^{2+}	1,8-naphthalimide derivative	Fluorimetric	2.5×10^{-6}	Fernández-Alonso et al. [22]
Amphiphilic copolymer	pH	Vinyl monomer	Fluorimetric	—	Han et al. [29]
Hydrophilic copolymer	pH	Pyridyl substituted benzimidazole derivative	Fluorimetric	—	Shen et al. [30]
Conjugated polymer	Cu^{2+}	Pendant benzimidazolyl moieties	Fluorimetric	—	Wu et al. [28]
Conjugated polymer	Fe^{3+} and PO_4^{3-}	Pendant benzimidazolyl moieties	Fluorimetric	3.38×10^{-6}	Saikia *et al* (2011) [27]
Coordination polymer	Fe^{3+}	Bis(benzimidazole) derivative	Fluorimetric	3.2×10^{-6}	Hao et al. [36]
Metal organic framework	Humidity and formaldehyde	Benzimidazolyl-attached bent organic ligand	Colorimetric	—	Yu et al. (2014) [32]
Metal organic framework	$Cr_2O_7^{2-}$	1,6-Bis(benzimidazol-1-yl)hexane ligand	Fluorimetric	2.16×10^{-6}	Li et al. [31]
Coordination polymer	Fe^{3+} ion and nitroaromatics	Benzimidazole ligand	Fluorimetric	3.70×10^{-7}	Zhou et al. [33]
Coordination polymer	Multi-analyte	Benzimidazole-appended tripodal tridentate ligand	Fluorimetric	—	Tripathi et al. [37]
Coordination polymer	pH	Benzimidazole-functionalized organic ligand	Phosphorescence	—	Yang et al. [38]
Coordination polymer	Fe^{3+}	Benzimidazole-functionalized organic ligand	Fluorimetric	2.53×10^{-6}	Zhao et al. [39]
Coordination polymer	Fe^{3+}	Benzimidazole-functionalized organic ligand	Fluorimetric	2.72×10^{-5}	Wei et al. [40]
Zr-UiO-66 nanocrystals	Fe^{3+}	Benzimidazole-functionalized organic ligand	Fluorimetric	—	Dong et al. [34]

Table 1.
Polymeric optical sensing materials based on benzimidazole derivatives.

can also be easily modified by tuning the ICT character of fluorescent molecules, which is often achieved by introducing electron donating groups (e.g. N,N-diethyl amino) and strong electron withdrawing moieties (e.g. -CN and –NO_2) on the opposite parts of molecular system (**Figure 1A**).

2.2 Covalently attached benzimidazole derivatives

As an alternative to dye-impregnation of polymers, fluorescent molecules can be covalently attached to polymeric materials, as demonstrated, for example, in a fluorescence solid sensor for the mercury detection based on a photocrosslinked membrane functionalised with (benzimidazolyl)methyl-piperazine derivative of 1,8-naphthalimide [22]. Benzimidazole, linked to a piperazine moiety by a methylene spacer, is responsible for the specific recognition of Hg^{2+} ions by forming a stable complex structure, that resulted in a strong fluorescence (**Figure 1B**). Materials developed by the covalent attachment of the sensing molecules usually have more advantages that those utilising physical entrapment, in which active molecules may easily leach out of the matrix. Stability and duration of covalently functionalised polymer materials are much better, and they even often provide improved analytical parameters of chemical sensor.

2.3 Luminescent polymers

Another approach to obtain fluorescent sensing materials is a clever design and synthesis of novel luminescent polymers. For instance, conjugated polymers are the constant trend in the development of novel functional materials [23, 24]. They effectively coordinate with many organic compounds or transition metals, which is very well conjoined with their excellent optical properties and exploited in optical chemical sensors. Detection methods are mostly relying on the fluorescence techniques, particularly the quenching effect ('superquenching') described by Stern-Volmer relationships. Fluorescent conjugated polymers also offer many advantages in regard to simple organic fluorophores, such as amplified sensitivity and the possibility of simple introduction of desired functional groups in order to achieve better interactions with the analyte. Benzimidazole is often found as a constituent of conjugated polymers [25–28]. Optical sensing ability of benzimidazole-based fluorescent polymers is demonstrated for the detection of pH [29, 30], metal ions [28] or inorganic anions [27], where benzimidazole moiety often plays a crucial

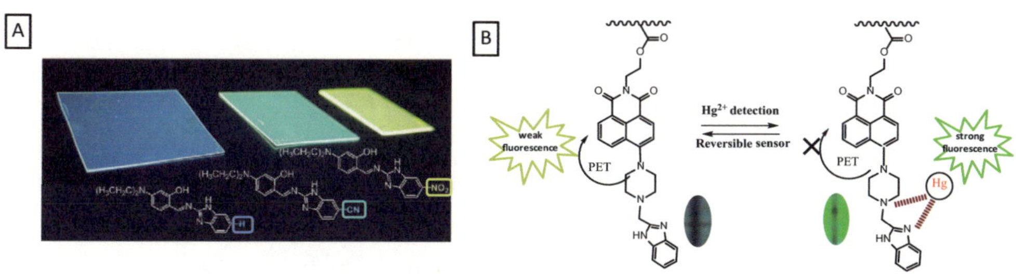

Figure 1.
(A) Fluorescent pH-sensitive bulk optodes based on immobilised Schiff base derivatives in plasticised PVC matrix. Tuneable fluorescent response of the optodes is a result of different substituents on the benzimidazole moiety. Reprinted from [21]. Copyright (2018), with permission from Elsevier. (B) Selective fluorescence solid sensor for Hg^{2+} based on N-(2-hydroxyethyl)-4-(4-(1Hbenzo[d]imidazol-2-yl)methyl) piperazine-1-yl)-1,8-naphthalimide, here presented by the author's courtesy. Fluorescence sensor undergoes fluorescence enhancement upon binding mercuric ion due to the inhibition of photo-induced electron transfer (PET) process from the piperazine to the naphthalimide moiety [22].

role in the sensing mechanism. For example, a copolymer built from *N*-(1-ethyl-2-(pyridin-4-yl)-1*H*benzo[*d*]imidazol-5-yl)methacrylamide and 2-hydroxyethyl methacrylate exhibits a pH sensitivity due to acid-base equilibria on the heteroatom of pyridyl-substituted benzimidazole moiety [30] (**Figure 2A**).

Besides conjugated polymers, benzimidazole-based materials can be developed as luminescent metal organic frameworks (MOFs) [31–35] or coordination polymers [36–42]. Such advanced functional materials have been extensively applied in the field of luminescence sensing due to their diverse structural characteristics and tunable pore sizes. For example, luminescence sensing of iron is achieved by a coordination polymer employing the linear 2,5-dichloroterephthalic acid ligand and the flexible bis(benzimidazole) derivatives. Ligand affords the capacity to strongly bind metal atoms, while bis(benzimidazole) derivatives can freely twist around two methylene -CH_2 groups with disparate angles to generate different conformations [36]. Luminescent MOFs have been exploited for the development of sensing materials for humidity and formaldehyde, such as a porous Cu(I)-MOF, constructed from CuI and 1-benzimidazolyl-3,5-bis(4-pyridyl)benzene (**Figure 3**) [32]. 3D cadmium metal-organic framework was demonstrated as sensing material for the detection of $Cr_2O_7^{2-}$ in water [31], while diamond-like coordination polymer exhibits selective emission quenching responses towards the Fe^{3+} ion and nitroaromatics [33].

Interesting to note is the emerging trend in the development of the so-called 'smart' materials, where the final product exhibit multistimuli-responsive photoluminescence sensing properties. Tripathi et al. developed Hg(II) coordination polymer with benzimidazole-appended tripodal tridentate ligand, 1,3,5-tris(benzimidazolylmethyl)benzene. Luminescent material is the first example of Hg(II) coordination polymer with multistimuli-responsive properties (**Figure 2B**). Luminescence quenching response is observed to a range of stimuli, including anions, solvents and nitroaromatic compounds [37].

2.4 Inorganic polymers

Inorganic polymers, such as networks of metal oxides obtained by sol-gel process, are also attractive substrates for immobilising sensing molecules. Sol-gel materials are very popular for the development of optical sensors, especially nanosized

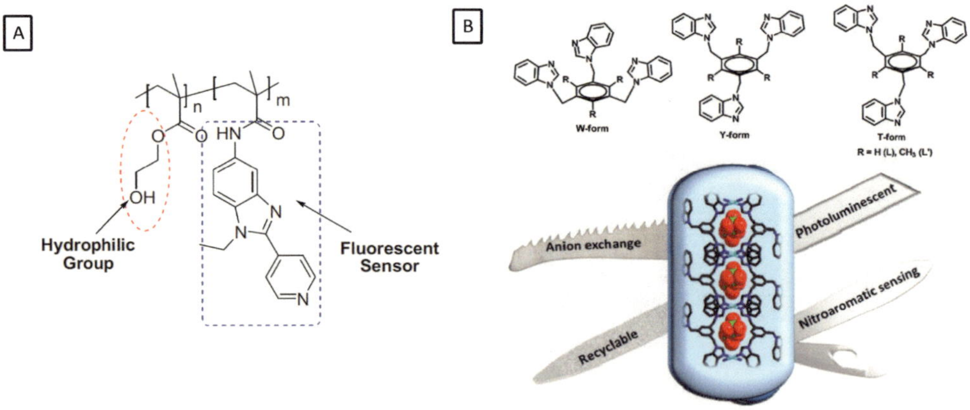

Figure 2.
(A) Chemical structure of pH-responsive copolymer of N-(1-ethyl-2-(pyridin-4-yl)-1Hbenzo[d]imidazol-5-yl) methacrylamide and 2-hydroxyethyl methacrylate, here presented by the author's courtesy. pH sensitivity is achieved by pyridyl substituted benzimidazole moiety [30]. (B) Representation of multistimuli-responsive 'smart' mercury(II) coordination polymer and different possible conformations of benzimidazole-based ligand in metal complexes. Adapted with permission from [37]. Copyright (2018) American Chemical Society.

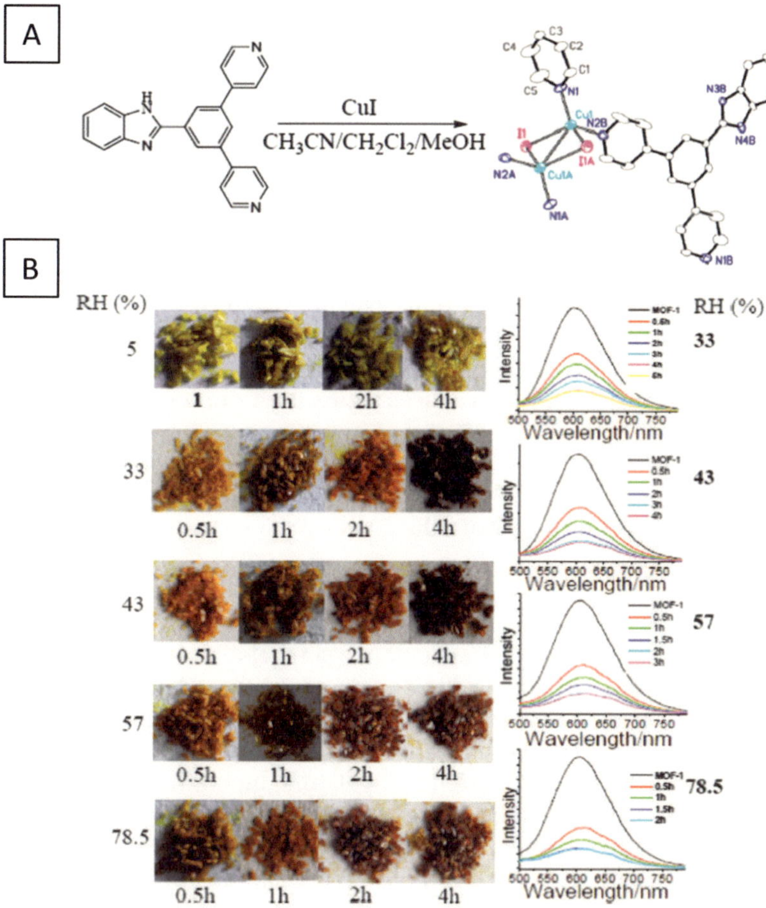

Figure 3.
(A) Highly sensitive naked eye colorimetric sensor for water and formaldehyde detection based on a porous Cu(I)-MOF constructed from CuI and 1-benzimidazolyl-3,5-bis(4-pyridyl)benzene. (B) The colour change of the bulk crystal samples of MOF in atmospheres with different relative humidity (RH 33–78.5%) and the corresponding solid-state emission spectra. Adapted with permission from [32]. Copyright (2013) Royal Society of Chemistry.

probes [43]. Basically, the sol-gel process is a method for the synthesis of ceramic or glass materials at low temperature, starting from the colloidal suspension ('sol'). Hydrolysis of alkoxy metal groups in the precursors followed by polycondensation results in a network structure ('gel'). Meantime, fluorescent indicators can be easily incorporated in sol-gel by impregnation, chemical or covalent immobilisation. These materials are porous, so that the analyte can freely diffuse. They are robust and biocompatible, which makes them suitable for intracellular sensing. Hoffman et al. have developed novel benzimidazole-based fluorescent materials using the sol-gel process [44]. Tetraethylorthosilicate (TEOS) was used as an inorganic precursor for the development of new silica hybrid materials. Although sol-gel chemistry is firmly embedded in the field of chemical sensors, there is a lack of benzimidazole-based sol-gel materials. The reason can be poor solubility and self-assembly properties of many benzimidazole derivatives, often inducing gelation process and thus, making the development of novel sol-gel materials, in a classical manner described above, a challenging task. However, the gelation of such compounds has been shown as an excellent method for preparing new sensing membranes, which will be discussed in further sections.

To conclude this section, we can highlight several facts. Literature shows that polymer-based materials are the most common substrates for the preparation of novel optical sensing platforms based on benzimidazole derivatives. Benzimidazole moiety retained its functional properties upon immobilisation in presented polymeric platforms. Although they are thoroughly explored and their potential for sensing applications is often emphasised, luminescent polymers that incorporate benzimidazole moiety are not adequately exploited in optical sensors. Even though polymer-based sensing materials are still relatively rare, a recent advance in developing benzimidazole-based ultralong-persistent room temperature phosphorescence (RTP) materials that exhibit reversible pH-responsive emission [38] represents a significant breakthrough of benzimidazole derivatives in materials science. Unfortunately, the biggest disadvantages of most polymer-based sensing materials are still very limited, such as selectivity, poor photostability and often leaching of indicator dyes.

3. Self-assembled sensing materials

3.1 Gels

Soft matter research and supramolecular organogels are one of the emerging scientific areas in the last decade. Functional materials based on supramolecular organogels are very attractive for the applications in tissue engineering, medical implants, controlled drug release, environmental studies etc. Small organic molecules have often been investigated as π-gelators, including benzimidazole derivatives [46, 47]. Utilising their fluorescence and self-assembling properties, benzimidazole-based gels are successfully demonstrated as novel functional materials. For example, a family of alkylpyridinylium benzimidazole derivatives was synthesised in order to examine its gelation properties [48], while several fluorescent π-gelators based on benzimidazole are presented as stimuli responsive systems and sensors [49–53]. Ghosh et al. presented sensing system for Ag^+ based on the cholesterol-appended benzimidazole. Benzimidazole moiety with conformational flexibility can exhibit different alignments upon metal ion chelation, while the cholesterol is likely oriented to exert hydrophobic-hydrophobic interaction for establishing cross-linked network for solvent trapping. The addition of Ag^+ ions to the solution of presented molecules in DMF:H_2O (1:1, v/v) at room temperature causes instant gelation and the change of colour, visible by a naked eye [51]. Another example of multi-analyte sensor array based on benzimidazole and acylhydrazone naphthol moities was demonstrated by Yao et al. [52]. The latter sensing system is able to detect many analytes such as CN^-, Al^{3+}, Fe^{3+} and L-Cys with a possibility for the selective identification of Fe^{3+} and Al^{3+} in the gel state (**Figure 6**).

3.2 Aggregation-induced emitters

Self-assembly of benzimidazole derivatives takes a great role in emerging mechanisms and designs of novel optical sensing materials. One of the research directions of the self-assembled molecules are the sensing materials based on the emissive (nano)aggregates. Aggregation of organic fluorophores is mostly investigated as an undesirable side effect in many biological or chemical applications due to fluorescence quenching. However, development of novel organic luminophores with aggregation-induced emission (AIE) changed the aspect of aggregation phenomena and the AIE was introduced as an analytical tool in a wide range of application, such as bioimaging, optoelectronics and chemosensors [54]. AIE or AIEE (aggregation-induced emission enhancement) can be observed in the so called 'poor' solvents, in crystalline or powder forms. With that in mind, benzimidazole-based fluorophores capable of emitting intense fluorescence in the aggregated form (AIE emitters) can also be classified as novel sensing materials [55–58]. For example, self-assembled nanoaggregates of benzimidazole-based acrylonitrile derivative are presented as sensing system for pH, based on aggregation-deaggregation mechanism and aggregation-induced emission (AIE) [59]. 2-Benzimidazolyl-substituted acrylonitrile dye exhibits fluorescence emission in the red, green or cyan spectral regions, depending on its protonation degree. The neutral form is capable of self-assembly in the aqueous environment (pH between 5 and 9), exhibiting stable red-orange fluorescence emission at 600 nm. Thus, due to the aggregation-induced emission (AIE), from the single molecular entity, tri-state system (RGB) is derived. The aggregation and emission are pH switchable and fully-reversible. Gogoi et al. presented a novel AIE system based on a benzimidazole derivative for the detection of pyrophosphate (PPi) (**Figure 4**) [56]. The benzimidazole moiety has a functional role in assembling aggregate structures and the recognition of Ppi. Molecules of benzimidazole derivative self-assemble in nanostructures when so-called 'bad' solvent, H_2O, is added into the THF solution ('good' solvent). Self-assembled nanostructures exhibit pronounced emission at λ = 530 nm. Their π-π stacking is affected by the Ppi presence, thus assembled aggregates are of different emission properties and sizes. Another example is offered by Singh et al. by the preparation of fluorescent aggregates for sensing chemical warfare agents (diethylchlorophosphate) from benzimidazolium-based receptors containing 2-mercaptobenzimidazole and 2-mercaptobenzthiazole as functional groups, using anionic surfactants [60]. Authors presented receptors with benzimidazolium moiety in the centre, as well as receptors with two fluorescent arms as binding- and signalling units that initially interacts with chemical warfare agents and captures the hydrolysed product of an organophosphate.

An AIE phenomenon is especially exploited in solid-state, where the concentration effect commonly causes fluorescence quenching. Having in mind that fluorophores emitting in the solid states are extremely rare, especially red ones, benzimidazole-based AIE molecular systems show a great prospect for future applications of pristine powder samples or crystals as solid-state sensors and 'smart' materials [61–65].

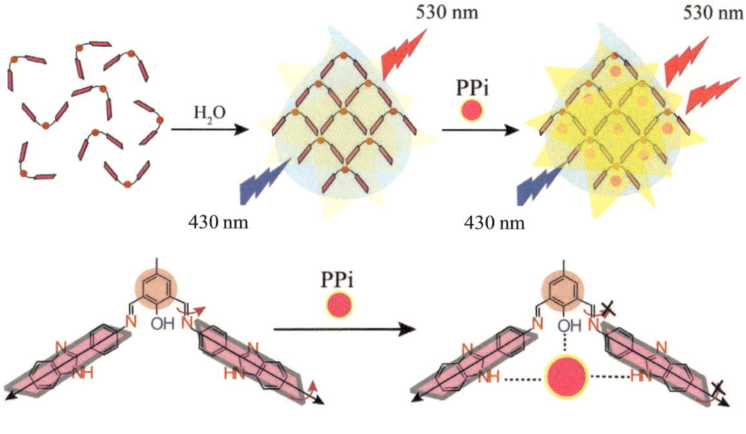

Figure 4.
Aggregation-induced emission of benzimidazole-based derivative and detection of pyrophosphate (PPi). Reprinted with permission from [56]. Copyright (2015) American Chemical Society.

In conclusion, due to its emerging and multidisciplinary character, further research on optical sensing applications of benzimidazole-based gel membranes and self-assembled structures is strongly encouraged. Gelation is proven and widely investigated effect in many benzimidazole-based compounds, yet not exploited enough for the preparation of novel chemical sensors. Meanwhile, research on the aggregation-induced emission phenomena has taken momentum in all areas of application. Beside the fact that certain benzimidazole-based derivatives exhibit AIE property, which is not often found within small heterocyclic molecular systems, optical chemical sensors based on this principle are rarely found. Self-assembled materials for optical sensing that incorporate benzimidazole unit are summarised in **Table 2**.

4. Nanomaterials for optical sensing

Nowadays, the term *nano* appears in all aspects of our life, technology and science, including the optical chemical sensors. In most general way, an optical nanosensor can be defined as a device smaller than 1 μm that is continuously tracking an analyte and simultaneously converting optical information into an analytically useful signal [66]. Fluorescence is the most commonly applied detection technique, due to its high sensitivity and relative simplicity of measurement [67]. Nanosensors can be macromolecular nanostructures, nano-sized polymer materials and sol-gels, multi-functional core-shell systems, multi-functional magnetic beads or nanosensors based on quantum dots or metal beads. We have previously mentioned the nanoaggregates formed by the self-assembly process. Although such type of nanomaterial can be classified as nanosensors, the emphasis in this section is placed on synthesis of nano-sized substrate materials functionalised with benzimidazole derivatives. Most commonly used method for the preparation of nanosensors is previously mentioned sol-gel process resulting in silica nanoparticles [68, 69]. Some other methods, such as precipitation, are often utilised for the preparation of polymer nanoparticles [70]. Research in the field of benzimidazole-based nanosensors is still in the early stages. Benzimidazole-based nanomaterials for optical sensing of metal ions are so far demonstrated as hybrid silica materials [44, 68, 71], ZnO nanoparticles decorated with benzimidazole-based organic ligand [72] or self-assembled nano hyperbranched polymer [73]. For example, Badiei et al. recently presented SBA-15 nanoporous silica functionalised with 2,6-bis(2-benzimidazolyl) pyridine for the selective recognition of mercury (**Figure 5**) [71]. Fluorescence intensity of the SBA-15 functionalized material quenched in the presence of Hg^{2+} ions, wherein the sensor is applicable in the physiological pH range of 6–8.

5. Other benzimidazole-based materials for optical sensing

A simple, fast and economic determination of target analyte, on-site and without a reference device is one of the key challenges of modern analytical chemistry. As mentioned in previous sections, the response to this challenge came forth in the form of optical chemical sensors. In addition, design and development of sensing materials as straightforward optical sensors enable countless possibilities of their applications, especially in modern technology where the emphasis is put on mobile, wearable and wireless devices. Simple, yet effective materials for colorimetric or fluorimetric detection of analytes can easily be achieved using filter paper or TLC plates. Paper substrates themselves are an attractive platform for the

DOI: http://dx.doi.org/10.5772/intechopen.85643

Material	Analyte	BI-based sensing molecule	Detection method	Limit of detection (mol L^{-1})	Ref.
Gel	Multi-analyte	Benzimidazole and acylhydrazone naphthol moities	Fluorimetric	—	Yao et al. [52]
Gel	Ag^+	Cholesterol appended benzimidazole	Colorimetric	4.31×10^{-5}	Ghosh et al. [51]
Gel	pH and anions	Benzimidazole moiety and four amide units	Colorimetric / Fluorimetric	—	Xue et al. [49]
Gel	Picric acid	Cholesterol-based anthraquinone-coupled imidazole	Colorimetric	4.30×10^{-6}	Mondal et al. [50]
Gel	Na_2S	Carboxylic acid functionalized benzimidazole	Fluorimetric	—	Yao et al. [53]
AIEgen	Pyrophosphate	Dipodal benzimidazole- functionalized sensor	Fluorimetric	1.67×10^{-9}	Gogoi et al. [56]
AIEgen	pH	Acrylonitrile derivative	Fluorimetric	—	Horak et al. [59]
AIEgen	Warfare agents	Benzimidazolium-based dipodal receptors	Fluorimetric	10×10^{-9}	Singh et al. [60]
Powder	F^- and COO^-	Benzimidazole derivative	Colorimetric	0.38×10^{-3}	Chaudhuri et al. [64]
SBA-15 nanoporous silica	Hg^{2+}	2,6-bis(2-benzimidazolyl) derivative	Fluorimetric	2.6×10^{-6}	Badiei et al. [71]
ZnO nanoparticles	Zn^{2+}	Benzimidazole-based organic ligand	Colorimetric	4.09×10^{-9}	Kaur et al. [72]
Nano hyperbranched polyester	Fe^{3+}	Benzimidazole end groups	Fluorimetric	—	Wang et al. [73]
Filter paper	TNT	Pyrene-substituted benzimidazole-isoquinolinones	Colorimetric	50×10^{-6}	Boonsri et al. [74]
Filter paper	CN^-	Acrylonitrile-embedded benzimidazole-anthraquinone	Colorimetric	37×10^{-9}	Kumar et al. [76]
Filter paper	Ni^{2+}	2-(2′-hydroxyphenyl)benzimidazole	Colorimetric	—	Dhaka et al. [77]
TLC plates	Acid/amine vapours	Carbazole-based benzimidazole derivatives	Colorimetric	—	Aich et al. [75]

Table 2.
Benzimidazole-based materials for optical sensing.

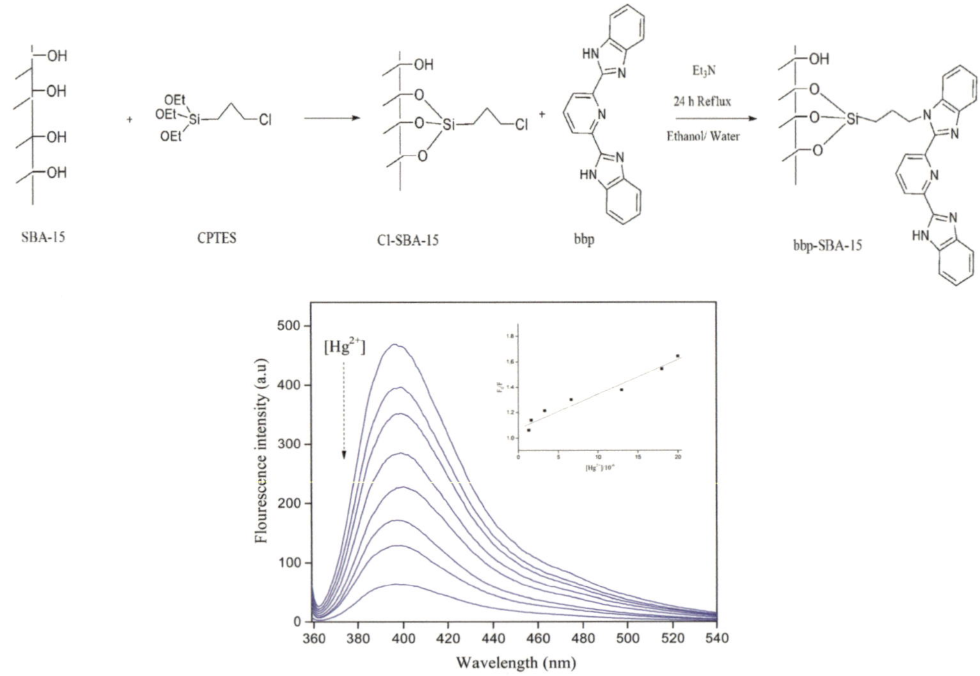

Figure 5.
Synthesis procedure of benzimidazole functionalized SBA-15 material and fluorescence emission of the aqueous suspended nanoparticles (0.4 g L^{-1}) upon titration of increasing amount of Hg^{2+} ions. Inset: Stern-Volmer plot, λ_{exc} = 353 nm. Reprinted with permission from [71]. Copyright (2018) Springer Nature.

use in a wide range of optical sensing, due to the possibility of a passive sample manipulation by capillary forces. So far, paper-based optical chemical sensors for neutral molecules, anions and cations, relying on benzimidazole derivatives as recognition element, have been successfully presented by several research groups. For example, Boonsri et al. demonstrated paper-based sensors for the trinitrotoluene (TNT) detection [74]. Sensing material prepared from pyrene-substituted benzimidazole-isoquinolinones can readily detect TNT in aqueous media by a naked-eye observation at concentrations as low as 50 μM. Optical sensing of acid/ amine vapours with three carbazole-based benzimidazole derivatives in the solid state was also demonstrated using TLC plates [75]. Plates were immersed with benzimidazole-based dyes and then exposed to trifluoroacetic acid (TFA) vapours for 1 minute. In following step, the TLC plates which were exposed with TFA vapours were further revealed to triethyl amine vapours and the restored colour was observed in each case (**Figure 6**).

Anion detection was demonstrated by the ratiometric detection of CN^- based on acrylonitrile embedded benzimidazole-anthraquinone coated on the filter paper [76]. Paper strips coated with the sensing molecule showed a distinct colour change from yellow-greenish to red under UV light in the presence of the CN^- ions. Dhaka et al. demonstrated a 'bare-eye' probe for the detection of Ni^{2+} based on 2-(2′-hydroxyphenyl)benzimidazole. Colourimetric sensing of Ni^{2+} was demonstrated on filter paper. Paper test strips exhibit distinct visual change from colourless to yellow-gold [77]. Other materials for optical sensing that incorporate benzimidazole unit are summarised in **Table 2**.

Besides optical sensing, paper-based materials coated with functional benzimidazole derivatives are also presented as 'smart', stimuli responsive materials with potential applications in security, optoelectronic or fluorescent imaging [62]. Simple sensing substrates such as paper and textile materials are perfectly suited

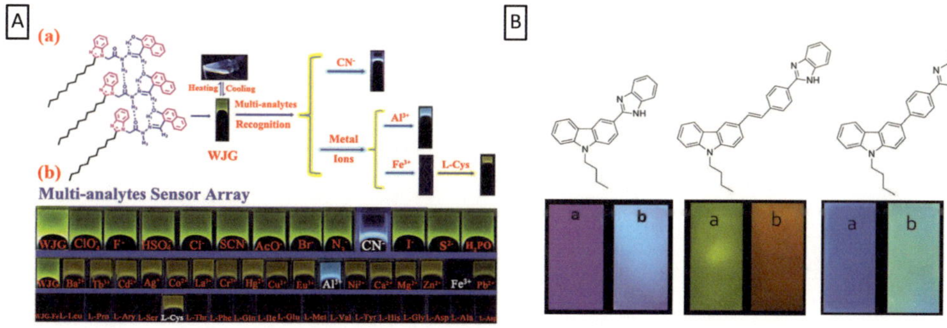

Figure 6.
(A) Proposed self-assembly mechanism of a supramolecular AIE gel and its multiple-stimuli responsive behaviour (a) and fluorescence responses of the multi-analyte sensor array to the presence of various anions, cations and amino acids. Reprinted with permission from [52]. Copyright (2018) Royal Society of Chemistry. (B) TLC plates immersed with carbazole benzimidazole-based dyes observed under the UV light (λ_{exc} = 365 nm) before (a) and after exposure to TFA vapours (b). Adapted with permission from [75]. Copyright (2016) Royal Society of Chemistry.

for applications in emerging mobile and wearable chemical sensors. Design and development of compatible 'sensing chemistries' that operate in the background of such devices is a constant challenge. The multifunctional nature of materials based on molecules such as benzimidazole can perfectly respond to this challenge.

6. Conclusion

Benzimidazole unit represents an important multifunctional building block in optical chemical sensors, with proven potential for the development of novel functional (nano)materials. Solid-state optical sensing systems incorporating benzimidazole derivatives are reviewed and discussed. Polymers are most commonly used substrates for the development of optical chemical sensors. Materials for optical sensing based on benzimidazole are also demonstrated as gels, sol-gel matrices, silica or polymer nanoparticles, (nano)aggregates and TLC or paper-based strips.

The role of benzimidazole moiety in optical sensing (nano)materials is important and crucial, since it maintains the function of the system and plays a key role in the formation of the analytical signal in the majority of chemical sensing systems reviewed here. Besides, the planar moiety significantly contributes to the conjugation of the chromo/fluorophore system. Although benzimidazole derivatives reviewed in the literature are mostly fluorescent sensors, several probes based on colourimetric switches are also demonstrated. It is very challenging to transfer the sensing chemistry from a solution to the solid state, which is successfully comprehended for the benzimidazole derivatives. It is even observed for some classes of chromophores with relatively unattractive sensing properties in aqueous solution (such as low quantum yield, decomposition upon protonation) to be drastically improved upon immobilisation in a polymer matrix.

Although examples of sensing materials presented in the literature show that benzimidazole derivatives can be successfully and easily applied in optical chemical sensors, they are yet insufficiently explored. Challenges in development of novel optical sensing (nano)materials are constantly emerging, since the scientific and industrial field of mobile and wearable sensors are experiencing great progress. Simple, fast and economic determination of target analyte, on-site and without reference device is a request that a multifunctional molecule such as benzimidazole can perfectly respond to.

Acknowledgements

This work was supported by the Croatian Science Foundation under grant number IP-2014-09-3386 entitled 'Design and synthesis of novel nitrogen-containing heterocyclic fluorophores and fluorescent nanomaterials for pH and metal-ion sensing' which is gratefully acknowledged.

Conflict of interest

Authors have no conflict of interest to declare.

Abbreviations

AIE	aggregation-induced emission
AIEE	aggregation-induced emission enhancement
DMF	dimethylformamide
ICT	intramolecular charge transfer
MOF	metal organic framework
NLO	non-linear optics
pHEMA	poly(2-hydroxyethyl methacrylate)
PPi	pyrophosphate
PVC	poly(vinyl chloride)
RGB	red, green and cyan spectrum
RTP	ultralong-persistent room temperature phosphorescence
SBA-15	porous silica
TEOS	tetraethylorthosilicate
TFA	trifluoroacetic acid
THF	tetrahydrofuran
TLC	thin layer chromatography
TNT	trinitrotoluene
UV	ultraviolet

Author details

Ema Horak[1]*, Robert Vianello[1] and Ivana Murković Steinberg[2]

1 Computational Organic Chemistry and Biochemistry Group, Ruđer Bošković Institute, Zagreb, Croatia

2 Department of General and Inorganic Chemistry, Faculty of Chemical Engineering and Technology, Zagreb, Croatia

*Address all correspondence to: Ema.Horak@irb.hr

References

[1] Keri RS, Hiremathad A, Budagumpi S, Nagaraja BM. Comprehensive review in current developments of benzimidazole-based medicinal chemistry. Chemical Biology and Drug Design. 2015;**86**(1):799-845

[2] Hung WY, Chi LC, Chen WJ, Chen YM, Chou SH, Wong KT. A new benzimidazole/carbazole hybrid bipolar material for highly efficient deep-blue electrofluorescence, yellow-green electrophosphorescence, and two-color-based white OLEDs. Journal of Materials Chemistry. 2010;**20**(45):10113-10119

[3] Vijayan N, Babu RR, Gopalakrishnan R, Ramasamy P, Harrison WTA. Growth and characterization of benzimidazole single crystals: A nonlinear optical material. Journal of Crystal Growth. 2004;**262**(1-4):490-498

[4] Manoharan S, Anandan S. Cyanovinyl substituted benzimidazole based (D-*π*-A) organic dyes for fabrication of dye sensitized solar cells. Dyes and Pigments. 2014;**105**:223-231

[5] Saltan GM, Dincalp H, Kiran M, Zafer C, Erbas SC. Novel organic dyes based on phenyl-substituted benzimidazole for dye sensitized solar cells. Materials Chemistry and Physics. 2015;**163**:387-393

[6] Hranjec M, Horak E, Babic D, Plavljanin S, Srdovic Z, Steinberg IM, et al. Fluorescent benzimidazo[1,2-a] quinolines: Synthesis, spectroscopic and computational studies of protonation equilibria and metal ion sensitivity. New Journal of Chemistry. 2017;**41**(1):358-371

[7] Perin N, Nhili R, Ester K, Laine W, Karminski-Zamola G, Kralj M, et al. Synthesis, antiproliferative activity and DNA binding properties of novel 5-Aminobenzimidazo [1,2-*a*] quinoline-6-carbonitriles. European Journal of Medicinal Chemistry. 2014;**80**:218-227

[8] Perin N, Uzelac L, Piantanida I, Karminski-Zamola G, Kralj M, Hranjec M. Novel biologically active nitro and amino substituted benzimidazo [1,2-*a*] quinolines. Bioorganic and Medicinal Chemistry. 2011;**19**(21):6329-6339

[9] Kulhanek J, Bures F, Pytela O, Mikysek T, Ludvik J. Imidazole as a donor/acceptor unit in charge-transfer chromophores with extended π-linkers. Chemistry-An Asian Journal. 2011;**6**(6):1604-1612

[10] Molina P, Tarraga A, Oton F. Imidazole derivatives: A comprehensive survey of their recognition properties. Organic and Biomolecular Chemistry. 2012;**10**(9):1711-1724

[11] Wolfbeis OS. Materials for fluorescence-based optical chemical sensors. Journal of Materials Chemistry. 2005;**15**(27-28):2657-2669

[12] Basabe-Desmonts L, Reinhoudt DN, Crego-Calama M. Design of fluorescent materials for chemical sensing. Chemical Society Reviews. 2007;**36**(6):993-1017

[13] Horak E, Kassal P, Murković Steinberg I. Benzimidazole as a structural unit in fluorescent chemical sensors: The hidden properties of a multifunctional heterocyclic scaffold. Supramolecular Chemistry. 2018;**30**:838-857

[14] Mohr GJ. Polymers for optical sensors. In: Optical Chemical Sensors. Vol. 2006. Dordrecht: Springer Netherlands; 2006. pp. 297-321

[15] Mistlberger G, Crespo GA, Bakker E. Ionophore-based optical sensors.

In: Cooks RG, Pemberton JE, editors. Annual Review of Analytical Chemistry, Vol 7. Annual review of analytical chemistry. 2014. pp. 483-512

[16] Tomar PK, Chandra S, Singh I, Kumar A, Malik A, Singh A. Development of a new copper(II) ion-selective PVC membrane electrode based on tris(2-benzimidazolylmethyl) amine. Journal of the Indian Chemical Society. 2011;**88**(11):1739-1744

[17] Yan ZN, Zhang SY, Wang HX, Kang YX. Preparation and analytical application of new Cr^{3+}-selective membrane electrodes based on acylhydrazone-containing benzimidazole derivatives. Journal of the Iranian Chemical Society. 2016;**13**(3):411-420

[18] Firooz AR, Ensafi AA, Karimi K, Khalifeh R. Specific sensing of mercury(II) ions by an optical sensor based on a recently synthesized ionophore. Sensors and Actuators B: Chemical. 2013;**185**:84-90

[19] Firooz AR, Ensafi AA, Kazemifard N, Sharghi H. A highly sensitive and selective bulk optode based on benzimidazol derivative as an ionophore and ETH5294 for the determination of ultra trace amount of silver ions. Talanta. 2012;**101**:171-176

[20] Horak E, Vianello R, Hranjec M, Krištafor S, Zamola GK, Steinberg IM. Benzimidazole acrylonitriles as multifunctional push-pull chromophores: Spectral characterisation, protonation equilibria and nanoaggregation in aqueous solutions. Spectrochimica Acta Part A: Molecular and Biomolecular Spectroscopy. 2017;**178**:225-233

[21] Horak E, Kassal P, Hranjec M, Steinberg IM. Benzimidazole functionalised Schiff bases: Novel pH sensitive fluorescence turn-on chromoionophores for ion-selective optodes. Sensors and Actuators B: Chemical. 2018;**258**:415-423

[22] Fernandez-Alonso S, Corrales T, Pablos JL, Catalina F. A switchable fluorescence solid sensor for Hg^{2+} detection in aqueous media based on a photocrosslinked membrane functionalized with (benzimidazolyl)methyl-piperazine derivative of 1,8-naphthalimide. Sensors and Actuators B-Chemical. 2018;**270**:256-262

[23] Anantha-Iyengar G, Shanmugasundaram K, Nallal M, Lee K-P, Whitcombe MJ, Lakshmi D, et al. Functionalized conjugated polymers for sensing and molecular imprinting applications. Progress in Polymer Science. 2019;**88**:1-129

[24] Alvarez A, Costa-Fernández JM, Pereiro R, Sanz-Medel A, Salinas-Castillo A. Fluorescent conjugated polymers for chemical and biochemical sensing. TrAC Trends in Analytical Chemistry. 2011;**30**(9):1513-1525

[25] Altarawneh S, Nahar L, Arachchige IU, El-Ballouli AO, Hallal KM, Kaafarani BR, et al. Highly porous and photoluminescent pyrene-quinoxaline-derived benzimidazole-linked polymers. Journal of Materials Chemistry A. 2015;**3**(6):3006-3010

[26] Choi HW, Kim YS, Yang NC, Suh DH. Synthesis of a new conjugated polymer based on benzimidazole and its sensory properties using the fluorescence-quenching effect. Journal of Applied Polymer Science. 2004;**91**(2):900-904

[27] Saikia G, Iyer PK. A remarkable superquenching and superdequenching sensor for the selective and noninvasive detection of inorganic phosphates in saliva. Macromolecules. 2011;**44**(10):3753-3758

[28] Wu CS, Liu CT, Chen Y. Multifunctional copolyfluorene containing pendant benzimidazolyl groups: Applications in chemical sensors and electroluminescent devices. Polymer Chemistry. 2012;**3**(12):3308-3317

[29] Han B, Zhou NC, Zhang W, Cheng ZP, Zhu J, Zhu XL. Fluorescence emission of amphiphilic copolymers bearing benzimidazole groups: Stimuli-responsive behaviors in aqueous solution. Journal of Polymer Science Part A: Polymer Chemistry. 2013;**51**(20):4459-4466

[30] Shen LJ, Zhao P, Zhu WH. A ratiometric hydrophilic fluorescent copolymer sensor based on benzimidazole chromophore for microbioreactors. Dyes and Pigments. 2011;**89**(3):236-240

[31] Li JX, Liu D, Hao ZC, Cui GH. An unusual (3,4,5)-connected 3D cadmium metal-organic framework as a luminescent sensor for detection of $Cr_2O_7^{2-}$ in water. Inorganic Chemistry Communications. 2018;**97**:79-82

[32] Yu Y, Zhang XM, Ma JP, Liu QK, Wang P, Dong YB. Cu(I)-MOF: Naked-eye colorimetric sensor for humidity and formaldehyde in single-crystal-to-single-crystal fashion. Chemical Communications. 2014;**50**(12):1444-1446

[33] Zhou L, Zhao K, Hu YJ, Feng XC, Shi PD, Zheng HG. A bifunctional photoluminescent metal-organic framework for detection of Fe^{3+} ion and nitroaromatics. Inorganic Chemistry Communications. 2018;**89**:68-72

[34] Dong Y, Zhang H, Lei F, Liang M, Qian X, Shen P, et al. Benzimidazole-functionalized Zr-UiO-66 nanocrystals for luminescent sensing of Fe^{3+} in water. Journal of Solid State Chemistry. 2017;**245**:160-163

[35] Li S-L, Lan Y-Q, Ma J-C, Ma J-F, Su Z-M. Metal–organic frameworks based on different benzimidazole derivatives: Effect of length and substituent groups of the ligands on the structures. Crystal Growth and Design. 2010;**10**(3):1161-1170

[36] Hao SY, Hou SX, Hao ZC, Cui GH. A new three-dimensional bis(benzimidazole)-based cadmium(II) coordination polymer. Spectrochimica Acta Part A-Molecular and Biomolecular Spectroscopy. 2018;**189**:613-620

[37] Tripathi S, Bardhan D, Chand DK. Multistimuli-responsive hydrolytically stable "smart" mercury(II) coordination polymer. Inorganic Chemistry. 2018;**57**(18):11369-11381

[38] Yang YS, Wang KZ, Yan DP. Ultralong persistent room temperature phosphorescence of metal coordination polymers exhibiting reversible pH-responsive emission. Acs Applied Materials and Interfaces. 2016;**8**(24):15489-15496

[39] Zhao XX, Liu D, Li YH, Cui GH. Bifunctional silver(I) coordination polymer exhibiting selective adsorptive of Congo red and luminescent sensing for ferric ion. Polyhedron. 2018;**156**:80-88

[40] Wei XJ, Li YH, Qin ZB, Cui GH. Two zinc(II) coordination polymers for selective luminescence sensing of iron(III) ions and photocatalytic degradation of methylene blue. Journal of Molecular Structure. 2019;**1175**:253-260

[41] Hao JM, Yu BY, Van Hecke K, Cui GH. A series of d(10) metal coordination polymers based on a flexible bis(2-methylbenzimidazole) ligand and different carboxylates: Synthesis, structures, photoluminescence and catalytic properties. CrystEngComm. 2015;**17**(11):2279-2293

[42] Wang XF, Du CC, Zhou SB, Wang DZ. Six complexes based on bis(imidazole/benzimidazole-1-yl) pyridazine ligands: Syntheses, structures and properties. Journal of Molecular Structure. 2017;**1128**:103-110

[43] Jeronimo PCA, Araujo AN, Montenegro MCBSM. Optical sensors and biosensors based on sol-gel films. Talanta. 2007;**72**(1):13-27

[44] Hoffmann HS, Stefani V, Benvenutti EV, Costa TMH, Gallas MR. Fluorescent silica hybrid materials containing benzimidazole dyes obtained by sol-gel method and high pressure processing. Materials Chemistry and Physics. 2011;**126**(1-2):97-101

[45] Babu SS, Praveen VK, Ajayaghosh A. Functional π-gelators and their applications. Chemical Reviews. 2014;**114**(4):1973-2129

[46] Ma XX, Xie JS, Tang N, Wu JC. AIE-caused luminescence of a thermally-responsive supramolecular organogel. New Journal of Chemistry. 2016;**40**(8):6584-6587

[47] Yu H, Kawanishi H, Koshima H. Preparation and photophysical properties of benzimidazole-based gels. Journal of Photochemistry and Photobiology A: Chemistry. 2006;**178**(1):62-69

[48] Shen X, Jiao T, Zhang Q, Guo H, Lv Y, Zhou J, et al. Nanostructures and self-assembly of organogels via benzimidazole/benzothiazole imide derivatives with different alkyl substituent chains. Journal of Nanomaterials. 2013;**2013**:8

[49] Xue P, Lu R, Jia J, Takafuji M, Ihara H. A smart gelator as a chemosensor: Application to integrated logic gates in solution, gel, and film. Chemistry – A European Journal. 2012;**18**(12):3549-3558

[50] Mondal S, Ghosh K. Anthraquinone derived cholesterol linked imidazole gelator in visual sensing of picric acid. Chemistry Select. 2017;**2**(17):4800-4806

[51] Ghosh K, Panja S, Bhattacharya S. Visual sensing of Ag^+ ions through gelation of cholesterol-appended benzimidazole and associated ion conducting behaviour. Chemistry Select. 2017;**2**(3):959-966

[52] Yao H, Wang J, Song S-S, Fan Y-Q, Guan X-W, Zhou Q, et al. A novel supramolecular AIE gel acts as a multi-analyte sensor array. New Journal of Chemistry. 2018;**42**(22):18059-18065

[53] Yao H, Wu HP, Chang J, Lin Q, Wei TB, Zhang YM. A carboxylic acid functionalized benzimidazole-based supramolecular gel with multi-stimuli responsive properties. New Journal of Chemistry. 2016;**40**(6):4940-4944

[54] Hong Y, Lam JWY, Tang BZ. Aggregation-induced emission. Chemical Society Reviews. 2011;**40**(11):5361-5388

[55] Cao YL, Yang MD, Wang Y, Zhou HP, Zheng J, Zhang XZ, et al. Aggregation-induced and crystallization-enhanced emissions with time-dependence of a new Schiff-base family based on benzimidazole. Journal of Materials Chemistry C. 2014;**2**(19):3686-3694

[56] Gogoi A, Mukherjee S, Ramesh A, Das G. Aggregation-induced emission active metal-free chemosensing platform for highly selective turn-on sensing and bioimaging of pyrophosphate anion. Analytical Chemistry. 2015;**87**(13):6974-6979

[57] Malakar A, Kumar M, Reddy A, Biswal HT, Mandal BB, Krishnamoorthy G. Aggregation induced enhanced emission of 2-(2 '- hydroxyphenyl) benzimidazole. Photochemical

and Photobiological Sciences. 2016;**15**(7):937-948

[58] Wu Z, Sun JB, Zhang ZQ, Gong P, Xue PC, Lu R. Organogelation of cyanovinylcarbazole with terminal benzimidazole: AIE and response for gaseous acid. RSC Advances. 2016;**6**(99):97293-97301

[59] Horak E, Hranjec M, Vianello R, Steinberg IM. Reversible pH switchable aggregation-induced emission of self-assembled benzimidazole-based acrylonitrile dye in aqueous solution. Dyes and Pigments. 2017;**142**:108-115

[60] Singh A, Raj P, Singh N. Benzimidazolium-based self-assembled fluorescent aggregates for sensing and catalytic degradation of diethylchlorophosphate. Acs Applied Materials and Interfaces. 2016;**8**(42):28641-28651

[61] Zhan Y, Wei Q, Zhao J, Zhang X. Reversible mechanofluorochromism and acidochromism using a cyanostyrylbenzimidazole derivative with aggregation-induced emission. RSC Advances. 2017;**7**(77):48777-48784

[62] Horak E, Robić M, Šimanović A, Mandić V, Vianello R, Hranjec M, et al. Tuneable solid-state emitters based on benzimidazole derivatives: Aggregation induced red emission and mechanochromism of D-π-a fluorophores. Dyes and Pigments. 2019;**162**:688-696

[63] Benelhadj K, Massue J, Retailleau P, Ulrich G, Ziessel R. 2-(2 '-Hydroxyphenyl)benzimidazole and 9,10-Phenanthroimidazole chelates and borate complexes: Solution- and solid-state emitters. Organic Letters. 2013;**15**(12):2918-2921

[64] Chaudhuri T, Mondal A, Mukhopadhyay C. Benzimidazole: A solid state colorimetric chemosensor for fluoride and acetate. Journal of Molecular Liquids. 2018;**251**:35-39

[65] Maeda C, Todaka T, Ueda T, Ema T. Color-Tunable solid-state fluorescence emission from Carbazole-based BODIPYs. Chemistry-A European Journal. 2016;**22**(22):7508-7513

[66] Borisov SM, Klimant I. Optical nanosensors–Smart tools in bioanalytics. The Analyst. 2008;**133**(10):1302-1307

[67] Wolfbeis OS. An overview of nanoparticles commonly used in fluorescent bioimaging. Chemical Society Reviews. 2015;**44**(14):4743-4768

[68] Montalti M, Rampazzo E, Zaccheroni N, Prodi L. Luminescent chemosensors based on silica nanoparticles for the detection of ionic species. New Journal of Chemistry. 2013;**37**(1):28-34

[69] Song X, Li F, Ma J, Jia N, Xu J, Shen H. Synthesis of fluorescent silica nanoparticles and their applications as fluorescence probes. Journal of Fluorescence. 2011;**21**(3):1205-1212

[70] Borisov SM, Mayr T, Mistlberger G, Waich K, Koren K, Chojnacki P, et al. Precipitation as a simple and versatile method for preparation of optical nanochemosensors. Talanta. 2009;**79**(5):1322-1330

[71] Badiei A, Razavi BV, Goldooz H, Ziarani GM, Faridbod F, Ganjali MR. A novel fluorescent chemosensor assembled with 2,6-Bis(2-Benzimidazolyl)pyridine-functionalized Nanoporous silica-type SBA-15 for recognition of Hg^{2+} ion in aqueous media. International Journal of Environmental Research. 2018;**12**(1):109-115

[72] Kaur N, Raj P, Kaur N, Kim DY, Singh N. Supramolecular hybrid of ZnO nanoparticles with benzimidazole

based organic ligand for the recognition of Zn^{2+} ions in semi-aqueous media. Journal of Photochemistry and Photobiology A: Chemistry. 2017;**347**:41-48

[73] Wang XX, Zeng FY, Ma ZY, Jiang YL, Han QR, Wang BX. Self-assembly of benzimidazole-ended nano hyperbranched polyester and its host-guest response. Materials Letters. 2016;**173**:191-194

[74] Boonsri M, Vongnam K, Namuangruk S, Sukwattanasinitt M, Rashatasakhon P. Pyrenyl benzimidazole-isoquinolinones: Aggregation-induced emission enhancement property and application as TNT fluorescent sensor. Sensors and Actuators B: Chemical. 2017;**248**:665-672

[75] Aich K, Das S, Goswami S, Quah CK, Sarkar D, Mondal TK, et al. Carbazole-benzimidazole based dyes for acid responsive ratiometric emissive switches. New Journal of Chemistry. 2016;**40**(8):6907-6915

[76] Kumar G, Gupta N, Paul K, Luxami V. Acrylonitrile embedded benzimidazole-anthraquinone based chromofluorescent sensor for ratiometric detection of CN^{-} ions in bovine serum albumin. Sensors and Actuators B-Chemical. 2018;**267**:549-558

[77] Dhaka G, Kaur N, Singh J. Spectral studies on benzimidazole-based “bare-eye” probe for the detection of Ni^{2+}: Application as a solid state sensor. Inorganica Chimica Acta. 2017;**464**:18-22

Chapter 10

Benzimidazole as Solid Electrolyte Material for Fuel Cells

Daniel Herranz and Pilar Ocón

Abstract

This chapter is focused in the application of benzimidazole, mainly in the form of poly[2,2′-(*m*-phenylene)-5,5′-bisbenzimidazole] (PBI) and poly(2,5-benzimidazole) (ABPBI), in the fuel cell technology. A short introduction is given of the fuel cell principles, explaining both the theory and the high importance of this technology. PBI and ABPBI are used in a certain type of fuel cells: the polymer electrolyte fuel cells and are key materials in the composition of some of the electrolyte membranes used. Commercially available membranes composed of PBI are indicated in order to give an overview of their potential performance. The synthesis of the polymers is explained. Moreover, the preparation of the different kinds of membranes, both in proton exchange membrane fuel cells (PEMFCs) and anion exchange membrane fuel cells (AEMFCs) is studied. A deep description is given about the properties that make this family of compounds so interesting for the fuel cell technology as well as an how these polymers have been characterized with the corresponding analysis. The comparison with other ion exchange membranes is also discussed. Special attention will be given to the state of the art of different kinds of PBI/ABPBI fuel cell electrolyte membranes, in which our group and others are working nowadays.

Keywords: polybenzimidazole, electrolyte, fuel cells, proton exchange membrane, anion exchange membrane

1. Introduction

Benzimidazole and its family can be used in the energy world easily in the form of polymers, since these materials have the possibility to create designed structures for many applications. The fuel cells and electrolyzers are emerging technologies with wonderful potential. In these technologies, an electrolyte is needed to separate two electrodes where electrochemical reactions occur. The separation must be physical and electrical, but the electrolyte allows the ionic conduction of ions in order to close the circuit (so the current goes through the external circuit and can be used) and to make possible the continuity of the reactions at the electrodes. Here is where benzimidazoles (in the form of polybenzimidazole, e.g.) play a key role, in the conformation of a solid polymer electrolyte membrane, alone or with other chemical materials.

But, what is a fuel cell? What do we understand for membranes in this field? Fuel cells are electrochemical devices that convert directly the chemical energy of the reagents into electrical energy and side-products via an electrochemical reaction. This process allows theoretical efficiencies as high as 80% [1], which is a wonderful advantage compared to the thermal machines limited thermodynamically by the Carnot cycle. There are many types of fuel cells, the most relevant are alkaline fuel cells (AFCs), polymer electrolyte membrane fuel cells (PEMFCs), phosphoric acid fuel cells (PAFCs), molten carbonate fuel cells (MCFCs), and solid oxide fuel cells (SOFCs).

Polymer electrolyte membrane fuel cells have as principal characteristics the low operation temperature (<120°C), high power density, and easy scale-up, making them a promising technology for power generation. Their main application fields are backup power, portable power, distributed generation, and transportation [1]. It is relevant to note the role of transition energy technology, since they can play an important function in the near future in order to overcome the fossil fuel depletion and mitigate the climate change. The reason is that fuels like hydrogen or alcohols, which are produced by unsustainable ways, could be produced with renewable energies. An example of this is the actual production of hydrogen mainly from catalytic reforming of methane and just some from electrolysis [2]. The hydrogen can be produced from electrolysis powered with electricity coming from renewables. This should be done when production is higher than the demand, allowing to store chemically the energy and later use it when needed with a PEMFC; this is known as the "hydrogen economy system." It is also possible to accumulate energy in short-chain alcohols like methanol or ethanol and use them to power PEMFCs [3, 4], mainly used in the portable applications. A great advantage of this technology is the low pollution associated with the process. For example, when hydrogen is used as fuel, the only products are electricity and water. The potential of PEMFCs is really promising but still drawbacks as high cost (mainly from the expensive catalysts based in Pt) and low durability have to be overcome for a general commercialization [1].

In PEMFCs one of the most important components is the polymeric ion exchange membrane (IEM) that works as an electrolyte. It has to be an electrical insulator to force the produced electrons to go through the external circuit, it also has to avoid the mixture of the reagents supplied in anode and cathode, and it is responsible of the adequate ionic conductivity of the ions traveling through it. Depending on the ion movement, two types of IEMs can be distinguished: anion exchange membranes (AEMs), where the ionic charge carriers are the hydroxide ions (OH^-) that travel from cathode to anode, and cation exchange membranes (CEMs) where generally the proton ion (H^+) moves from anode to cathode in the fuel cell. For that reason, the last ones are also called proton exchange membranes (PEMs). The AEMs are used in alkaline media and the others in acid media. The proton exchange membrane fuel cells (PEMFCs) have been historically more used because of the discovery of the Nafion® membrane that has good ionic conductivity and durability and has been the standard so far [5]. The higher mobility of the H^+ ion compared to OH^- in aqueous media has also been a relevant factor [6]. The alkaline media in the other hand does not have a standard membrane and presents relevant advantages that have produced high interest in the last years. Some of them are the faster electrochemical kinetics in the alkaline media, possible absence of noble metals as catalysts, minimized corrosion problems, and cogeneration of electricity and valuable chemicals [7].

Independently of the media, membranes are expected to have good ionic conductivity, long-term chemical and electrochemical stability, adequate mechanical strength, good moisture control, low fuel or oxygen crossover, and production costs compatible with intended application [5, 6].

In the FCs, the active materials (fuel and oxidant) are continuously fed and extracted. The fuel cell, **Figure 1**, is made up of two electrodes: the anode, where the

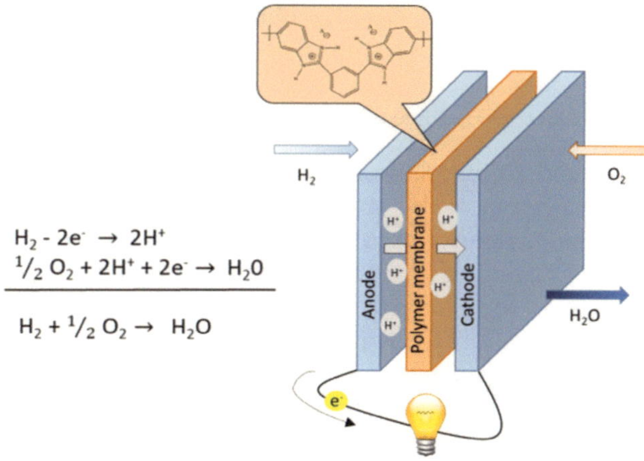

Figure 1.
Polymer exchange membrane fuel cell working with H_2 and O_2.

fuel is oxidized, and the cathode, where the oxidant (O_2) is reduced. It also involves an electrolyte, which acts as an ionic conductor and electrical insulator. The electrons obtained in the anode are addressed directly to the cathode through the external circuit, generating an electric current directly usable. In addition, the protons produced in the anode go through the electrolyte, up to the cathode to reduce O_2, generating water as the only product of the reaction. The reaction is exothermic and has a value of $\Delta H^0_r = -285.83$ kJ/mol for H_2O (l) and − 241.862 kJ/mol for H_2O (v). Although this is a spontaneous reaction, it needs to be catalyzed to be operational, since the kinetics of the process is too slow otherwise.

At atmospheric pressure, the maximum potential difference obtained by the fuel cell will be determined by the difference of energy between the initial and final state of the system. The Gibbs free energy variation of the process, ΔG, can be calculated from the operation temperature (T) and changes with both enthalpy (ΔH) and entropy (ΔS) of the reaction. Under standard conditions

$$\Delta G^0 = \Delta S^0 - T\,\Delta S^0 \tag{1}$$

and the maximum potential difference, obtained in the fuel cell, $E^0_{theoric}$, will be

$$E^0 = \frac{-\Delta G^0}{nF} = 1.23V \tag{2}$$

where n is the number of electrons exchanged and F is Faraday's constant. At 298 K and 1 atm, $\Delta G^0 = -237.340$ J/mol and therefore $E^0 = 1.23$ V. For an operating temperature of 80°C, the values of ΔH and ΔS change, but slightly, and the decrease in ΔG will be mainly due to the temperature, resulting in a theoretical potential difference of 1.18 V approximately. However, in practice this potential, called the open circuit potential, is significantly lower than this potential value, usually less than 1 V. This suggests that some losses appear in the fuel cell even when no external current is generated. The potential difference of the fuel cell in operation, that is, when the current is passing through the system, $E_{fuel\ cell\ (I)}$, will be given by the sum of thermodynamic or reversible value (I = 0), minus the anode and cathode activation overvoltage and the ohmic losses or overvoltage. The electrode kinetics was represented by the Butler-Volmer equation, the mass transport process was described by the multicomponent Stefan-Maxwell equations and Fick's law, and the ionic and electronic resistances are described by Ohm's law. The E $_{\text{fuel cell (I)}}$ value could be obtained by

$$E_{fuel\ cell\ (I)} = E_{Reversible\ (I=0)} - \eta_{activation} - \eta_{ohmic} \tag{3}$$

The losses considered are in relation to the activation overvoltages, and they are dependent on the kinetics of the processes involved and therefore directly related to the goodness of catalyst used for the process. Thus, $\eta_{\text{activation}}$ is related directed with both the oxidation kinetic reaction and the reduction kinetic reaction of the reagent involved in the catalysts surface materials. The $\eta_{\text{activation}}$ for an H_2/O_2 fed in PEMFC will come mainly determined by the slow kinetics of oxygen reduction reaction (ORR) on the catalyst material in comparison to H_2 oxidation, while $\eta_{\text{activation (transport)}}$ is the consequence of material transport. This overpotential considers the combination of the flow of reactants and products in the fuel cell. The polarization from concentration gradients occurs when a reactant is rapidly consumed at the electrode by the electrochemical reaction so that gradients are established. The η_{ohmic} = iR will be due to the combination of resistors provided by internal/external electrical contacts and ionic resistance due to ion motion through the membrane. Therefore, the fuel cell when current is not zero has an $E_{\text{fuel cell(I)}}$ expression like this:

$$E_{fuel\ cell\ (I)} = E_{rever\ (I=0)} - \frac{RT}{\alpha_c} ln\left(\frac{i}{i_{0,c}}\right) - \frac{RT}{\alpha_a} ln\left(\frac{i}{i_{0,a}}\right) - \frac{RT}{\alpha_c} ln\left(\frac{i_{I,c}}{i_{I,c} - i}\right) - \frac{RT}{\alpha_a} ln\left(\frac{i_{I,a}}{i_{I,a} - i}\right) \tag{4}$$

$$E_{rev\ (I=0)} = 1.229 - \left(8.5 \times 10^{-4}\right)(T - 298.15) + \left(4.308 \times 10^{-5}\right)T[\ln(P_{H2}) + 0.5\ln(P_{O2})] \tag{5}$$

being i_{oc}, α_c, i_{Lc} and i_{oa}, α_a, i_{La} the exchange current density, transfer coefficient, and limit current density of the cathodic and anodic processes, respectively [8]. The polarization curve of the device can be found in **Figure 2**, where the different losses mentioned above are indicated.

It was previously stated that the ion exchange polymer membrane is electrically insulator and practically impermeable to reactant gases, but some small amount of mainly H_2 will crossover from anode to cathode. Hydrogen that permeates through the membrane does not participate in the electrochemical reaction on the anode

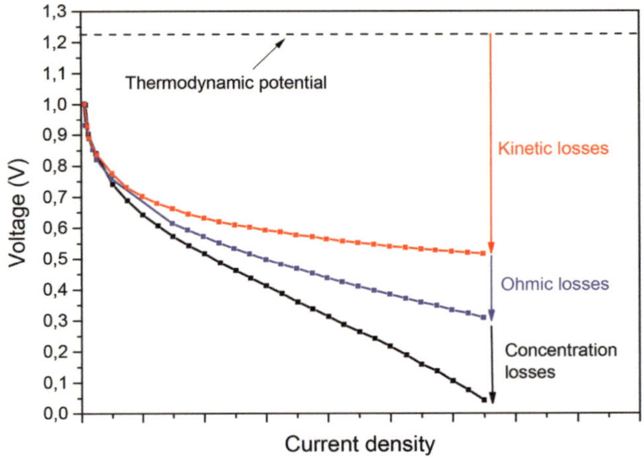

Figure 2.
Polarization curves with voltage losses of a fuel cell.

side. Each hydrogen molecule on the cathode side reacts with oxygen on the surface of the catalyst resulting in two fewer electrons in the generated current that travels through the external circuit and thus in a reduction of cathode and the overall fuel cell potential. These losses are not big in fuel cell operation, but when the fuel cell is at open circuit potential or at very low current densities, this situation may have a dramatic effect on fuel cell potential. At least, all these losses have to be taken into account when the device works and have a lot to do with good fuel cell performance.

2. Synthesis of polybenzimidazole materials

Polybenzimidazoles are synthesized by the repetitive reaction of aromatic amino groups with carboxyl groups using a 1:2 molar ratio by the process of step-grow polymerization [9]. Usually the monomer reagents are a diacid and a tetra-amine, like the example in **Figure 3**. There are many polybenzimidazoles but the ones that have presented better application and have been more studied are poly(2,2′-(*m*-phenylene)-5,5′-bibenzimidazole), known as PBI, and poly(2,5-benzimidazole), known as ABPBI. Both were first synthesized by Vogel and Marvel in 1961 [10]. For PBI the synthesis was a two-step process with an intermediate prepolymer that prevented the production of high molecular weight polymer. Cho et al. [11, 12] discovered a process with 3,3′,4,4′-tetraaminobiphenyl (TAB) and isophthalic acid (IPA) to do the synthesis in a single step obtaining high molecular weight, in the presence of catalyzers and at temperatures higher than 350°C. It is important to know the molecular weight of the polymer, which is obtained by the measurement of the inherent viscosity (IV, in dLg^{-1}) of the polymer dissolved in concentrated sulfuric acid. For membrane application, usually casted from solution, it is interesting to have high molecular weight in order to achieve mechanically stable membranes that can support higher doping and thus obtain better ionic conductivity. The previously described method of Vogel and Marvel and Cho et al. can be classified in the heterogeneous molten/solid state synthesis [13, 14]. The other synthesis method used is the homogeneous solution synthesis, using solvents as polyphosphoric acid (PPA) [15]; this method allows to use moderate temperature and more stable monomers and is excellent to synthesize linear high molecular weight polymers at laboratory or small batch scale. These advantages make this synthesis method the most commonly used. Another example of solvent is Eaton's reagent, a mixture of phosphorus pentoxide (P2O5) and methanesulfonic acid (MSA) proposed by Eaton et al. [16], which has low viscosity making it suitable for the homogeneous solution synthesis and the acid washing after it [17, 18]. A shorter reaction time with high molecular weight has been obtained using homogeneous solution microwave-assisted synthesis recently, both for PBI and ABPBI [14].

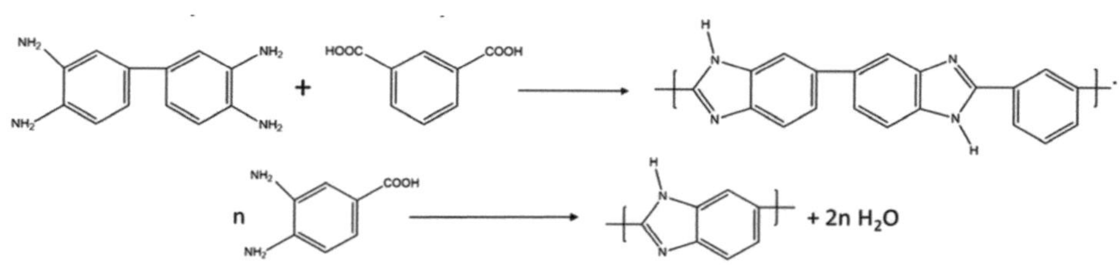

Figure 3.
Example synthesis of (top) poly(2,2′-(m-phenylene)-5,5′-bibenzimidazole), abbreviated as PBI, and (bottom) poly(2,5-benzimidazole) (ABPBI).

ABPBI is synthesized from a single monomer, (3,4-diaminobenzoic acid) (DABA), which as the advantages of being less expensive, commercially available, and non-carcinogenic. The scheme is shown in **Figure 3**. Different syntheses have been done by the homogeneous solution method in PPA or Eaton's reagent, and inherent viscosity values as high as 7.33 have been reached, as reported by Li et al. by using recrystallized DABA [19]. This is essential for the direct casting of ABPBI membranes since it has been suggested by Asensio and Gómez-Romero that values of at least 2.3 dL g^{-1} are necessary to cast good membranes [13].

In the case of ABPBI, since there is only a monomer, its purity is not as critical as in PBI; however, the use of high purity monomer produces polymers of high molecular weight [20]. Since polybenzimidazoles have to be doped in order to become ionic conductors, two methods are used to prepare the membranes: direct casting from the polymerization solution, as the work developed by Asensio et al. [21], or dissolving the previously synthesized polymer and then doing the casting of the membrane. The casting process consists in the formation of a thin film by the deposition of the polymer by evaporation of the solvent in the solution. To solubilize PBI or ABPBI, usually strong bases or acids are needed; only a few organic solvents can also do it; one of them is the N,N-dimethylacetamide (DMAc) [13, 22]. There is also an alternative way to cast ABPBI membranes from a mixture of NaOH and ethanol [23].

3. Properties of the materials and characterization

The structure of polybenzimidazoles has a good degree of flexibility and chemical and thermal resistance compared to other polymers with more single bonds in their main chain between aromatic units. The presence of aromatic units in the main chain to have higher thermal stability than the aliphatic analogs is also important [9]. In order to characterize polybenzimidazoles, one of the most important parameters is the molecular weight of the polymer, which will be highly related to the final membranes properties. The common way to obtain the molecular weight is by measurement of the intrinsic viscosity of the polymer (η_{IV}) at a certain temperature (normally 25–30°C). From the plotting of the specific viscosity (η_{sp}) as function of the polymer concentration, the intrinsic viscosity is calculated extrapolating to zero concentration. A simpler measurement process was proposed to do the calculation with a single-point method using Eq. (6), where C is the polymer concentration in a concentrated acid like 96 wt% H_2SO_4:

$$\eta_{IV} = (\eta_{SP} + 3\ln(1 + \eta_{SP}))/4C \quad (6)$$

The protocol test is to calculate the η_{sp} of a polymer solution 5 g L^{-1} in concentrated sulfuric acid at 30°C using an Ubbelohde viscometer. From the η_{IV} value, the average molecular weight is calculated with the Mark-Houwink- Sakurada expression:

$$\eta_{IV} = K * M_W^{\alpha} \quad (7)$$

where the Mark-Houwink constants depend on the molecular weight range and distribution. Values often used for this constants are $K = 1.94 \times 10^{-4}$ dL g^{-1}, and $\alpha = 0.791$, obtained from Buckley et al. by light scattering measurements. Other solvents as formic acid or MSA can also be used to measure the viscosity of polybenzimidazoles [14].

There are various techniques in order to investigate the structure of polybenzimidazoles. Nuclear magnetic resonance is very powerful for pure organic compounds or the repeating unit of a polymer. Solvents that can be used include deuterated dimethyl sulfoxide (DMSO- d_6) and deuterated sulfuric acid (D_2SO_4). The most commonly used to record ^{1}H-NMR spectra is DMSO-d_6 because with D_2SO_4, the fast exchange interaction with the proton in the imine of the imidazole rings (-NH-) causes the chemical shift of that hydrogen to be often indiscernible [24]. ^{1}H-NMR PBI characteristic signals in DMSO-d_6 are at 13.2 (2H), 9.1 (1H), 8.3 (2H), and 8.0–7.6 (7H) ppm, the first of them attributed to the imidazole protons and the others to the aromatic protons [25, 26]. IR and Raman spectroscopy are also used, mainly to identify different functional groups and obtain or corroborate the chemical structure of the polymers [24, 27, 28]. In PBI, the IR spectrum region from 2000 to 4000 cm^{-1} is interesting since N–H stretching modes occur in this range, showing three typical bands at 3415, 3145, and 3063 cm^{-1}. The broad band around 3145 cm^{-1} has been attributed to the stretching vibrations of N–H groups self-associated by hydrogen bonds, and the peak at 3145 cm^{-1} is assigned to the N–H groups stretching vibration. In the region from 1630 to 1500 cm^{-1}, the peaks observed come from the vibration of C=C and C=N bonds [27]. In the Raman spectra of PBI, the most significant absorption band comes from the benzene ring vibration and is located around 1000 cm^{-1} [28]. For the measurement of the Raman spectra, it is relevant to use an excitation wavelength of 785 nm (red laser) since it gives much less fluorescence than the 532 nm (green laser) [29]. Because the structure and functional groups are the same, ABPBI presents the same IR peaks than PBI, as reported by Asensio et al. [30]. They also investigated the bands appearing when the polymer membrane is doped with phosphoric acid: in the N–H stretching zone, they found the evolution of nitrogen protonation by the acid, and in the medium and high doped samples, the broad band of N^+–H vibration becomes stronger, while the nonassociated imidazole protons decreases. In polybenzimidazoles doped with alkaline media for anion conductivity purposes, the structure changes are also clearly identified. Aili et al. [31] investigated PBI with different degrees of KOH doping and found that in the IR spectra, at KOH concentrations higher than 15 wt.%, the N–H stretching band at 3415 cm^{-1} disappear as well as the broad band around 3100 cm^{-1} of shelf-associated hydrogen bonded N–H groups. They concluded that the IR data indicated the predominance of the deprotonated form of PBI with KOH concentrations of the bulk solution around 15–20 wt.%. In the ^{1}H-NMR spectrum the signal at 13.3 ppm of the N–H proton disappeared at high bulk KOH concentration, and most signals from the aromatic protons showed upfield shift compared to pristine PBI, indicating complete ionization. This full ionization of the polymer releases the extensive intermolecular hydrogen bonding allowing for high swelling behavior and water and KOH uptake and therefore enhanced ion conductivity. This study corroborates the knowledge that the introduction of species that interact with imidazole groups by hydrogen bonding decreases the intermolecular polybenzimidazole cohesion, causing a strong plasticizing effect observed in the great decay of the tensile strength and enhanced elongation at break when the doping level increases, especially when full ionization of the polymer is reached. Using an even higher concentration doping solution, they found that a higher crystallinity structure was obtained, as observed by XRD, mechanical test, and swelling behavior measurements. X-ray photoelectron spectroscopy (XPS) is also a helpful technique for the characterization of polybenzimidazoles, concretely for the capacity to distinguish the oxidation states of the elements present and allow their quantification in the surface of the membrane [29]. Other fundamental measurements usually performed on the synthesized membranes are the determination of the ionic conductivity, the swelling behavior with water and in acidic/alkaline media, or the thermogravimetric analysis (TGA). In conclusion, a full set of characterization analysis have been studied and are used to identify and test the properties of the synthesized polybenzimidazoles and the membranes prepared with them.

4. Commercial availability

There have been different companies relevant in the fuel cell membrane field, probably the most known one is DuPont for developing the Nafion® membrane made of a sulfonated tetrafluoroethylene-based fluoropolymer-copolymer with excellent thermal and mechanical stability as well as high proton conductivity in low-temperature fuel cells. Companies like Solvay, Gore, and others have also commercialized membranes with this chemistry. This membrane has been the standard for fuel cells used in low-temperature and acidic media, but at temperatures higher than 100°C, Nafion® performance drops dramatically due to the lower hydration level. It is in these conditions where membranes made of polybenzimidazoles have shown good performance and promising applicability, and production for commercialization has occurred. BASF Fuel Cell (formerly PEMEAS Fuel Cell), a part of one of the larger chemistry industries, has developed a product line about a membrane electrode assembly (MEA) based in a PBI membrane, Celtec® [32, 33]. These MEAs optimal operation conditions are between 120 and 180°C, doped in phosphoric acid. They have shown relevant advantages working as high temperature PMFCs, like high tolerance to fuel gas impurities such as CO (up to 3%), H_2S (up to 10 ppm), NH_3, or methanol, no humidification required, far simpler system due to elimination of water, and a less complex reformer technology. In addition, several advantages can be obtained for the electrocatalysis, but it is necessary to be especially careful at the high stability toward corrosion needed to ensure long fuel cell lifetimes, apart from high activity for the oxidation of the fuels and the oxygen reduction reaction. Other companies that commercialize PBI- and PBI-based membranes are "PBI Performance Products" with their Celazole® PBI PEM [22, 34] and Danish Power Systems with their Dapozol® membranes and MEAs [35]. Membranes based on PBI are of high applicability as it can be observed, both for the fuel cell technology in development and also for other applications as carbon capture, pervaporation dehydration processes, or electrochemical hydrogen separation, among others.

5. Proton exchange membrane fuel cells (PEMFCs)

Polybenzimidazole (PBI) as ionic exchange membrane can be used as proton exchange if the material is doped with phosphoric acid (H_3PO_4), sulfuric acid (H_2SO_4), and nitric acid (HNO_3) solvent media. The PBI has benzimidazole units in the polymer chain and bears the pKa = 5.5 that is responsible for the weak acid character, and they have excellent oxidative and thermal stability [36]. The acid molecules penetrate the membranes during doping process, due to the acid-base interaction between them and gradually swelling of PBI membrane. Therefore, PBI can be easily doped with different types of strong acids, which act as predominant protonation through the PBI membranes.

In these circumstances, the material can work as solid electrolyte in a fuel cell in temperature range between 100 and 200°C, overcoming the dehydration problem that the Nafion® membrane has in operation condition at around 100°C and in consequence the dramatically reduction of its proton conductivity, presenting a near zero electro-osmotic drag [37]. High temperature makes HT-PEMFC more tolerant to impurities in feed gases (CO, e.g.) and simplifies elimination of waste heat with a simpler cooling system. If the fuel cell is working with reformed natural gas as a power source, the device does not require humidification of reactants due to the simple water management; that is why all these features greatly simplify design of HT-PEMFC stack [38].

In the PBI/H_3PO_4 system, the polybenzimidazole acts not only as a matrix polymer but also as proton acceptor [39]. For HT-PEMFCs, PBI/H_3PO_4 is considered a reasonably successful solid electrolyte because the excellent conductivity and thermochemical stability. Phosphoric acid has been widely employed as an anhydrous proton conductor because of its high proton conductivity, low cost, and thermal stability. At temperatures above 150°C, the dehydration of the acid occurs and yields pyrophosphoric acid or higher oligomers, which exhibit worse proton conductivity. On the other hand, the long-running operation leads to the release and dilution of H_3PO_4 from the membranes, which results in a loss of the acid into the fuel cell gas/vapor exhaust streams, the decrease of membrane ionic conductivity, and thus a lower fuel cell performance occurs. The high proton conductivity of the membranes was proved only when the polymer holds a large excess of phosphoric acid [40]. The optimum doping level was around 5 moles H_3PO_4 per PBI repeat unit, where a compromise between conductivity and mechanical properties was achieved.

A thick membrane is not usually advantageous because it is mainly responsible for the large ohmic polarization and modest power performance of HT-MEA. However, approx. 100 μm has been implemented with the intention of improving their mechanical properties [41]. The acid doping is an essential process, but it softens the PBI membrane, causing membrane ripping in MEA fabrication. The mechanical stability of the doped PBI membrane can be improved by lowering the H_3PO_4 doping level; however, the proton conductivity is reduced [42].

The problems of HT-PEMFCs operating at temperatures up to 100°C are not solved yet and demonstrate the necessity of research on new and more satisfactory alternatives. In this context, the ionic liquids (ILs) have been used as nonaqueous and low-volatility proton carriers in replacement of aqueous electrolytes. The protic ILs for example are able to transport protons due to their acid-base character and their capability to form complex or intermolecular hydrogen bonds [43] even in nonaqueous conditions. This type of materials tries to overcome the formation of unstable materials in the operating conditions and then to improve the performance of the PEMFC at high temperatures. The first research team working in this subject was Watanabe and colleagues, who identified the potential electroactive use of ILs in fuel cell reactions [44]. Sometimes, polymer phase substrate and the IL result in nonhomogeneous and unmanageable membranes when both components are integrated together. In general, ILs and polymers dissolved in a common solvent and later are casted as a film. In this way hybrid membranes are obtained, and the materials may be studied once the solvent has been removed. PBI-based hybrid membranes holding ILs are examples of this methodology. Greenbaum et al. [45] demonstrated that the composite gel-type membranes obtained from H_3PO_4 and aprotic hydrophilic IL, namely, 1-propyl-3-methylimidazolium dihydrogen phosphate [PMI][H_2PO_4] and PBI, can be operated as ion exchange membrane up to 150°C in a PEMFC. The composite membranes were homogeneous and both chemically and thermally stable with wide temperature range. Nevertheless, phase separation occurred when mixing the 1-ethyl-3-methylimidazolium triflate [EMI][Tf] or 1-ethyl-3-methylimidazolium bis(trifluoromethanesulfonyl)imide [EMI][TFSI] ILs with H_3PO_4 and PBI, resulting in homogeneous membranes. Schauer et al. [46] investigated the use of aprotic ionic liquid 1-butyl-3-methylimidazoliumtrifluoromethanesulfonate [BMIM][TfO] and protic ionic liquid 1-ethylimidazoliumtrifluoromethanesulfonate [EIM][TfO] to prepare membranes with several different polymers: a polybenzimidazole derivative with benzofuranone (PBI-O-Ph), Udel®-type polysulfone (Udel® PSU), and poly(vinylidene fluoride-co-hexafluoropropene) fluoroelastomer. The proton conductivity of the membranes was a function of the temperature and the ionic liquid amount in the membrane and the polymeric matrix itself. For PBI-O-Ph-based membranes, the conductivity was very low up to 90°C. Wang et al. [47] studied the PBI/IL composite membranes where the IL was 1-hexyl-3-methylimidazolium trifluoromethanesulfonate [HMI][Tf], an organosoluble fluorine ionic liquid. The ionic conductivity reached a value as high as 1.6×10^{-2} S cm^{-1} at 250°C under anhydrous conditions, and the results depended on temperature and IL content. The IL [HMI][Tf] works simultaneously as plasticizer and ion carrier. On the other hand, the major drawback related to the IL addition is a loss of membranes' mechanical properties, resulting in a good solid electrolyte to carry out the functions of HT-PEMFC at temperature > 200°C.

In many cases imidazolium salts are the most investigated as ILs in these applications; composite membranes with good specific conductivity have been found for their application as electrolytes in PEMFCs; however low performances (maximum power densities of around 1 mW cm^{-2} [48]) have been obtained.

Another example of composite hybrid membranes is the use of PBI as matrix and the diethlyaminebisulfate/sulfate IL, [DE][SH], in different compositional ratios, PBI/[DE][SH_x], as was published by Ocón et al. [49]. In this case, the composite membranes were obtained using a solution casting method. The interaction between the IL and the PBI was analyzed by FTIR spectroscopy. The imine group from the imidazole ring of PBI composite membranes showed no evidence of protonation, and consequently, the interaction between the IL and PBI was weak, remaining free inside of the PBI structure and allowing for the ionic conduction. The mechanical properties and tensile stress of pristine PBI was deteriorated dramatically on increasing the IL content, despite the fact that the conductivity values were very acceptable for the described application. For demanding fuel cell operation conditions, such as 200°C, and low humidity conditions, the PBI/[DE][SH_x] membranes exhibited acceptable ionic conductivity values, higher than 0.01 S cm^{-1}. In addition to high proton conductivity in anhydrous environment, which is an indispensable condition for potential HT-PEMFC membrane candidates, other requisites must also be fulfilled: barrier to the reagent gases, thermal and dimensional stability under operating conditions, electrochemical stability under reducing and oxidizing potentials, and compatibility with the electrocatalyst. In this particular case, low open-circuit voltage (OCV) of the cell, 0.8 V, was obtained. This suggests a mixed potential, although no crossover was detected in the experiments. The authors suggested that kinetic complication could show up like additional oxidation and reduction reactions simultaneously with the corresponding oxygen reduction reaction (ORR) and hydrogen oxidation reaction (HOR), respectively; furthermore, the poisoning effect of the H_2S generated at the anode should not be ignored.

On the other side, the beneficial effect on the decrease of the IL viscosity was observed in the performance of the fuel cell. The optimum performance was obtained with no limiting current, being the maximum current density ca. 70 mA cm^{-2} and 13.5 mW cm^{-2}, using 100% relative humidity at 80°C. At temperature higher than 80°C, the system starts to dehydrate, whereas the IL viscosity increases and the proton diffusion was hindered. The performance at 150°C wasn't good showing clear evidences of the system dehydration at temperatures beyond 80°C. The migration of the IL from anode to cathode was demonstrated in *postmortem* analysis of PBI/[DE][SH_x] composite-based electrodes. The IL went out of the composite membrane, and in consequence the cell resistivity increased by a factor of six times after polarization measurements.

It is necessary to keep in mind that the requirements of cell lifetime vary for different applications, that is, 5000 h for cars, 20,000 h for buses, and 40,000 h for stationary application with continuous operation [43]. This means that the development of ionic exchange membranes with a long operating life is a challenge to develop.

6. Anion exchange membrane fuel cells (AEMFCs)

Many electrochemical systems use ion exchange membranes, such as fuel cells, electrolyzers, or redox flow batteries. Traditionally cation exchange membranes have been used in these systems due to the idea that anion exchange membranes had too low conductivity and stability. However, in the last years, many advances have been made, and anion exchange membranes (AEMs) are demonstrating to have performances comparable to acid ones, showing promising application in several technologies [2]. These membranes conduct negatively charged ions like OH^- or Cl^- and usually have positive-charged groups in the polymer structure, which could be directly present in the polymer backbone or more commonly fixed to it by extended side chains of varying lengths and chemistries. Varcoe et al. [2] investigated a deep review about the different chemistries of polymer backbones and head groups and their current state of research. The use of alkaline media, compared to acid media, has some advantages like the better electrochemical kinetics of the oxygen reduction reaction (ORR). This allows the possibility of using non-noble metals in the electrocatalysts reducing the fuel cell system cost. Other advantages are the minimized corrosion problems and the cogeneration of electricity and valuable chemicals [7, 50]. Compared to classical alkaline fuel cells (AFCs) where the electrolyte is in aqueous phase, the use of AEMs solves the carbonation problems and the difficulties of the liquid electrolyte management. The fuels commonly used in anion exchange membrane fuel cells (AEMFCs) are hydrogen and alcohols. Hydrogen is the common fuel in commercialization and research and gives the higher power densities. On the other hand, alcohols like methanol or ethanol have the advantages of easier handle, store, and transport and can be acquired from abundant biomass, which is environmentally friendly considering the process is carbon-neutral.

Among all the polymers available and tested for AEMFCs, polybenzimidazoles have demonstrated good applicability, and the most commonly used and studied are PBI and ABPBI. Some of their advantages remain in the properties previously described, as excellent thermal stability, which allows to use them at higher temperatures, superior mechanical properties that can withstand the performance conditions, and the presence of amine and imine groups which form strong hydrogen bonding interactions and can be further functionalized. The great stability properties have also encouraged many studies combining polybenzimidazoles with other polymers, creating blend or crosslinked membranes with excellent performances. Membranes based on polybenzimidazoles alone or with other polymers have also demonstrated low alcohols crossover, making them adequate electrolytes in alcohol fuel cells. In the alkaline media, the pristine form of PBI can be equilibrated in aqueous solutions of alkali metal hydroxides forming homogeneous systems with the hydroxide salt and water dissolved in the polymer matrix. These materials have shown high ion conductivity and great chemical stability at low alkali concentrations and have been tested as anion-conducting electrolytes in fuel cells with hydrogen or alcohol and in water electrolyzers. In order to understand the physical and chemical properties of polybenzimidazoles in alkaline media, Aili et al. have made a study with thin films of PBI in aqueous KOH solution with concentrations from 0 to 50 wt.% [31]. They observed by the EDS cross-sectional maps that the dissolved KOH is evenly distributed in the electrolyte membrane. The polymer has strong water affinity through hydrogen bonding with the imidazole groups, absorbing around the water molecules per repeating unit (r.u.), and KOH forms various hydrated complexes when dissolved in water. The degree of ionization of the polymer is determined by the position of the acid-base equilibrium presented in **Figure 4**. They observed that it depends on the KOH concentration as was expected, increasing the KOH content per PBI r.u. with the higher concentration of the bulk solution, reaching 2.6 KOH molecules/r.u. at bulk concentration of 25 wt.%.

A similar trend was observed for the water molecules, reaching more than 20 H_2O molecules/r.u. at KOH concentration around 20–25 wt.% in the bulk solution. In the polymer phase, the number of water molecules per KOH decreased while increasing the bulk solution concentration, showing a concentrating effect of KOH in the polymer. They did the measurements by titration and gravimetrically, getting consistent results that corroborate previous knowledge. They also observed the anisotropic swelling behavior of the polymer at different KOH concentrations that had been previously reported and performed X-ray diffraction (XRD) measurements to explain it. The explanation they found was that the increasing of surface area and thickness up to 15 wt.% concentration was due to the uptake of water and KOH, but further increasing the concentration leads to full ionization of the polymer, breaking many of the hydrogen bonds and separating the layered structure. This separation is easier in the interlayer dimension than in the intra-layer one, causing high thickness increase and area decrease. When KOH bulk solution concentration reached 50 wt.%, sharp peaks appeared in the XRD and were attributed to a crystalline phase of a poly(potassiumbenzimidazolide) hydrate with a symmetric and highly regular structure with crystallite size in the range of 70–120 nm. These crystalline peaks were vanished after washing in water until neutral pH. They also observed that the previously described effect of the introduction of water and KOH that disturbs the polymer hydrogen bonding of imidazole groups affected the mechanical properties, causing great decay in the tensile strength and enhanced elongation at break. When full ionization of the polymer was reached, at 20–25 wt. %, more than 200% elongation at break and 0.3 GPa elastic modulus were obtained, which compared with the 80% elongation at break and 3.0 GPa in pure water, showing the great differences. The IR measurements showed clearly that the chemical environment of the benzimidazole moieties changed greatly from the

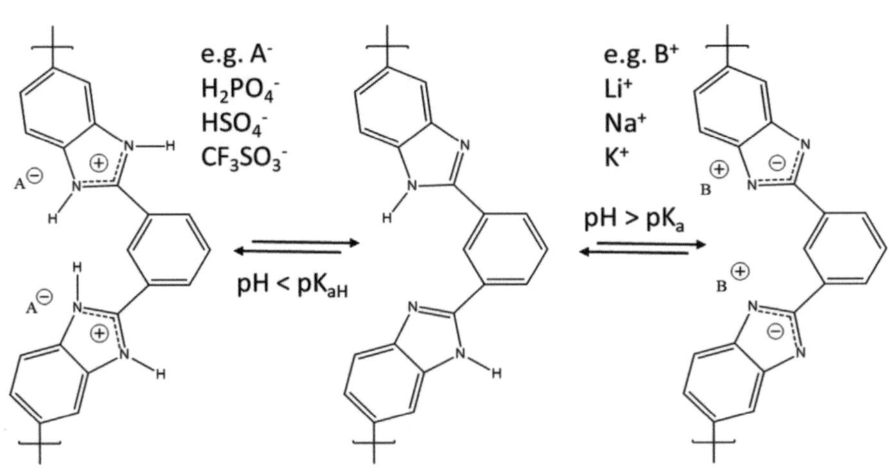

Figure 4.
Scheme showing the amphoteric nature of PBI in acidic (left) and alkaline (right) environments.

dissociation of the acidic proton. The result was that the deprotonated form of PBI predominates when the KOH concentration of the bulk solution is around 15–20%.

In order to discuss the different membranes based on polybenzimidazoles, the classification of anion exchange membranes made by Merle et al. will be useful [6]. Membranes are classified in three main groups: heterogeneous membranes, interpenetrating polymer networks, and homogeneous membranes. The heterogeneous membranes are composed by an anion exchange material embedded in an inert compound and can be divided in ion-solvating polymers if the inert compound is a salt or hybrid membranes in it is an inorganic segment. Polybenzimidazoles alone or blended with other polymers would fall into the category of ion-solvating polymers. The interpenetrating polymer network is a combination of two polymers in which one or both are synthesized or crosslinked in the presence of the other without any covalent bonds between them. The homogeneous membranes are composed only by the anion exchange material, forming a one-phase system, where the cationic charges are covalently bonded to the polymer backbone. Mobile counter ions are associated with the ionic sites to preserve the electroneutrality of the polymer. Examples of the cationic sites are the quaternary ammonium (QA) groups commonly used in AEMs. Depending on the production method and the starting materials, homogeneous membranes are divided into three types: (co) polymerization of monomers, modification into a polymer, and modification on a preformed film.

Alkali-doped PBI was investigated by Xing et al. for use in AEMFCs [51]. They obtained very interesting results, like conductivity as high as 9×10^{-2} S cm^{-1} at 25° C, higher than 2×10^{-2} S cm^{-1} of a H_2SO_4-doped PBI membrane, or the similar performance in hydrogen/oxygen fuel cells with alkali-doped PBI membrane and Nafion®117 membrane. Since that pioneering work, extensive attention has been paid to the alkali-doped PBI membranes, and thus great progress has been made. However, relevant issues are still remaining such as alkali leakage, fuel permeability, and mechanical stability. The single-cell performance of alkali-doped PBIs has been extensively studied with various fuels [52], such as hydrogen, methanol, ethanol, ethylene glycol, glycerol, formate, and borohydrides.

Using hydrogen as fuel, Zarrin et al. [53] have developed a stable and highly ion-conductive porous membrane doped with KOH. They found enhanced ionic conductivity by introducing the porosity in the membrane and obtained around twice better cell performance and conductivity compared with a commercial Fumapem® FAA membrane. Moreover, the KOH-doped PBI membrane maintained the ionic conductivity after 14 days of stability test, far more than the 3 h of the commercial one. The peak power density obtained with the porous PBI membrane of porosity 0.7 was 72 mW cm^{-2}, better than the 41 and 45 mW cm^{-2} obtained with a dense PBI membrane and the commercial FAA membrane, respectively. This better performance was demonstrated to be ascribed to the fact that the porous structure offered a higher ion transport rate through the membrane. One of the previously mentioned issues is the gradual alkali leakage during the cell operation. To solve it Zeng et al. [54] synthesized a sandwiched porous PBI membrane doped with KOH. The pore-forming method rendered numerous sponge-like walls and interconnected macropores, improving the interaction between the PBI and the doping alkali, indicating that both anionic conductivity and alkali retention could be enhanced by this method. Using this sandwiched porous PBI membrane doped with KOH in an AEMFC, they obtained an open-circuit voltage (OCV) of 1.0 V and a peak power density of 544 mW cm^{-2} at 90°C, which was higher than using the conventional membrane structure. They also investigated the durability of the fuel cell at a constant current density of 700 mW cm^{-2} and found that the conventional fuel cell had a dramatic voltage drop after short operation time, which was ascribed to the progressive release of the alkali solution. On the other hand, the sandwiched porous membranes performed with improved stability; the voltages reduced gradually to 0.1 V and remained there for another 25 h approximately. They explained that the performance enhancement was attributed to the retarding in the release of the alkali solution from the sponge-shaped wall, maintaining the high conductivity of the membrane. However, finally the leakage occurred, but as the authors indicated, the membrane could be reused after doping with KOH solution again.

Another approach was that used by Lu et al. [55]. They used PBI to react with poly(vinylbenzyl chloride) (PVBC), a polymer commonly used by other groups as for example Varcoe et al. in their grafted PTFE membranes [56, 57]. PVBC has the advantage of reacting with the imidazole rings of PBI creating a crosslinking connection with remaining -CH_2Cl groups unreacted that can be later functionalized as desired. For the functionalization of these groups, they decided to use the diamine 1,4-diazabicyclo (2.2.2) octane (DABCO), a very stable amine in alkaline media especially when only one of the two nitrogen is quaternized as previously reported [2, 6]. This method had the advantage that quaternization is done in the already casted membrane so it can be ensured that only one of the nitrogens react with PVBC obtaining the stability desired. Thanks to the good mechanical properties of PBI, they obtained membranes with good flexibility and strength both in dry conditions and saturated in water as well as high hydroxide conductivity (>25 mS cm^{-1} at room temperature) and superior chemical stability in alkaline environment. They tested the membrane in the H_2/O_2 fuel cell obtaining a peak power density of 230 mW cm^{-2} at 50°C and performed stability test, which showed high durability both in the constant current and continuous open-circuit voltage.

In addition to being used as an anion exchange membrane, alkali-doped PBI can work as ionomer, serving as ion-conductive pathway in the catalyst layer as well as a binder. Matsumoto et al. [58] developed a well-structured electrocatalyst for AEMFCs composed of carbon nanotubes (CNT), KOH-doped PBI ionomer, and platinum nanoparticles. This allowed them to obtain highly effective diffusivity and improved electrochemical activity, and they obtained a peak power density of 256 mW cm^{-2} at 50°C when tested in a H_2/O_2 fuel cell.

For fuel cells running on methanol, Hou et al. [59] tested a direct methanol fuel cell with a KOH-doped PBI membrane and observed that when a mixed solution of 2.0 M methanol and 2.0 M KOH was used as fuel, the OCV was around 1.0 V, and the peak power density was 31 mW cm^{-2} at 90°C. Wu et al. [60] prepared a membrane of KOH-doped PBI with CNT nanocomposites and obtained maximum power densities of 67 mW cm^{-2} and 104 mW cm^{-2} at 60 and 90°C, respectively, with a fuel composition of 2.0 M methanol + 6.0 M KOH and humidified oxygen. Li et al. [61] worked with pristine PBI membrane synthesized by solution casting method and treated it separately with 2.0 M H_3PO_4 and 6.0 M KOH to prepare a PEM and an AEM, respectively. They also studied several parameters of the structure design and operating parameters. They found that the conductivity of the KOH-doped PBI membrane was higher than the phosphoric acid membrane, 21.6 and 7.9 mS cm^{-1}, respectively. They also obtained a higher peak power density with the KOH-doped PBI membrane, 117.9 mW cm^{-2} at 90°C, than with the acid one, 46.5 mW cm^{-2}. They even reached a peak power density of 158.9 mW cm^{-2} at 90°C when using free-microporous layer electrodes and tripled the fuel flow rate.

In fuel cells running on ethanol, Hou et al. [62] developed a KOH-doped PBI membrane and found that with fuel composition of 2.0 M ethanol +2.0 M KOH, they obtained OCV of 0.92 V and maximum power density of 42.9 mW cm^{-2} at 75° C and 0.97 V and 60.9 mW cm^{-2} at 90°C. Modestov et al. [63] fabricated a membrane electrode assembly (MEA) employing non-platinum electrocatalysts and a KOH-doped membrane. In the anode they used a mixed solution of 3.0 M KOH

+2.0 M ethanol as fuel, while in the cathode they used air flow. With these conditions and at temperature of 80°C, a peak power density of 100 mW cm^{-2} was obtained at a voltage of 0.4 V. It was also found that by operating the fuel cell with pure oxygen, the current density was improved by 10%. Also using ethanol as fuel, recently Herranz et al. [29] tested the fuel cell performance of membranes synthesized with PBI and poly(vinyl alcohol) (PVA) with different weight ratios. PVA alcohol groups interacted with PBI by hydrogen bonding as well as allowing enhanced conductivity of the hydroxyl anion through the membranes. The increasing content in the PVA blend membrane leads to higher conductivities but if excessive could bring structural problems since PBI demonstrated to be essential for the membrane integrity. PVA:PBI 4:1 membrane obtained the best performance with a peak power density of 76 mW cm^{-2} at 90°C, 50% higher than a pristine KOH-doped PBI tested in the same conditions.

ABPBI has also been widely investigated for AEMs synthesis and application. Luo et al. [64] synthesized ABPBI and prepared the pristine membranes by the solution casting method. They studied the conductivity of the membranes at various alkali doping levels. They found high conductivity values for the membranes as 2.3×10^{-2} S cm^{-1} at 25°C and 7.3×10^{-2} S cm^{-1} at 100°C in the ABPBI membrane with alkali doping level of 0.37. They also founded the membranes have great thermal stability and excellent chemical stability, demonstrated by maintaining the conductivity values in alkaline media at 100°C for more than 1000 h.

Other alcohols and fuels have also been tested in AEMFCs using polybenzimidazoles in the membrane structure, showing promising results [65, 66]. Overall, the applicability and interest of benzimidazoles as AEMs are actual and will continue to increase due to their excellent properties.

7. Conclusions

Polybenzimidazoles have been deeply studied in the last decades, and great advancements have been done in their synthesis, making them economical materials with excellent thermal and mechanical properties as well as high chemical resistance in acidic and alkaline media. Their special structure with imidazole moieties and high intermolecular hydrogen bonding make them excellent materials to be used and ion exchange membranes for fuel cells. They can be used alone or in combination with other polymers or compounds, like the ionic liquids, as has been demonstrated many times. With them, it is possible to reach performances similar to other fuel cells and allow the application at higher temperatures, with all the benefits that implies. In the acidic media temperatures in the range of 120–200° C are used with good performances and easier water management, but still issues like structural stability with high doping level have to be solved. In order to help with the conductivity, ionic liquids have been investigated because of their nonaqueous and low-volatility properties as proton carriers. Interesting developments have been done but further research is necessary. In the alkaline media, their application has also attracted great interest. The ionization of the structure has been clearly identified at certain doping levels and the plasticizing effects it has. Pristine polybenzimidazole membranes have been directly doped with alkali solutions obtaining very good conductivity values, and other strategies like crosslinking with other polymers or synthesis of blend membranes have reported also promising results. The fuel cell performance is not yet as good as in the acidic media, but good results around 100 mW cm^{-2} have been obtained. Commercialization of membranes and MEAs based on PBI shows the potential they have, and research continues nowadays to develop them even more and better understand the possibilities of these wonderful materials in the fuel cell technology and the energy applications.

Acknowledgements

The authors want to acknowledge the Spanish Ministry of Economy Industry and Competitiveness (MINECO) project ENE2016-77055-C3-1-R and to Madrid Regional Research Council (CAM) project P2018/EMT-4344 (BIOTRES-CM).

Conflict of interest

The authors declare that they have no conflict of interest.

Nomenclature

PBI	Poly[2,2′-(*m*-phenylene)-5,5′-bisbenzimidazole]
ABPBI	Poly(2,5-benzimidazole)
PEMFCs	Proton exchange membrane fuel cells. Also used for general polymer electrolyte membrane fuel cells
AEMFCs	Anion exchange membrane fuel cells
IEM	Ion exchange membrane
AEMs/CEMs	Anion/cation exchange membranes
ORR	Oxygen reduction reaction
IV	Inherent viscosity
PPA	Polyphosphoric acid
MEA	Membrane electrode assembly
ILs	Ionic liquids
OCV	Open-circuit voltage
QA	Quaternary ammonium

Author details

Daniel Herranz and Pilar Ocón*
Department of Applied Physic Chemistry, University Autonomous of Madrid, Madrid, Spain

*Address all correspondence to: pilar.ocon@uam.es

References

[1] Wang Y, Chen KS, Mishler J, Cho SC, Adroher XC. A review of polymer electrolyte membrane fuel cells: Technology, applications, and needs on fundamental research. Applied Energy. 2011;**88**:981-1007. DOI: https://doi.org/10.1016/j.apenergy.2010.09.030

[2] Varcoe JR, Atanassov P, Dekel DR, Herring AM, Hickner MA, Kohl PA, et al. Anion-exchange membranes in electrochemical energy systems. Energy & Environmental Science. 2014;**7**:3135-3191. DOI: https://doi.org/10.1039/b000000x

[3] Corti HR, Gonzalez ER. Direct Alcohol Fuel Cells. Dordrecht: Springer; 2014. DOI: https://doi.org/10.1007/978-94-007-7708-8

[4] Zakaria Z, Kamarudin SK, Timmiati SN. Membranes for direct ethanol fuel cells: An overview. Applied Energy. 2016;**163**:334-342. DOI: https://doi.org/10.1016/j.apenergy.2015.10.124

[5] Smitha B, Sridhar S, Khan AA. Solid polymer electrolyte membranes for fuel cell applications—A review. Journal of Membrane Science. 2005;**259**:10-26. DOI: https://doi.org/10.1016/j.memsci.2005.01.035

[6] Merle G, Wessling M, Nijmeijer K. Anion exchange membranes for alkaline fuel cells: A review. Journal of Membrane Science. 2011;**377**:1-35. DOI: https://doi.org/10.1016/j.memsci.2011.04.043

[7] Pan ZF, An L, Zhao TS, Tang ZK. Advances and challenges in alkaline anion exchange membrane fuel cells. Progress in Energy and Combustion Science. 2018;**66**:141-175. DOI: https://doi.org/10.1016/j.pecs.2018.01.001

[8] O'Hayre RP, Cha S-W, Colella W, Prinz FB. editors. In: Fuel Cell Fundamentals. Hoboken, New Jersey: John Wiley & Sons; 2006. DOI: https://doi.org/10.1002/9781119191766

[9] Ebewele RO. Polymer Science and Technology. Vol. 74. Boca Raton, New York: CRC Press LLC; 1985. DOI: https://doi.org/10.1016/0025-5416(85)90434-3

[10] Vogel H, Marvel CS. Polybenzimidazoles, new thermally stable polymers. Journal of Polymer Science. 1961;**50**:511-539. DOI: https://doi.org/10.1002/pol.1961.1205015419

[11] Choe E. Catalysts for the preparation of Polybenzimidazoles. Journal of Applied Polymer Science. 1994;**53**:497-506. DOI: https://doi.org/10.1002/app.1994.070530504

[12] Choe EW. Single-stage melt polymerization process for the production of high molecular weight polybenzimidazole. US patent. 1982;**4**(312):976

[13] Li Q, Jensen JO, Savinell RF, Bjerrum NJ. High temperature proton exchange membranes based on polybenzimidazoles for fuel cells. Progress in Polymer Science. 2009;**34**:449-477. DOI: https://doi.org/10.1016/j.progpolymsci.2008.12.003

[14] Yang J, He R, Aili D. Synthesis of Polybenzimidazoles. High Temperature Polymer Electrolyte Membrane Fuel Cells. Switzerland: Springer; 2016. DOI: https://doi.org/10.1007/978-3-319-17082-4_7

[15] Iwakura Y, Uno K, Imai Y. Polyphenylenebenzimidazoles. Journal of Polymer Science. 1964;**2**:2605-2615. DOI: https://doi.org/10.1002/pol.1964.100020611

[16] Eaton PE, Carlson GR, Lee JT. Phosphorus pentoxide-methanesulfonic

acid. Convenient alternative to polyphosphoric acid. The Journal of Organic Chemistry. 1973;**38**:4071-4073. DOI: https://doi.org/10.1021/jo00987a028

[17] Kim H, Cho SY, An SJ, Eun YC, Kim J, Yoon H, et al. Synthesis of poly (2,5-benzimidazole) for use as a Fuel-cell membrane. Macromolecular Rapid Communications. 2004;**25**:894-897. DOI: https://doi.org/10.1002/marc.200300288

[18] Jouanneau J, Mercier R, Gonon L, Gebel G. Synthesis of sulfonated polybenzimidazoles from functionalized monomers: Preparation of ionic conducting membranes. Macromolecules. 2007;**40**:983-990. DOI: https://doi.org/10.1021/ma0614139

[19] JS W, MH L. S RF. High temperature membranes. In: W V, A L, HA G, editors. Handbook of Fuel Cells. Vol. 3. United States: John Wiley & Sons Ltd. 2003. pp. 436-446

[20] Asensio JA, Borro S, Gómez-Romero P. Polymer electrolyte fuel cells based on phosphoric acid-impregnated poly(2,5-benzimidazole) membranes. Journal of the Electrochemical Society. 2004;**151**:304-310. DOI: https://doi.org/10.1149/1.1640628

[21] Asensio JA, Borrós S, Gómez-Romero P. Proton-conducting membranes based on poly (2,5-benzimidazole) (ABPBI) and phosphoric acid prepared by direct acid casting. Journal of Membrane Science. 2004;**241**:89-93. DOI: https://doi.org/10.1016/j.memsci.2004.03.044

[22] Fishel KJ, Gulledge AL, Pingitore AT, Hoffman JP, Steckle WP, Benicewicz BC. Solution polymerization of polybenzimidazole. Journal of Polymer Science, Part A: Polymer Chemistry. 2016;**54**:1795-1802. DOI: https://doi.org/10.1002/pola.28041

[23] Litt M, Ameri R, Wang Y, Savinell R, Wainwright J. Polybenzimidazoles/phosphoric acid solid polymer electrolytes: Mechanical and electrical properties. Materials Research Society Symposium Proceedings. 1999;**548**:313-323. DOI: https://doi.org/10.1557/PROC-548-313

[24] He RH, Sun BY, Yang JS, Che QT. Synthesis of poly[2,2′-(m-phenylene)-5,5′-bibenzimidazole] and poly(2,5-benzimidazole) by microwave irradiation. Chemical Research in Chinese Universities. 2009;**25**:585-589

[25] Yang J, He R, Che Q, Gao X, Shi L. A copolymer of poly[2,2′-(m-phenylene)-5,5′-bibenzimidazole] and poly(2,5-benzimidazole) for high-temperature proton-conducting membranes. Polymer International. 2010;**59**:1695-1700. DOI: https://doi.org/10.1002/pi.2906

[26] Conti F, Willbold S, Mammi S, Korte C, Lehnert W, Stolten D. Carbon NMR investigation of the polybenzimidazole–dimethylacetamide interactions in membranes for fuel cells. New Journal of Chemistry. 2013;**37**:152. DOI: https://doi.org/10.1039/c2nj40728k

[27] Musto P, Karasz FE, MacKnight WJ. Hydrogen bonding in polybenzimidazole/polyimide systems: A Fourier-transform infra-red investigation using low-molecular-weight monofunctional probes. Polymer. 1989;**30**:1012-1021. DOI: https://doi.org/10.1016/0032-3861(89)90072-4

[28] Li Q, He R, Berg RW, Hjuler HA, Bjerrum NJ. Water uptake and acid doping of polybenzimidazoles as electrolyte membranes for fuel cells. Solid State Ionics. 2004;**168**:177-185. DOI: https://doi.org/10.1016/j.ssi.2004.02.013

[29] Herranz D, Escudero-Cid R, Montiel M, Palacio C, Fatás E, Ocón P. Poly (vinyl alcohol) and poly

(benzimidazole) blend membranes for high performance alkaline direct ethanol fuel cells. Renewable Energy. 2018;**127**:883-895. DOI: https://doi.org/10.1007/978-3-642-20487-6

[30] Asensio JA, Borrós S, Gómez-Romero P. Proton-conducting polymers based on benzimidazoles and sulfonated benzimidazoles. Journal of Polymer Science, Part A: Polymer Chemistry. 2002;**40**:3703-3710. DOI: https://doi.org/10.1002/pola.10451

[31] Aili D, Jankova K, Han J, Bjerrum NJ, Jensen JO, Li Q. Understanding ternary poly(potassium benzimidazolide)-based polymer electrolytes. Polymer. 2016;**84**:304-310. DOI: https://doi.org/10.1016/j.polymer.2016.01.011

[32] Mader J, Xiao L, Schmidt TJ, Fuel B, Ave V. Polybenzimidazole/acid complexes as high-temperature membranes. Advances in Polymer Science. 2008;**216**:63-124. DOI: https://doi.org/10.1007/12

[33] BASF Proton-Conductive Membrane. Available from: https://www.basf.com/global/en/who-we-are/organization/locations/europe/german-companies/BASF_New-Business-GmbH/our-solutions/proton-conductive-membrane.html [Accessed: February 10, 2019]

[34] PBI Products. Celazole PBI. Available from: https://pbipolymer.com/markets/membrane/ [Accessed: February 10, 2019]

[35] Danish Power Systems High Temperature PEM Fuel Cells. Available from: http://daposy.com/fuel-cells [Accessed: February 10, 2019]

[36] Wainright JS, Wang J-T, Weng D, Savinell RF, Litt M. Acid-doped Polybenzimidazoles: A new polymer electrolyte. Journal of the Electrochemical Society. 1995;**142**:L121-L123. DOI: https://doi.org/10.1149/1.2044337

[37] Weng D, Wainright JS, Landau U, Savinell RF. Electro-osmotic drag coefficient of water and methanol in polymer electrolytes at elevated temperatures. Journal of the Electrochemical Society. 1996;**143**:1260-1263. DOI: https://doi.org/10.1149/1.1836626

[38] Chandan A, Hattenberger M, El-Kharouf A, Du S, Dhir A, Self V, et al. High temperature (HT) polymer electrolyte membrane fuel cells (PEMFC)—A review. Journal of Power Sources. 2013;**231**:264-278. DOI: https://doi.org/10.1016/j.jpowsour.2012.11.126

[39] Kreuer KD, Fuchs A, Ise M, Spaeth M, Maier J. Imidazole and pyrazole-based proton conducting polymers and liquids. Electrochimica Acta. 1998;**43**:1281-1288. DOI: https://doi.org/10.1016/S0013-4686(97)10031-7

[40] Ma Y-L, Wainright JS, Litt MH, Savinell RF. Conductivity of PBI membranes for high-temperature polymer electrolyte fuel cells. Journal of the Electrochemical Society. 2004;**151**:A8-A16. DOI: https://doi.org/10.1149/1.1630037

[41] Vielstich W, Lamm A, Gasteiger HA, editors. In: Handbook of Fuel Cells: Fundamentals, Technology, Applications. United States: John Wiley & Sons, Ltd.; 2009

[42] Aili D, Allward T, Alfaro SM, Hartmann-Thompson C, Steenberg T, Hjuler HA, et al. Polybenzimidazole and sulfonated polyhedral oligosilsesquioxane composite membranes for high temperature polymer electrolyte membrane fuel cells. Electrochimica Acta. 2014;**140**:182-190. DOI: https://doi.org/10.1016/j.electacta.2014.03.047

[43] Wu J, Yuan XZ, Martin JJ, Wang H, Zhang J, Shen J, et al. A review of PEM

fuel cell durability: Degradation mechanisms and mitigation strategies. Journal of Power Sources. 2008;**184**: 104-119. DOI: https://doi.org/10.1016/j.jpowsour.2008.06.006

[44] Susan MABH, Noda A, Mitsushima S, Watanabe M. Brønsted acid–base ionic liquids and their use as new materials for anhydrous proton conductors. Chemical Communications. 2003;**3**:938-939. DOI: https://doi.org/10.1039/b300959a

[45] Ye H, Huang J, Xu JJ, Kodiweera NKAC, Jayakody JRP, Greenbaum SG. New membranes based on ionic liquids for PEM fuel cells at elevated temperatures. Journal of Power Sources. 2008;**178**:651-660. DOI: https://doi.org/10.1016/j.jpowsour.2007.07.074

[46] Schauer J, Sikora A, Plíšková M, Mališ J, Mazúr P, Paidar M, et al. Ion-conductive polymer membranes containing 1-butyl-3-methylimidazolium trifluoromethanesulfonate and 1-ethylimidazolium trifluoromethane-sulfonate. Journal of Membrane Science. 2011;**367**:332-339. DOI: https://doi.org/10.1016/j.memsci.2010.11.018

[47] Wang JTW, Hsu SLC. Enhanced high-temperature polymer electrolyte membrane for fuel cells based on polybenzimidazole and ionic liquids. Electrochimica Acta. 2011;**56**:2842-2846. DOI: https://doi.org/10.1016/j.electacta.2010.12.069

[48] Che Q, He R, Yang J, Feng L, Savinell RF. Phosphoric acid doped high temperature proton exchange membranes based on sulfonated polyetheretherketone incorporated with ionic liquids. Electrochemistry Communications. 2010;**12**:647-649. DOI: https://doi.org/10.1016/j.elecom.2010.02.021

[49] Mamlouk M, Ocon P, Scott K. Preparation and characterization of polybenzimidzaole/diethylamine hydrogen sulphate for medium temperature proton exchange membrane fuel cells. Journal of Power Sources. 2014;**245**:915-926. DOI: https://doi.org/10.1016/j.jpowsour.2013.07.050

[50] Vijayakumar V, Nam SY. Recent advancements in applications of alkaline anion exchange membranes for polymer electrolyte fuel cells. Journal of Industrial and Engineering Chemistry. 2019;**70**:70-86. DOI: https://doi.org/10.1016/j.jiec.2018.10.026

[51] Xing B, Savadogo O. Hydrogen/oxygen polymer electrolyte membrane fuel cells (PEMFCs) based on alkaline-doped polybenzimidazole (PBI). Electrochemistry Communications. 2000;**2**:697-702. DOI: https://doi.org/10.1016/S1388-2481(00)00107-7

[52] Wu QX, Pan ZF, An L. Recent advances in alkali-doped polybenzimidazole membranes for fuel cell applications. Renewable and Sustainable Energy Reviews. 2018;**89**: 168-183. DOI: https://doi.org/10.1016/j.rser.2018.03.024

[53] Zarrin H, Jiang G, Lam GY-Y, Fowler M, Chen Z. High performance porous polybenzimidazole membrane for alkaline fuel cells. International Journal of Hydrogen Energy. 2014;**39**: 18405-18415. DOI: https://doi.org/10.1016/j.ijhydene.2014.08.134

[54] Zeng L, Zhao TS, An L, Zhao G, Yan XH. A high-performance sandwiched-porous polybenzimidazole membrane with enhanced alkaline retention for anion exchange membrane fuel cells. Energy & Environmental Science. 2015;**8**:2768-2774. DOI: https://doi.org/10.1039/c5ee02047f

[55] Lu W, Zhang G, Li J, Hao J, Wei F, Li W, et al. Polybenzimidazole-crosslinked poly(vinylbenzyl chloride) with quaternary 1,4-diazabicyclo (2.2.2) octane groups as high-performance anion exchange membrane for fuel cells.

Journal of Power Sources. 2015;**296**:204-214. DOI: https://doi.org/10.1016/j.jpowsour.2015.07.048

[56] Herman H, Slade RCT, Varcoe JR. The radiation-grafting of vinylbenzyl chloride onto poly(hexafluoropropylene-co-tetrafluoroethylene) films with subsequent conversion to alkaline anion-exchange membranes: Optimisation of the experimental conditions and characterisation. Journal of Membrane Science. 2003;**218**:147-163. DOI: https://doi.org/10.1016/S0376-7388(03)00167-4

[57] Poynton SD, Slade RCT, Omasta TJ, Mustain WE, Escudero-Cid R, Ocón P, et al. Preparation of radiation-grafted powders for use as anion exchange ionomers in alkaline polymer electrolyte fuel cells. Journal of Materials Chemistry A. 2014;**2**:5124-5130. DOI: https://doi.org/10.1039/c4ta00558a

[58] Matsumoto K, Fujigaya T, Yanagi H, Nakashima N. Very high performance alkali anion-exchange membrane fuel cells. Advanced Functional Materials. 2011;**21**:1089-1094. DOI: https://doi.org/10.1002/adfm.201001806

[59] Hou H, Sun G, He R, Sun B, Jin W, Liu H, et al. Alkali doped polybenzimidazole membrane for alkaline direct methanol fuel cell. International Journal of Hydrogen Energy. 2008;**33**:7172-7176. DOI: https://doi.org/10.1016/j.ijhydene.2008.09.023

[60] Wu JF, Lo CF, Li LY, Li HY, Chang CM, Liao KS, et al. Thermally stable polybenzimidazole/carbon nano-tube composites for alkaline direct methanol fuel cell applications. Journal of Power Sources. 2014;**246**:39-48. DOI: https://doi.org/10.1016/j.jpowsour.2013.05.171

[61] Li L-Y, Yu B-C, Shih C-M, Lue SJ. Polybenzimidazole membranes for direct methanol fuel cell: Acid-doped or alkali-doped? Journal of Power Sources. 2015;**287**:386-395. DOI: https://doi.org/10.1016/j.jpowsour.2015.04.018

[62] Hou H, Sun G, He R, Wu Z, Sun B. Alkali doped polybenzimidazole membrane for high performance alkaline direct ethanol fuel cell. Journal of Power Sources. 2008;**182**:95-99. DOI: https://doi.org/10.1016/j.jpowsour.2008.04.010

[63] Modestov AD, Tarasevich MR, Leykin AY, Filimonov VY. MEA for alkaline direct ethanol fuel cell with alkali doped PBI membrane and non-platinum electrodes. Journal of Power Sources. 2009;**188**:502-506. DOI: https://doi.org/10.1016/j.jpowsour.2008.11.118

[64] Luo H, Vaivars G, Agboola B, Mu S, Mathe M. Anion exchange membrane based on alkali doped poly(2,5-benzimidazole) for fuel cell. Solid State Ionics. 2012;**208**:52-55. DOI: https://doi.org/10.1016/j.ssi.2011.11.029

[65] Couto RN, Linares JJ. KOH-doped polybenzimidazole for alkaline direct glycerol fuel cells. Journal of Membrane Science. 2015;**486**:239-247. DOI: https://doi.org/10.1016/j.memsci.2015.03.031

[66] Zeng L, Zhao TS, An L, Zhao G, Yan XH. Physicochemical properties of alkaline doped polybenzimidazole membranes for anion exchange membrane fuel cells. Journal of Membrane Science. 2015;**493**:340-348. DOI: https://doi.org/10.1016/j.memsci.2015.06.013

Chapter 11

Supramolecular Assembly of Benzimidazole Derivatives and Applications

Ana Beloqui

Abstract

Herein, we focus on the chemical and physical properties of benzimidazole and its derivatives used for the synthesis of supramolecular materials. The design and modification of benzimidazole opens the scope of the diversity of structures (different sizes and morphologies) that can be built. The synthesized materials include not only small coordination complexes but also isolated crystals, metal-organic frameworks, metal-coordination polymers, smart nanocontainers, and more advanced macrostructures such as microflowers and nanowires. These supramolecular structures are based on noncovalent interactions, mostly on metal coordination chemistry and π-π stacking interactions. Moreover, the same molecule, due to its chemical structure, can undergo both sorts of interactions in order to induce the self-assembly into supramolecular materials. In this process, as it is shown in this chapter, the conditions used for the assembly determine the final structure and morphology of the fabricated macromolecule. Finally, we show most recent applications of these materials in the field of sensing, photoluminescence, fuel cell, and fabrication of new nanostructures.

Keywords: self-assembly, supramolecular interactions, metal-imidazole coordination, π-π stacking interactions

1. Introduction

Benzimidazole and its derivatives are mostly known by their role in therapeutic drugs and by their pharmacological activities, for example, antimicrobial, analgesic, and anti-inflammatory [1]. Moreover, they are part of essential biomolecules as vitamin B_{12} [2]. Thus, the biological activity of benzimidazole and its derivatives is unquestionable. However, there is a growing research interest in using benzimidazole derivatives for their assembly into supramolecular structures for technological applications. This implies the formation of well-defined complex bond through noncovalent interactions. In this regard, the interest on benzimidazole molecule is twofold (**Figure 1**). On the one hand, benzimidazole is a popular N-donor ligand that is often used in coordination chemistry, meaning that it can through metal- or small-molecule coordination to the assembly of molecules. Indeed, the imidazole ring is commonly found as part of essential components of biological products,

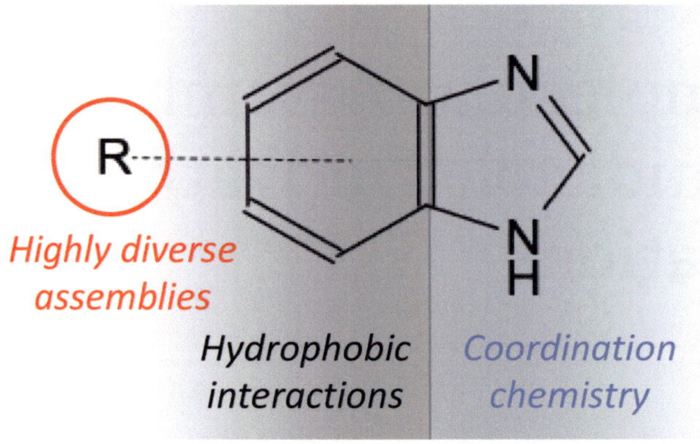

Figure 1.
The physicochemical nature of benzimidazole allows the assembly through different chemistries such as hydrophobic interactions (mainly through the benzyl group) and small-molecule coordination (mainly through the imidazole ring). Benzimidazole derivatives (different "R") will lead to the synthesis of assemblies of diverse composition and thus morphology and properties.

Type of material	Assembly	Ligand	Application	Reference
Metalogel	Metal-metal, hydrophobic interactions	2,6-Bis(benzimidazol-2′-yl)pyridine	Nanocontainer for small molecules	[4–6]
Macrocycles	Anion binding coordination	N-methylbenzimidazole	Tautomer switch	[7]
Coordination polymer	Metal coordination, Ag(I), Cu(II), lanthanides (III), hydrophobic interaction	N,N′-bisoctadecyl-2-(1*H*-benzimidazole-2-carbonyl)-L-glutamic amide	Chiral materials, photoluminiscence	[8]
Metal-organic crystals	Metal coordination, Cu(II), Zn(II)	1-Benzylamonium	Photoluminiscence	[9]
Metal-organic frameworks	Metal coordination, Cd(II)	2-Pyridin-3-yl-1*H*-benzoimidazole	Photoluminiscence	[10]
		2-Pyridin-4-yl-1*H*-benzoimidazole		
Metal-organic frameworks-metalogel	Metal coordination, Ag(I)	2-Heptadecylbenzimidazole	Dye adsorption	[11]
Coordination polymer	Metal coordination, Cd(II)	1*H*-benzimidazole-5-carboxylic acid	Fluorescence	[12]
Metal-organic frameworks	Metal coordination, Cd(II)	1,1′-(1,5-Pentanediyl) bis-1*H*-benzimidazole	Photoluminescence	[13]
Coordination polymer	Metal coordination, lanthanides (III)	Tris(benzimidazole-2-ylmethyl) amine	Adsorbent materials	[14]
Coordination polymer	Metal coordination, Zn(II) and Cd(II)	3-(1*H*-benzoim-dazol-2-yl) propanonic acid	—	[15]
Coordination polymer	Metal coordination, Co(II)	1,1-(1,4-Butanediyl) bis-1*H*-benzimidazole	—	[16]
Coordination polymer	Metal coordination, Cd(II)	1,3-Bis(5,6-dimethylbenzimidazole) propane	Photoluminescence	[17]
		1,4-Bis(5,6-dimethylbenzimidazole) butane		
		1,6-bis(5,6-dimethylbenzimidazole) hexane		

Type of material	Assembly	Ligand	Application	Reference
Macrocycles and large structures	Metal coordination, Co(II)	1,1′-(1,4-butanediyl) bis(benzimidazole)	Thermostable polymers	[18]
Helical coordination polymers	Metal coordination, Cd(II), Zn(II)	1,1-(1,4-Butanediyl) bis-1*H*-benzimidazole	Electrochemistry	[19]
Self-assembled polymer	Hydrophobic interactions	4-(1*H*-Benzimidazole-2-yl)-benzoic acid	Integration in functional devices	[20]
Metal-organic macrocycles	Metal coordination-Pd(II)-anion binding	di-Benzimidazole	Nanocontainers	[21]
Conjugates of calix-6-arenes	Covalent	(Tris)imidazole	Cu sensor	[22]
Liquid polymers	Hydrogen bonding	40-(6-(Benzimidazolethio) hexoxy)-biphenyl-4-yl 4-(alkoxy) benzoate	Liquid crystals for electronic conduction	[23]
Coordination polymer	Metal coordination, Zn(II)	2,6-bis(1′-methylbenzimidazolyl) pyridine	Self-healing material	[24]

Table 1.
Examples of materials synthesized through supramolecular assembly of benzimidazole derivatives.

such as histidine (in proteins), purine, histamine, and nucleic acids. In the specific case of proteins, it is common to find the imidazole ring in coordination with metal cations, which are essential for their biological function. On the other hand, the benzyl ring of the benzimidazole can undergo physical interactions (hydrophobic, π-π interactions) with other planar benzyl groups or hydrophobic moieties. Moreover, the controlled intra- and intermolecular π-π interactions can lead the supramolecular assembly into large structures.

Supramolecular assemblies are emergent structures basically driven by physical interactions [3]. It consists in a controlled multilevel organization process, from the assembly of discrete elementary molecular units via noncovalent interaction, to the further assembly of those into complex functional structures. The interaction forces operate under entropic constraints, looking for energy minimization. Selected organic and/or inorganic molecules can determine the chemical and structural composition of the eventually formed material. Therefore, this is a versatile methodology for the fabrication of nano-microstructures of defined size, morphology, and properties.

In this chapter, we show different approaches that are currently utilized for the controlled assembly of benzimidazole and its derivatives for the formation of large and ordered structures. Derivatives of imidazole are specifically designed for the assembly into structures with diverse size and morphology [4]. We will go through the methodologies that allow the fabrication of isolated crystals, metal-coordinated polymers, metal-organic frameworks, helical structures, smart nanocontainers, and advanced structures such as microflowers or nanowires. Moreover, benzimidazole can be combined with biological macromolecules (proteins, nucleic acid) to trigger their assembly. Finally, we show that benzimidazole derivatives, besides the key role they have shown in the self-assembly of macromolecules, are used in a reasonably broad range of technological applications such as sensing, photoluminescence, fuel cell, and fabrication of new nanostructures. A list of examples in which benzimidazole molecule is used for the assembly into supramolecular structures are summarized in **Table 1**.

2. Supramolecular self-assembly through coordination chemistry

The combination of organic and inorganic compounds to raise supramolecular structures is an on-growing research field. This approach broadens the applications and the nature of the morphologies and chemistries that can be applied for. Of particular importance is the assembly of metal (inorganic)-organic structures based on coordination chemistry. A good design of the ligands and metal cations and the selection of the appropriate synthesis method can lead to the formation of well-defined crystals, frameworks, or polymers as it is discussed below. Furthermore, the anion binding chemistry is another field of application of coordination chemistry. Small molecules such as anions coordinate to ligands to form stable complexes with interesting properties.

In this section, we focus on the ability of the imidazole ring to coordinate to small molecules such as anions and metal ions for the formation of discrete complexes and large supramolecular structures.

2.1 Supramolecular assembly with small molecules

The field of the coordination chemistry compiles a broad range of interactions, from classical Werner transition metal complexes, clusters, and organometallics, to host-guest complexes and supramolecular complexes (e.g., crown ethers or cryptands) [25]. Usually, in most of the examples found in the literature, benzimidazole derivatives are coordinated through metal-cation interaction. However, the anion binding chemistry has an important impact within the coordination chemistry field and, as it is going to be shown below, in the formation of discrete complexes and large supramolecular assemblies with benzimidazole derivatives.

Anionic species are generally larger than metal cations and thus might require greater size of the ligands. Moreover, the coordination of anions is usually saturated and therefore they only interact with ligands via weak forces, that is, hydrogen bonding and van der Waals interactions. Additionally, the relatively narrow pH window in which many anions exist determines the stability of the synthesized complexes. Anions play significant roles in biology, as receptors or cofactors, and in a broad number of applications such as sensing, crystal engineering, transmembrane transport, or anion-based catalysis [26].

Imidazole and benzimidazole derivatives have been used as ligands in anion binding-coordinated complex formation. In the case of imidazole molecule, both the nitrogen atoms within the imidazole ring are both covalently bond to sp3-hybridized carbon atoms. In these systems, the positively charged imidazolium group works as a hydrogen bond donor, interacting with the coordinated anion through a combination of hydrogen bonding and electrostatic interactions [7]. Furthermore, when using benzimidazole units, those can be employed as NH hydrogen bond donors and, in this case, a tautomerism process may affect the nature of the hydrogen bond presented to an anionic guest. As example, the N-methylbenzimidazole-based ligands selectively interact with the dihydrogen phosphate ion, acting as both a hydrogen bond donor and acceptor. Hence, several receptors containing benzimidazole derivatives have been reported as colorimetric, fluorescent, and electrochemically active sensors (**Table 1**).

2.2 Supramolecular assembly with metal cations

As a component of vitamin 1benzimidazole moiety exhibits good coordination ability with various transitional metal ions, such as Mn(II), Fe(II), Co(II), Ni(II), Cd(II), Hg(II), Pd(II), Cu(II), Zn(II), Ag(I), and Pb(II). In addition, it has been shown to coordinate with rare earth metal ions (lanthanides) [8]. The metal coordination of the imidazole ring is used in nature for the hierarchical assembly of biopolymers, for example, mussel byssus or worm jaws, and metalloproteins, in which the metal cations can not only show a structural function, but also contribute in their functionality, for example, catalysis in enzymes.

The metal coordination chemistry is an advantageous and widely used approach for assembly. The metal-ligand bonds are usually stronger that anion binding coordination; they are highly directional and kinetically labile. This means that, from one side, the symmetry and stereochemical preference is usually imposed by the metal cation and that this process is governed by a thermodynamic control, which is usually dependent on pH variations. Finally, metal ions bring along their intrinsic reactivity (as Lewis acidity, redox reactivity), which can be transferred to the assembled material.

The aforementioned advantages of metal coordination have been exploited to arrange small organic ligands into well-defined assemblies, organized arrays, that is, metal-organic frameworks (MOFs) and polymers. These materials have found applications in the host-guest chemistry, sensing, storage/separation, and catalysis.

2.2.1 Metal-organic macrocycles

Benzimidazole unit has been used for the formation of complexes using a very simple synthesis approach [27]. The assemblage into well-defined crystals is driven through the spontaneous metal coordination assembly at room temperature. As example, the combination of $CuCl_2{\cdot}6H_2O$ salt in a mixture of water/MeOH solution with benzimidazole in a molar ratio of 1:4 [9, 28] derives into the formation of dark blue crystals that consisted in a complex with formula $[Cu(bim)_4Cl_2]{\cdot}2H_2O$. Similar results are obtained by mixing $ZnCl_2$ in a DMF solution for the synthesis of crystals of $[Zn(bim)_2Cl_2]{\cdot}3H_2O$.

Nevertheless, most of the cases require the derivatization of the benzimidazole unit to allow its assembly into larger arrays, frameworks, and polymers as it is explained below.

2.2.2 Synthesis of metal-organic frameworks

The exploration of metal-organic frameworks (MOFs) has received much attention because of their well-defined architectures and wide range of potential applications in different fields. The assembly of transition metal cations such as Zn(II) and Cd(II) with multidentate nitrogen-containing ligands has produced various MOFs with fascinating structures and luminescent and catalytic properties [10, 11, 29]. The selection of chelating or bridging organic linkers often favors a structure-specific assembly, guiding the eventual morphology of the formed macromolecule. The factors that govern the formation of such complexes are complicated and include not only the nature of the metal ions and the ligand structure but also anion-directed interactions, hydrogen bonds, van der Wall forces, and employed reaction conditions.

The design and prediction of MOFs with potential properties is still a challenge to date [12]. Usually, the synthesis of MOFs starts from stiff bridging ligands via relatively strong dative bonds. Nevertheless, it has been proven that the contribution of hydrogen-bonded interactions leads to highly stable and porous architectures. Thus, of extreme importance is the design of ligands that can eventually undergo stabilization interactions.

Benzimidazole derivatives have been used for the synthesis Cd-based MOFs. Cd-containing structures, both discrete assemblies and infinite molecular frameworks, have been released from and characterized due to their useful properties in catalysis, luminescent materials, NLO materials, phase transformation, and host-guest chemistry. The use of 1,1′-(1,5-pentanediyl)bis-1*H*-benzimidazole as ligand formed a supramolecular structure that showed photoluminescence properties, which could be modulated by the influence of different counterions [13]. This points out the influence of anions in the arrangement of the coordination molecules and thus the final structure of the macromolecules.

Changing from transition metals to lanthanides leads to new applications and properties. The benzimidazole derivative (tris(benzimidazole-2-ylmethyl)amine, ntb) has been utilized as ligand for the synthesis of lanthanide (Ln: Nd^{3+}, Eu^{3+}, Gd^{3+}, and Er^{3+}) coordination monomers ($[ln(ntb)(NO_3)_3]$) that are further assembled via hydrogen bonding into three-dimensional (3D) frameworks [14]. Synthesis and crystallization conditions controlled the eventual morphology of the materials for each lanthanide used (monoclinic, hexagonal and cubic crystals). The Eu^{3+} and Nd^{3+} derivatives showed solid-state photoluminiscence in the near-infrared and visible region. In this case, the use of benzimidazole derivatives for the synthesis of porous coordination frameworks is advantageous for the supraorganization of structures through hydrogen-bonded frameworks, which is known to provide highly stable porous structures.

2.2.3 Fabrication of metal coordination polymers

There is a strong controversy on the use of the terms metal-organic frameworks and metal coordination polymers to assign the arrangement of an array of ligands through noncovalent coordination using metal cations [30]. As the term, metal-organic framework is very much appropriate to use for three-dimensional networks, the formation of one-dimensional and two-dimensional extended structures such as layers is named metal coordination polymers.

As the structures of coordination polymers are strongly influenced by the organic ligands and metal ions [15, 16, 31], it is important to choose suitable ligands and metal ions under appropriate synthetic conditions in order to synthesize coordination complexes with interesting structures. The flexibility of the ligand is a key parameter to direct the assembly into polymers instead of frameworks [17].

In this field, the synthesis of flexible divergent ligands is preferred. As example, the introduction of butane moieties to benzimidazole units provides the required flexibility for the fabrication of metal-coordinated polymers using Co(II). Hence, the ligand 1,1′-(1,4-butanediyl)bis(benzimidazole) (L) can be used for the fabrication of Co polymers (L1, $[CoL_2(H_2O)_2](NO_3)_2{\cdot}8H_2O$; and L2, $[CoL(H_2O)_2(CH_3CO_2)_2]H_2O$) Those polymers were obtained from the same ligand just varying slightly the synthesis conditions [18]. This variation leads to different composition and morphology for each of the polymers. While L1 forms infinite networks, the coordination of Co(II) in L2 leads to the formation of an infinite zigzag two-dimensional polymeric structure. Furthermore, same ligand L in presence of Cd(II) ions led to the fabrication of one-dimensional helical chain polymeric structures [19].

The assembly of these structures usually is sensitive to pH values and the protonation states of the ligands, as in the case of carboxylates and nitrogen donor groups. In order to have a better understanding of the effect of the pH value on the aromatic nitrogen-donor ligands, Li et al. [12] studied the reaction system using the benzimidazole derivative H_2bic (1*H*-benzimidazole-5-carboxylic acid) as ligand. In this case, they used Cd(II) for the assembly of the metal-coordinated polymer. The assembly was performed at different pH values, leading to different morphologies and compositions of the polymer. Hence, at pH 5.0, a two-dimensional supramolecular assembly consisting in stacked one-dimensional chains was obtained. However, when the pH was raised to 6.5, a rhombus network structure was obtained. Finally, at a pH of 7.2, a three-dimensional architecture based on binuclear cadmium units was retrieved.

3. Supramolecular self-assembly through π-π stacking interactions

As aforementioned, coordination chemistry is not the only noncovalent interaction used for the assembly of supramolecular structures. Indeed, nanostructures can be formed through interactions such as hydrogen bond, pi-pi stacking, metal coordination, or electrostatic interactions. In the specific case of benzimidazole derivatives, the heteroaromatic benzimidazole moiety introduces both pi-pi attacking and coordination unit. Benzimidazole derivatives have shown gelification and formation of structures in absence of metal cations or anions and completely different structures in presence of metal cations [8, 20]. As example, N,N′-bisocyadecyl-2-(1*H*-benzimidazle-2-carbonyl)-L-glutamic amide, BzLG) ligand was assembled through a supramolecular gelation method named low-molecular-weight organogels (LMWGs) in various organic solvents or in water. Once the gel is formed, the solvent is removed and nanostructures can be obtained with relatively uniform structures and large quantity. BzLG gelifies in several organic solvents, including cyclohexane, toluene, acetonitrile, ethanol, and dimethyl formamide. Obtained structures relied upon the solvent used in the gelification process: from nanotubes in dimethyl formamide and acetone to nanofibers in acetonitrile and cyclohexane.

When BzLG was used to coordinate with transitional metal ions and lanthanide ions, completely different structures were obtained. Nanotube flowers were obtained upon addition of $Eu(NO_3)_3$ and $Tb(NO_3)_3$, while in the case of $Cu(NO_3)_2$, microflower structures were observed. The latter structures were very similar to the nanoflower structures formed from bovine serum albumin (BSA) in phosphate buffer and $Cu(SO_4)_2$ metal salt reported in the fabrication of protein nanoflowers and nanosponges [32, 33].

4. Benzimidazole conjugates as smart nanocontainers

The design of stimuli-responsive materials is a growing research field. These materials are capable of altering their chemical and/or physical properties upon exposure to an external stimulus, such as temperature, humidity, light, or pH. Thus, they have been applied to biomedical applications as drug delivery vectors or as degradable biocompatible containers.

The ability of benzimidazole to coordinate metal cations can be used for the detection and study of metal cations in smart biodevices and organic nanocages. Here, we show two examples, benzimidazole conjugates of cyclodextrin and calix-6-arene, in which benzimidazole units are coupled to arranged molecules that can act as containers of small molecules, releasing those molecules under specific stimuli.

4.1 Cyclodextrin conjugates

Cyclodextrins (β-CD) are cyclic biomacromolecules typically containing six to eight glucose subunits bound through α-1,4 glycosidic bonds. They show toroid-like structures, with two external rims, one larger and the other smaller, that expose the secondary and primary hydroxyl groups of the glucose subunits, respectively. Importantly, the interior of the toroid is hydrophobic, being able to host hydrophobic molecules.

The chemical nature of benzimidazole makes it a good ligand to interact with cyclodextrin (β-CD) oligomers through hydrophobic interactions. Moreover, the physicochemical properties of benzimidazole molecules can be altered with the pH [34]. Therefore, under neutral pH condition, benzimidazole interacts with the inner part of the β-CD, blocking the diffusion through the macromolecule. As the pH is lowered, the dissociation constant between benzimidazole and β-CD decreases and benzimidazole is released to the medium. Using this approach, β-CD pH-responsive nanovalves have been fabricated: as the pH decreases, the external rings are opened, and the cargo release occurs. More complex structures were designed for sensing glucose or lactose using this system [35, 36].

4.1.1 Assembly into cyclodextrin-like architectures

The versatility and the high ability of benzimidazole derivatives to assemble are here demonstrated [4–6]. Through the controlled coordination to metal cations, that is, Pd(II), it is possible to self-assemble macrocyclic containers that mimic β-CD [21]. Moreover, these nanocontainers have the ability to bind anion guests and induce the transformation in the morphology and compositional unit of the nanocontainer. The hydrogen bonding between the inner surface of the macrocycles and the bound guests induced the fit—transformation properties of the assembled material, as observed in nature. Hence, the in situ anion-adaptative self-assembly gives rise to Pd_nL_{2n} species for n:3, 6, 7. As example, Sun et al. demonstrated the assembly of BzI-based ligands using square-planar palladium(II) ions into well-defined hydrogen-bonding pockets that will find applications in molecular sensing and catalysis.

4.2 Calix-6-arene conjugates

The benefits of the combination of benzimidazole molecules with nanocontainer scaffolds are also evidenced in this example in which benzimidazole is attached to calix-6-arenes [22]. Calix-6-arenes are organic macrocycles composed of (derivatives of) phenol subunits. Due to its hydrophobic cavity, they can be used to host smaller hydrophobic molecules or ions. They have been extensively used as a molecular platform to host catalytic units.

In benzimidazole-calixarene conjugates, the benzimidazole moieties are localized hanging out from the small rim of the macrocycle, pointing toward the environment. The coordination of metal cations to the imidazole ring of the benzimidazole molecule mimics the hydrophobic environment of the copper site in proteins and enzymes [37, 38]. The coordination of metal cations, that is, Cu(II), triggers a detectable modification in the structure of the conjugate, therefore being this a very sensitive method for the detection of metal ions.

5. Recent applications of assembled materials and new perspectives

As aforementioned, the assembly of benzimidazole and its derivatives has been utilized for the fabrication of stable materials with applications in several fields [39–42], some of them collected in **Table 1**: sensor fabrication [22], drug delivery systems [35], fuel cell design, biomedicine [43], conductivity in liquid crystals [23], or the fabrication of nanostructures [17].

Additionally, new applications of benzimidazole-derived materials are being raised. Thanks to the intrinsic properties of polymeric structures and the optical properties of some of the macroassemblies described above (luminescence, phosphorescence, or fluorescence), there is a growing interest in using these structures as healable and/or writable materials. These polymeric assemblies respond to external stimuli, for example, heat, presence of ions, with the surface rearrangement, and entanglement of the chains. The use of benzimidazole moieties in these systems, thanks to their chromophore rings, allows the use of an optical irradiation to heal the material [24]. Moreover, the ability of these polymers to bind and interact with different anions shows that their optical properties are used for advanced application in data recording and security protection [44].

6. Conclusions

In this chapter, we show the use of benzimidazole and its derivatives as important ligands that are currently being used for their supramolecular assembly into large structures with interesting properties and applications. Due to the physicochemical properties of benzimidazole, different chemistries and interactions can guide the assembly into very stable materials. Among them, the metal coordination to the imidazole ring seems to be the most exploited one for the formation of metal macromolecules, metal coordination polymers, and metal-organic frameworks. The selected benzimidazole derivatives and metal cation, together with the utilized synthesis conditions, will lead to the formation of materials with very diverse size and morphology. Thus, the establishment of a solid knowledge on the prediction of the assembled structures would contribute to the advancement of material science with strong strategic implications for the on-demand synthesis of smart responsive materials. However, in spite of the large efforts that are being done, currently there is no method to predict the composition and structure of the eventually synthesized materials.

Acknowledgements

This project has received funding from the Spanish Ministry of Economy and Competitiveness (MINECO) and FEDER funds in the frame of "Plan Nacional—Retos para la Sociedad" call with the grant reference MAT2017-88808-R. This work was performed under the Maria de Maeztu Units of Excellence Programme—MDM-2016-0618. A.B. thanks Diputación de Guipúzcoa for current funding in the frame of Gipuzkoa Fellows program.

Author details

Ana Beloqui
CIC nanoGUNE, Donostia-San Sebastian, Spain

*Address all correspondence to: a.beloqui@nanogune.eu

References

[1] Boiani M, González M. Imidazole and benzimidazole derivatives as chemotherapeutic agents. Mini Reviews in Medicinal Chemistry. 2005;**5**(4): 409-424. DOI: 10.2174/1389557053544047

[2] Weissbach H, Barker HA. Isolation and properties of B12 coenzymes containing benzimidazole or dimethylbenzimidazole. Proceedings of the National Academy of Sciences of the United States of America. 1959;**45**(4): 521-525. DOI: 10.1073/pnas.45.4.521

[3] Davis AV, Yeh RM, Raymond KN. Supramolecular assembly dynamics. Proceedings of the National Academy of Sciences. 2002;**99**(8): 4793-4796. DOI: 10.1073/pnas.052018299

[4] Jiang B, Zhang J, Zheng W, Chen LJ, Yin GQ, Wang YX, et al. Construction of alkynylplatinum(II) bzimpy-functionalized metallacycles and their hierarchical self-assembly behavior in solution promoted by Pt···Pt and π–π interactions. Chemistry—A European Journal. 2016;**22**(41):14664-14671. DOI: 10.1002/chem.201601682

[5] Datta S, Saha ML, Stang PJ. Hierarchical assemblies of supramolecular coordination complexes. Accounts of Chemical Research. 2018;**51**(9):2047-2063. DOI: 10.1021/acs.accounts.8b00233

[6] Zhang Y, Zhou Q-F, Huo G-F, Yin G-Q, Zhao X-L, Jiang B, et al. Hierarchical self-assembly of an alkynylplatinum(ll) bzimpy-functionalized metallacage via Pt···Pt and π–π interactions. Inorganic Chemistry. 2018;**57**(7):3516-3520. DOI: 10.1021/acs.inorgchem.7b02777

[7] Gale PA, Hiscock JR, Lalaoui N, Light ME, Wells NJ, Wenzel M. Benzimidazole-based anion receptors: Tautomeric switching and selectivity. Organic & Biomolecular Chemistry. 2012;**10**(30):5909-5915. DOI: 10.1039/ c1ob06800h

[8] Zhou X, Jin Q, Zhang L, Shen Z, Jiang L, Liu M. Self-assembly of hierarchical chiral nanostructures based on metal-benzimidazole interactions: Chiral nanofibers, nanotubes, and microtubular flowers. Small. 2016;**12**(34):4743-4752. DOI: 10.1002/smll.201600842

[9] Bibi S, Mohammad S, Manan NSA, Ahmad J, Kamboh MA, Khor SM, et al. Synthesis, characterization, photoluminescence, and electrochemical studies of novel mononuclear Cu(II) and Zn(II) complexes with the 1-benzylimidazolium ligand. Journal of Molecular Structure. 2017;**1141**:31-38. DOI: 10.1016/j.molstruc.2017.03.072

[10] Wang JH, Tang GM, Qin TX, Wang YT, Cui YZ, Ng SW. Structural and luminescent properties of a series of Cd(II) pyridyl benzimidazole complexes that exhibit extended three-dimensional hydrogen bonded networks. Journal of Coordination Chemistry. 2017;**70**(7):11189. DOI: 10.1080/00958972.2017.1299143

[11] Zhang YM, You XM, Yao H, Guo Y, Zhang P, Shi BB, et al. A silver-induced metal-organic gel based on biscarboxyl-functionalised benzimidazole derivative: Stimuli responsive and dye sorption. Supramolecular Chemistry. 2014;**26**(1):39-47. DOI: 10.1080/10610278.2013.822968

[12] Guo Z, Cao R, Li X, Yuan D, Bi W, Zhu X, et al. A series of cadmium(II) coordination polymers synthesized at different pH values. European Journal of Inorganic Chemistry. 2007;**5**:742-748. DOI: 10.1002/ejic.200600844

[13] Xiao B, Hou H, Fan Y, Tang M. Impact of counteranion on the self-assembly of Cd(II)-containing MOFs: Syntheses, structures and photoluminescent properties. Inorganica Chim Acta. 2007;**360**(9):3019-3025. DOI: 10.1016/j.ica.2007.02.038

[14] Jiang JJ, Pan M, Liu JM, Wang W, Su CY. Assembly of robust and porous hydrogen-bonded coordination frameworks: Isomorphism, polymorphism, and selective adsorption. Inorganic Chemistry. 2010;**49**(21): 10166-10173. DOI: 10.1021/ic1014384

[15] Li XY, Peng YQ, Li J, Fu WW, Liu Y, Li YM. Assembly of ZnII and CdII coordination polymers with different dimensionalities based on the semi-flexible 3-(1*H*-benzimidazol-2-yl)propanoic acid ligand. Acta Crystallographica Section E: Crystallographic Communications. 2018;**74**(1):28-33. DOI: 10.1107/S2056989017017534

[16] Wang XL, Hou LL, Zhang JW, Liu GC, Mu B, Lin HY. Two different dimensional bbbm-based cobalt(II) coordination polymers tuned by benzenedicarboxylates: Assembly, structures and properties. Inorganica Chim Acta. 2013;**397**:88-93. DOI: 10.1016/j.ica.2012.11.024

[17] Jiao CH, He CH, Geng JC, Cui GH. Syntheses, structures, and photoluminescence of three cadmium(II) coordination polymers with flexible bis(benzimidazole) ligands. Journal of Coordination Chemistry. 2012;**65**(16):2852-2861. DOI: 10.1080/00958972.2012.706283

[18] Agarwal RA, Aijaz A, Ahmad M, Sañudo EC, Xu Q, Bharadwaj PK. Two new coordination polymers with Co(II) and Mn(II): Selective gas adsorption and magnetic studies. Crystal Growth & Design. 2012;**12**(6):2999-3005. DOI: 10.1021/cg300217v

[19] Liu S, Yang Y, Qi Y, Meng X, Hou H. New Cd(II), Zn(II) helical coordination polymers constructed from hybrid ligands 4-ferrocenylbutyrate and 1,1-(1,4-butanediyl)bis-1*H*-benzimidazole. Journal of Molecular Structure. 2010;**975**(1-3):154-159. DOI: 10.1016/j.molstruc.2010.04.013

[20] Wang X, Zeng F, Ma Z, Jiang Y, Han Q, Wang B. Self-assembly of benzimidazole-ended nano hyperbranched polyester and its Host-guest response. Materials Letters. 2016;**173**:191-194. DOI: 10.1016/j.matlet.2016.01.129

[21] Zhang T, Zhou LP, Guo XQ, Cai LX, Sun QF. Adaptive self-assembly and induced-fit transformations of anion-binding metal-organic macrocycles. Nature Communications. 2017;**8**:15898. DOI: 10.1038/ncomms15898

[22] Izzet G, Akdas H, Hucher N, Giorgi M, Prangé T, Reinaud O. Supramolecular assemblies with calix-6-arenes and copper ions: From dinuclear to trinuclear linear arrangements of hydroxo—Cu(II) complexes. Inorganic Chemistry. 2006;**45**(3):1069-1077. DOI: 10.1021/ic051221e

[23] Tan S, Wei B, Liang T, Yang X, Wu Y. Anhydrous proton conduction in liquid crystals containing benzimidazole moieties. RSC Advances. 2016;**6**(40):34038-34042. DOI: 10.1039/C6RA03375J

[24] Burnworth M, Tang L, Kumpfer JR, Duncan AJ, Beyer FL, Fiore GL, et al. Optically healable supramolecular polymers. Nature. 2011;**472**(7343): 334-337. DOI: 10.1038/nature09963

[25] Bowman-James K. Alfred Werner revisited: The coordination chemistry of anions. Accounts of Chemical Research. 2005;**38**(8):671-678. DOI: 10.1021/ar040071t

[26] Zhao J, Yang D, Yang XJ, Wu B. Anion coordination chemistry: From recognition to supramolecular assembly. Coordination Chemistry Reviews. 2019;**378**:415-444. DOI: 10.1016/j.ccr.2018.01.002

[27] Shaker SA, Khaledi H, Cheah SC, Ali HM. New Mn(II), Ni(II), Cd(II), Pb(II) complexes with 2-methylbenzimidazole and other ligands. Synthesis, spectroscopic characterization, crystal structure, magnetic susceptibility and biological activity studies. Arabian Journal of Chemistry. 2016;**9**:S1943-S1950. DOI: 10.1016/j.arabjc.2012.06.013

[28] Pettinari C, Marchetti F, Cingolani A, Troyanov SI, Drozdov A. Ligation properties of N-substituted imidazoles: Synthesis, spectroscopic and structural investigation, and behaviour in solution of zinc(II) and cadmium(II) complexes. Polyhedron. 1998;**17**(10):1677-1691. DOI: 10.1016/S0277-5387(97)00455-5

[29] Xiao B, Yang LJ, Xiao HY, Fang SM. HgII-containing MOFs constructed from bis-benzimidazole-based ligand with highly selective anion-exchange capacity. Journal of Coordination Chemistry. 2011;**64**(24):4408-4420. DOI: 10.1080/00958972.2011.639875

[30] Biradha K, Ramanan A, Vittal JJ. Coordination polymers versus metal-organic frameworks. Crystal Growth & Design. 2009;**9**(7):2969-2970. DOI: 10.1021/cg801381p

[31] Bentz KC, Cohen SM. Supramolecular metallopolymers: From linear materials to infinite networks. Angewandte Chemie International Edition. 2018;**57**(46):14992-15001. DOI: 10.1002/anie.201806912

[32] Rodriguez-Abetxuko A, Morant-Miñana MC, López-Gallego F, Yate L, Seifert A, Knez M, et al. Imidazole-grafted nanogels for the fabrication of organic–inorganic protein hybrids. Advanced Functional Materials. 2018;**28**(35). DOI: 10.1002/adfm.201803115

[33] Ge J, Lei J, Zare RN. Protein–inorganic hybrid nanoflowers. Nature Nanotechnology. 2012;**7**(7):428-432. DOI: 10.1038/nnano.2012.80

[34] Yi S, Zheng J, Lv P, Zhang D, Zheng X, Zhang Y, et al. Controlled drug release from cyclodextrin-gated mesoporous silica nanoparticles based on switchable host-guest interactions. Bioconjugate Chemistry. 2018;**29**(9):2884-2891. DOI: 10.1021/acs.bioconjchem.8b00416

[35] Díez P, Sánchez A, Gamella M, Martínez-Ruíz P, Aznar E, De La Torre C, et al. Toward the design of smart delivery systems controlled by integrated enzyme-based biocomputing ensembles. Journal of the American Chemical Society. 2014;**136**(25):9116-9123. DOI: 10.1021/ja503578b

[36] Llopis-Lorente A, Díez P, de la Torre C, Sánchez A, Sancenón F, Aznar E, et al. Enzyme-controlled nanodevice for acetylcholine-triggered cargo delivery based on janus Au–mesoporous silica nanoparticles. Chemistry—A European Journal. 2017;**23**(18):4276-4281. DOI: 10.1002/chem.201700603

[37] Holm RH, Kennepohl P, Solomon EI. Preface: Bioinorganic enzymology. Chemical Reviews. 1996;**96**(7):2237-2238. DOI: 10.1021/cr9604144

[38] Lewis EA, Tolman WB. Reactivity of dioxygen-copper systems. Chemical Reviews. 2004;**104**(2):1047-1076. DOI: 10.1021/cr020633r

[39] Ricco R, Pfeiffer C, Sumida K, Sumby CJ, Falcaro P, Furukawa S, et al. Emerging applications of metal-organic frameworks. CrystEngComm. 2016;**18**(35):6532-6542. DOI: 10.1039/c6ce01030j

[40] Riccò R, Liang W, Li S, Gassensmith JJ, Caruso F, Doonan C, et al. Metal-organic frameworks for cell and virus biology: A perspective. ACS Nano. 2018;**12**:13-23. DOI: 10.1021/acsnano.7b08056

[41] Nath I, Chakraborty J, Verpoort F, McInerney JO, Cotton JA, Pisani D, et al. Metal organic frameworks mimicking natural enzymes: A structural and functional analogy. Chemical Society Reviews. 2016;**45**(15):4127-4170. DOI: 10.1039/c6cs00047a

[42] Furukawa H, Cordova KE, O'Keeffe M, Yaghi OM. The chemistry and applications of metal-organic frameworks. Science. 2013;**341**(6149):1230444. DOI: 10.1126/science.1230444

[43] Horcajada P, Gref R, Baati T, Allan PK, Maurin G, Couvreur P, et al. Metal-organic frameworks in biomedicine. Chemical Reviews. 2012;**112**(2):1232-1268. DOI: 10.1021/cr200256v

[44] Sun H, Liu S, Lin W, Zhang KY, Lv W, Huang X, et al. Smart responsive phosphorescent materials for data recording and security protection. Nature Communications. 2014;**5**(1):3601. DOI: 10.1038/ncomms4601